AF729427

Cell and Developmental Biology of Arabinogalactan-Proteins

Cell and Developmental Biology of Arabinogalactan-Proteins

Edited by

Eugene A. Nothnagel

University of California
Riverside, California

Antony Bacic

and

Adrienne E. Clarke

University of Melbourne
Parkville, Victoria, Australia

Kluwer Academic / Plenum Publishers

New York, Boston, Dordrecht, London, Moscow

Library of Congress Cataloging-in-Publication Data

Cell and developmental biology of arabinogalactan-proteins/edited by Eugene A. Nothnagel, Antony Bacic, and Adrienne E. Clarke.
p. cm.
"Proceedings of the Twentieth Symposium in Plant Physiology, held January 21–23, 1999, at the University of California, Riverside"—T.p. verso.
Includes bibliographical references.
ISBN 0-306-46469-1
1. Arabinogalactan—Congresses. I. Nothnagel, Eugene A. II. Bacic, A. (Antony) III. Clarke, A. E. (Adrienne Elizabeth) IV. Symposium in Plant Physiology (20th: 1999: University of California, Riverside)

QK898.A67 C46 2000
572'.68—dc21

00-058755

Proceedings of the Twentieth Symposium in Plant Physiology, held January 21–23, 1999, at the University of California, Riverside

ISBN: 0-306-46469-1

© 2000 Kluwer Academic/Plenum Publishers, New York
233 Spring Street, New York, N.Y. 10013

http://www.wkap.nl/

10 9 8 7 6 5 4 3 2 1

A C.I.P. record for this book is available from the Library of Congress.

All rights reserved

No part of this book may be reproduced, stored in a retrieval system, or transmitted in any form or by any means, electronic, mechanical, photocopying, microfilming, recording, or otherwise, without written permission from the Publisher

Printed in the United States of America

Preface

This volume captures a sense of the impact that the study of arabinogalactan-proteins (AGPs) is having in plant physiology. We have moved from a very diffuse set of observations on the distribution of AGPs in plants to a point where we can see some threads of research direction and some critical issues to be resolved. In this context it is interesting to recall that the reviewer of a proposal seeking funding for the meeting held at the University of California at Riverside in January, 1999, noted that "it is difficult to get an idea of where the field is headed other than you can find AGPs wherever you look!" It is true that AGPs are ubiquitous. We are now seeing, however, that different AGPs may have different tissue and cellular locations and that some appear and disappear during development. With this distribution and developmental control of AGPs, it is not unreasonable to consider that they may well have a range of functions. Indeed, some AGPs may be critical to survival, and this could be one reason why many people have observed a very low frequency of plants resulting from antisense experiments using AGP polypeptide backbone genes. In response to the reviewer's comment that it is difficult to get an idea of where the field is headed, one approach is to consider the blockages to further understanding of both the form and function of AGPs and their reciprocal relationships.

FORM OF ARABINOGALACTAN-PROTEINS

The fact that there are so many closely related but not identical AGPs leads to the questions of "What is an AGP?" and "When does one AGP differ from another?"

In this context, both the structure of the polypeptide backbones and the glycosyl constituents need to be understood. Features of the polypeptide backbones needing improved understanding include their domain structure, their diversity, whether they are glycosylphosphatidylinositol-anchored or not, and whether the genes encoding the polypeptide backbones include introns or not. Some of the questions regarding the carbohydrate chains are "How many types of chains?", "Where are the points of glycosylation?", "What parts of the polypeptide backbone are covered or exposed?", "What is the diversity of the carbohydrate sequences?", "What is the overall shape of the molecule?", and "What parts of the polypeptide and the carbohydrate side chains might be accessible for binding to other molecules?"

POSSIBLE FUNCTIONS OF ARABINOGALACTAN-PROTEINS

Many different functions have been implied from observations of the involvement of AGPs in such diverse physiological effects as programmed cell death, cell division, arrest of growth (reversible), oxidative bursts/wounding, somatic embryogenesis, pollen tube growth, chilling protection, microsporogenesis, growth suppression, and xylem formation. Observations on many of these involvements of AGPs are detailed in this volume, but at present, no one function of a single AGP is understood in detail.

In furthering our understanding of both the form and the function of AGPs, it became clear during the meeting that there are several critical issues to be addressed -

(i) *The specificity of the Yariv reagent.* It is important to establish the exact identities of the molecules to which the Yariv reagent binds. Certainly the reagent precipitates AGPs from tissue extracts. It would be valuable, however, to know the key features of both the polysaccharide and the polypeptide components which are required for binding to the Yariv reagent.

(ii) *Sequence analysis of glycosyl chains.* Understanding the precise structure and arrangement of the glycosyl chains is critical to our ultimate understanding of both the form and the function of AGPs. A current roadblock is the lack of routine and facile methods for isolating the glycosyl chains and establishing the monosaccharide components, linkages, and overall sequence. Commercial availability of glycosidases with defined specificities for the linkages which commonly occur in AGPs would be extremely helpful. The panel of monoclonal antibodies currently available is a useful tool set, although this usefulness is limited by lack of detailed specificity studies for many of these antibodies.

(iii) *Availability of defined oligosaccharide fragments as reagents.* The fact that the specificity of the antibodies has been defined in only a few cases reflects the lack of defined oligosaccharides that can be used as reagents. Commercial availability of oligosaccharides of galactose and arabinose in the relevant linkages, for example, would enable rapid progress to be made in applying immunocytochemical techniques to the cellular location and function of different AGPs.

(iv) *Biosynthesis of AGPs.* Understanding how the various chains are formed and at what stage the glycosylation of the polypeptide backbone occurs will give us insight into the groups of AGPs with similar but distinct structural features.

(v) *The number of AGPs within a particular plant tissue.* We do not yet have a sense of the complete range of AGPs in plant tissues. There are salt-extractable, non-salt extractable, membrane-bound, and non-membrane-bound AGPs, but the complete range for any particular tissue or cell type and the differences between the members of this range are still not defined.

(vi) Arabidopsis *mutants.* The range of *Arabidopsis* mutants available provides an invaluable experimental tool for understanding the relationship between form and function of AGPs. Nonetheless, the precise details of the chemistry of the different AGPs will be required to get an understanding of the mutations in these complex proteoglycans.

A related topic discussed during the meeting was the potential commercial use of AGPs. There is evidence supporting the idea that they can be used to induce immunostimulation in animals, and there may be other medicinal uses. Industrial uses rely on their functionality as emulsifiers, for example, in the food and cosmetic industries. There are many opportunities for applying current knowledge of the chemistry of AGPs to these industrial applications, many of which are presently based on empirical observations.

All in all, we anticipate that by the time the next meeting is held, we will have created a much broader knowledge base of both the form and the function of AGPs. We can look forward to a further understanding of how the design of these complex and varied proteoglycans is adapted for their seemingly myriad functions.

Adrienne E. Clarke
University of Melbourne

Acknowledgements

As editors of this volume, we wish to thank each of the authors for their scholarly contribution and for their cooperation and patience as this volume was assembled. Prominent acknowledgement and our enormous gratitude go to Nancy Day and Laura Heraty of the Department of Botany and Plant Sciences at Riverside for their editorial assistance and many, many hours of word processing to produce the camera-ready copy of this volume. We also thank MaryAnn McCarra, our editor at Kluwer Academic/Plenum Publishers.

As organizers of the 20th Symposium in Plant Physiology at the University of California, Riverside, upon which this volume is based, we prominently acknowledge Symposium Coordinators Cindi McKernan and Susana Aparicio, whose expertise and tireless attention to every detail in Riverside were greatly appreciated by both the organizers and the attendees at the symposium. We similarly thank Joanne Noble for her coordination and assistance in Melbourne. Our thanks also go to many other members of the Department of Botany and Plant Sciences including Cherie Cooksey, Ann Montejano, and Susan Miller, who handled finances, accounting, and many other essential functions; Van Stout and Lee Gross, who supplied the display boards for the poster session; and Rick Miranda, Watt Pattanagul, Kristen Lennon, Jean-Claude Mollet, David Puthoff, Donna Dubay, and Ricardo Cesped, who provided ground transportation and other assistance.

We are sincerely grateful to the following governmental agencies, industrial corporations, and University of California programs and administrators for their financial contributions to the support of the symposium:

United States Department of Agriculture
(Award 98-35304-6921 from the Plant Growth and Development Program)
National Science Foundation
(Award 9808309 from the Integrative Plant Biology Program)
Pioneer Hi-Bred International, Inc.
Bestfoods Baking Company
GLYKO, Inc.
Dionex Corporation
Biosupplies Australia Pty. Ltd.
University of California BioSTAR Project
Michael Clegg, Dean, College of Natural and Agricultural Sciences, UCR
Elizabeth Lord, Chair, Department of Botany and Plant Sciences, UCR

Neither the symposium nor this volume would have been possible without the financial support of these sponsors.

A. E. Clarke — *University of Melbourne*
E. A. Nothnagel — *University of California, Riverside*
A. Bacic — *University of Melbourne*

Contents

SECTION 2: LOCALIZATION AND ACTION OF ARABINOGALACTAN-PROTEINS AT THE SUBCELLULAR AND CELLULAR LEVELS

SECTION 3: ARABINOGALACTAN-PROTEINS IN SOMATIC EMBRYOGENESIS

SECTION 4: ARABINOGALACTAN-PROTEINS IN REPRODUCTIVE DEVELOPMENT

SECTION 5: ARABINOGALACTAN-PROTEINS IN VEGETATIVE DEVELOPMENT

SECTION 6: MEDICALLY AND INDUSTRIALLY IMPORTANT ARABINOGALACTAN-PROTEINS AND RELATED MACROMOLECULES

SECTION 7: SHORT PAPERS AND ABSTRACTS

Abbreviations

η	intrinsic viscosity
AAL	*Aleuria aurantia* L-Fuc-specific lectin
AG	arabinogalactan, a pure polysaccharide
AGP	arabinogalactan-protein
ALANAP	2-[(1-napthalenylamino)carbonyl]benzoic acid
Ara	arabinose
α-L-Arafase	α-L-arabinofuranosidase
AraGalGal	L-Ara*f*α(1→3)D-Galβ(1→6)D-Gal
AVG	aminoethoxyvinylglycine
BAPTA	1,2-bis(2-aminophenoxy)ethane-N,N,N',N'-tetraacetic acid
BD	buoyant density
BLAST	basic local alignment search tool
BSA	bovine serum albumin
CAM	cell adhesion molecule
CBPRP	chitin-binding proline-rich protein
dBEST	non-redundant EST divisions
DEAE	diethylaminoethyl
3,4-dhp	3,4-dehydro-L-proline
DP	degree of polymerization
2,2'-dp	2,2'-dipyridyl
ECM	extracellular matrix
ELISA	enzyme-linked immunosorbent assay
EMS	ethyl methanesulfonate
EP3	extracellular matrix protein 3, an acidic endochitinase
EST	expressed sequence tag

f	furanose
F-AGPs	L-fucosylated AGPs
Fuc	fucose
FucAraGalGal	L-Fucα(1→2)L-Ara*f*α(1→3)D-Galβ(1→6)D-Gal
α-L-Fucase	α-L-fucosidase
GA_3	gibberellic acid
Gal	galactose
$(\alpha\text{-D-Gal})_3$ or αGalY	1,3,5-tri-(*p*-α-D-galactosyloxyphenylazo)-2,4,6-trihydroxybenzene; see also $(\beta\text{-D-Glc})_3$
$(\beta\text{-D-Gal})_3$, or βGalY	1,3,5-tri-(*p*-β-D-galactosyloxyphenylazo)-2,4,6-trihydroxybenzene; see also $(\beta\text{-D-Glc})_3$
GalA	galacturonic acid
β-1,3-Galase	exo-β-1,3-galactanase
β-1,6-Galase	endo-β-1,6-galactanase
Glc	glucose
$(\beta\text{-D-Glc})_3$ or βGlcY	1,3,5-tri-(*p*-β-D-glucosyloxyphenylazo)-2,4,6-trihydroxybenzene; also referred to as β-glucosyl Yariv phenylglycoside, reagent, or antigen; one of a class of several synthetic molecules used as probes of AGPs; the $(\beta\text{-D-Glc})_3$ and $(\beta\text{-D-Gal})_3$ members of this class bind and precipitate many types of AGPs, but $(\alpha\text{-D-Gal})_3$ and $(\beta\text{-D-Man})_3$ do not precipitate AGPs and are thus useful as negative controls; $(\beta\text{-D-Glc})_3$ is the most widely-used member of this class and is sometimes referred to as simply Yariv reagent or Yariv antigen
GlcA	glucuronic acid
β-GlcA-ase	β-glucuronidase
GlcN	glucosamine
GlcNAc	N-acetylglucosamine
GPC	gel permeation chromatography
GPI	glycosylphosphatidylinositol
GUS	β-glucuronidase
HIC	hydrophobic interaction chromatography
HPLC	high performance liquid chromatography
HRGP	hydroxyproline-rich glycoprotein
Hyp	*trans*-4-hydroxy-L-proline
IC_{50}	50% inhibitory concentration
lacto-*N*-tetraose	D-Galβ(1→3)D-GlcNAcβ(1→3)D-Galβ(1→4)D-Glc
LTP	lipid transfer protein
MAb	monoclonal antibody
mAGP	morphoregulatory AGP
Man	mannose
$(\beta\text{-D-Man})_3$ or βManY	1,3,5-tri-(*p*-β-D-mannosyloxyphenylazo)-2,4,6-trihydroxybenzene; see also $(\beta\text{-D-Glc})_3$
4-O-Me-GlcA	4-O-methyl-glucuronic acid

4-O-Me-GlcA(Ara)GalGal	4-O-Me-D-GlcAβ(1→6)[L-Ara*f*α(1→3)]D-Galβ(1→6)D-Gal
M_n	number-average molecular weight
M_r	relative molecular weight
M_w	weight-average molecular weight
MWCO	molecular weight cut-off
NMR	nuclear magnetic resonance
nsLTP	nonspecific lipid transfer protein
p	pyranose in sugar names, para in other chemicals
PA	pyridylamino
PAC	a domain found in Proline-rich proteins and Arabinogalactan proteins, and contains conserved Cysteine residues
PAGE	polyacrylamide gel electrophoresis
PBS	phosphate-buffered saline
PCD	programmed cell death
PCR	polymerase chain reaction
PELP	pistil-specific extensin-like protein
P/HRGP	proline/hydroxyproline-rich glycoprotein
PI-PLC	phosphatidylinositol-specific phospholipase C
PNP	*p*-nitrophenyl, *p*-nitrophenol
PRP	proline-rich protein
RG	rhamnogalacturonan
Rg	radius of gyration
Rh	hydrodynamic radius
Rha	rhamnose
RI	refractive index
SDS-PAGE	sodium dodecyl sulfate-polyacrylamide gel electrophoresis
SEM	scanning electron microscopy
SERK	somatic embryogenesis receptor kinase
TCA	trichloroacetic acid
T-DNA	segment of *Agrobacterium* tumor-inducing plasmid that integrates into host plant chromosomal DNA
TEM	transmission electron microscopy
TIBA	2,3,5-triiodobenzoic acid
TUNEL	terminal deoxynucleotidyl transferase-mediated dUTP nick-end labeling
UV	ultraviolet
WE-AGP	water-extractable arabinogalactan-peptide
WE-AX	water-extractable arabinoxylan
XG	xyloglucan
Xyl	xylose

Cell and Developmental Biology of Arabinogalactan-Proteins

Chapter 1

A Brief History of Arabinogalactan-Proteins

B.A. Stone[1] and K. Valenta[2]
[1]*Department of Biochemistry and* [2]*Borchardt Library, La Trobe University, Bundoora 3083, Australia*

In this paper key events in the development of our current understanding of the structure, functional roles and biosynthesis of arabinogalactan-proteins (AGPs) will be analyzed. In making this analysis, use has been made of citations in reviews (Smith and Montgomery 1959, Clarke *et al* 1979, Fincher *et al* 1983, Showalter 1993, Bacic *et al* 1996, Du *et al* 1996a, Knox 1995 1996, Nothnagel 1997, Sommer-Knudsen *et al* 1998). These were analyzed by a bibliographical system (Garfield 1979) based on the citation index that enabled key events, their chronology and relationships in the evolution of the AGP story to be identified. Unfortunately the citation networking program that has proved useful, for example, in defining how DNA theory was developed, was not accessible. Instead, milestones in AGP history are summarized in sectors as follows: structural elucidation of exudate gums and cell and tissue AG-peptides and -proteins; detection, cell and tissue location of AGPs; separation and analysis of AGPs; function of AGPs; and, finally, their biosynthesis.

1. EXUDATE GUM STRUCTURE

The history of AGPs begins at least 4,000 years ago, when exudate gums, in particular gum arabic, were gathered and shipped by the Egyptians (Caius and Radha 1939) for use as an adhesive for mineral pigments in paints. However, the first chemical characterization of an exudate gum was made by Gay-Lussac and Thenard (1810) in the course of their elemental analysis of vegetable and animal materials. In this study it was shown that the elementary formula $C_n(H_2O)_n$ could describe vegetable materials that were not acids or resins but were analogues of sucrose, viz starch, gum, milk sugar, fibres and the crystalline principle of manna, mannitol. This was the genesis of the name "hydrate de carbone" for this group of compounds. The exudate gum from *Prunus* was characterized by John (1812), and later, gum arabic was shown to be an acid (Neubauer 1854).

Kiliani (1880) first isolated arabinose from gum arabic, and later the analytical studies of O'Sullivan (1884 1890) in London on the constitution of gum arabic led to its basic chemical characterization. Gum arabic was one of the carbohydrate polymers investigated by newly developing methods for characterizing polysaccharides. These methods included, in particular, partial acid hydrolysis and methylation, developed by Haworth and Hirst in Birmingham, and periodate oxidation, developed by O'Colla in Ireland. These were applied to gum arabic by Fred Smith (1939 a, b 1940), Alistair Stephen (1951), J.K.N. Jones (1953) and Thomas Dillon *et al* (1954). The first three carbohydrate chemists were all students of Haworth and Hirst. Anderson *et al* (1966) described the organization of the galactan backbone and the arabinosyl substitution. These and other studies led Gerald Aspinall and associates (1969) to recognize two types of AGs, among which, gum arabic, with its 3,6-galactan backbone was classified as Type II.

The 1970s were marked by the prolific analytical and structural studies on gum arabic and other Type II exudate gums, from numerous sources, by D.M.W. Anderson and coworkers in Edinburgh. With this came the recognition that their structural detail had taxonomic significance (Anderson and Dea 1969). Most significantly, in the context of this AGP history, Anderson's amino acid analyses (Anderson *et al* 1972) showed clearly that proteins, rich in Ser, Thr, Hyp and Asp, were associated with the gum polysaccharides and that the protein content across the various gum sources was quite variable. In South Africa, Churms and Stephen (1984) subjected gum arabic to the Smith periodate degradation procedure, which revealed that the galactan backbone was apparently constructed on a regular repeating pattern.

The heterogeneous nature of gum arabic was clearly demonstrated with use of hydrophobic chromatography by Randall *et al* (1989), and later a gum arabic fraction was shown definitively to be a glycoprotein by Lamport and coworkers (Qi *et al* 1991).

2. STRUCTURES OF CELL AND TISSUE AG-PEPTIDES AND AG-PROTEINS

Although a number of examples of Type II AGs were known by the beginning of the 1970s, their definitive covalent association with proteins was not established until Fincher *et al* (1974) reported the isolation and purification of a saturated-ammonium sulfate-soluble AG-peptide from wheat endosperm. The peptide had a high Hyp, Ala and Ser content and was joined to a Hyp residue in the peptide by an alkali-stable linkage.

Contemporaneously, but quite independently and only a few kilometres away, Michael Jermyn (Fig 1) was engaged in a parallel investigation. Jermyn and Yeow (1975) applied the crucial observation of Yariv and associates in Israel that β-glycosyl phenylazo dyes, the now well-known "Yariv reagents", could complex with and precipitate not only gum arabic (Yariv *et al* 1967b) but also similar molecules from a whole range of embryophytes. This ground-breaking work established the ubiquity of AGPs and provided the basis for their detection. These developments formed the basis of much structural and analytical work that followed rapidly. Detailed structures of Yariv-positive polymers from plant tissue cultures (Anderson *et al* 1977) and tissues from whole plants (Gleeson and Clarke 1979) confirmed their characterization as AGPs.

Figure 1. Michael A. Jermyn (1920–1989).

Gradually more detail accumulated about AGP structures. The alkali-stable AG–protein linkage was shown to be between β-galactan and Hyp (Strahm *et al* 1981, McNamara and Stone 1981), and the Pro/Hyp-rich polypeptide component was demonstrated to be in a partial helix conformation (van Holst and Fincher 1984). Significant differences were found in the fine structures of AGPs from different organs of the same plant (Nakamura *et al* 1984, Tsumuraya *et al* 1984), and as in gum arabic, the AG framework was shown to be regular (Bacic *et al* 1987). A significant difference between the structure of a classical AGP and a fraction from gum arabic was recognized by Qi *et al* (1991).

Information on the protein of AGPs was gathered much more slowly. Partial amino acid sequence information using the Edman procedure was forthcoming from

AGPs deglycosylated by trifluoromethanesulfonic acid or hydrofluoric acid. However, this at first did not generate primers of use in a cloning approach to defining polypeptide sequences because of the high content of Pro and Ala and the consequent problems with the complementary DNA. The problem was overcome by Shao-Lim Mau and Richard Simpson (Mau *et al* 1995), who generated peptides by proteolysis of hydrofluoric acid-deglycosylated polypeptides with thermolysin at elevated temperatures. Their sequences led to useful primers for a PCR approach. The first sequences based on cDNA clones were published (Du *et al* 1994 1996b, Chen *et al* 1993 1994, Mau *et al* 1995) from the laboratory of Adrienne Clarke in Melbourne. Genomic clones followed (Li and Showalter 1996).

The most recent structural milestone was the report by Bacic and colleagues (Youl *et al* 1998) of the discovery, using contemporary mass spectrometric technology, of a hitherto unrecognized glycosylphosphatidylinositol anchor motif on certain AGPs.

3. DETECTION OF AGPS

Without suitable specific reagents for the detection of AGPs, the subject would not have developed. All the reagents in use have their genesis in immunochemistry. Heidelberger's library of antisera, specific for different epitopes on bacterial surface polysaccharides, in particular those from pneumococci, found applications in many polysaccharide detection and structural determination strategies. Thus, the cross-reactions between pneumococcal Type II and IV antisera and epitopes on gum arabic were recognised by Heidelberger *et al* (1956), and the epitopes were proposed to be terminal α-glucuronosyl and terminal β-galactosyl residues, respectively.

The first Yariv reagents were synthesized at the Albert Einstein College of Medicine in New York (Yariv *et al* 1962) in an attempt to generate artificial monosaccharide antigens. These were analogues of the phenylazoglycosides used by Goebel and Avery in the 1940s to couple glycosides of known structure to proteins for use as antigens. Yariv *et al* (1967a) found that the α-L-fucosyl reagent precipitated a fucose-binding lectin from *Lotus tetragonolobus* seeds. An extension to the testing of other seed and plant extracts with other β-glycosyl reagents (Yariv *et al* 1967b) was presumably obvious at this time when many lectins were being discovered and characterized. Jermyn and Yeow (1975) observed that the β-glucosyl Yariv reagent interacted with extracts from a wide variety of plants to precipitate materials that they termed "β-lectins," but which we now recognize as AGPs.

Later, specific carbohydrate-binding proteins were introduced and included the spontaneously occurring, monoclonal myeloma protein J539 that recognizes (1→6)-β-oligogalactosyl epitopes (Glaudemans 1974) and the lectin from the giant clam (*Tridacna maxima*) (Baldo *et al* 1978) that recognizes β-galactan residues. Following the development of techniques for generating monoclonal antibodies by Milstein and Kohler, an antibody that specifically detected terminal arabinosyl residues on AGPs was developed by Marilyn Anderson and colleagues (1984). Later Paul Knox and associates (Stacey *et al* 1990) at the John Innes Institute, United Kingdom, produced the now widely used JIM series of monoclonals that have been applied with spectacular results for the tissue location of AGPs, as well as in functional studies.

4. CELL AND TISSUE LOCATION

The cell and tissue location of AGPs has been determined chiefly on the basis of Yariv staining and MAb reactivity. The AGPs were shown to occur intercellularly, intracellularly and in secretions (Anderson *et al* 1977, Clarke *et al* 1978, Gleeson and Clarke 1980), and ubiquitously in media from suspension-cultured cells (Aspinall *et al* 1969, Anderson *et al* 1977). The AGPs were also recognized to be membrane-associated (Larkin 1977 1978), to be present in the periplasmic space (Samson *et al* 1984) and to be components of cell walls (Serpe and Nothnagel 1995). Organ-specific AGPs were reported by van Holst and Clarke (1986).

5. SEPARATION AND ANALYSIS

Apart from their presence in exudate gums and tissue culture media, AGPs are generally found in very small amounts, and the development of methodology for small-scale separation and analysis was appropriate. Separation by protein gel electrophoresis (Laemmli 1970) is often satisfactory, but alternative techniques were developed. These included radial gel diffusion and rocket electrophoresis combined with Yariv reagent detection (van Holst and Clarke 1986).

For preparative separation of AGPs from other proteins, density gradient centrifugation in CsCl (Fincher *et al* 1974) and affinity chromatography on J539 columns (Andrew and Stone 1983) offered practical possibilities. Later, hydrophobic chromatography was shown to allow resolution of several fractions from gum arabic (Randall *et al* 1989).

6. FUNCTION

The functions of AGPs in plants has exercised the minds of biochemists and cell biologists since the earliest discoveries. The sticky exudate gums were hypothesized to seal wounds (Smith and Montgomery 1959), and nutrient properties were attributed to stylar AGPs (Labarca and Loewus 1972). The heterogeneity of the peripheral monosaccharides noted in the review by Clarke *et al* (1979) was taken to imply a fundamental informational role. However, the same review concluded with the reflection that AGPs might simply be part of the "ballast" of living organisms or "the garbage bin of plant metabolism"!!!

The notion of the involvement of AGPs in cell differentiation was first proposed by Basile (1979 1980) and has been increasingly supported by further publications from his laboratory and that of others using the JIM series of monoclonal antibodies. A role for AGPs in somatic embryogenesis (Stacey *et al* 1990) and pattern formation in roots (Knox *et al* 1989) has been proposed. The AGPs have also been implicated in cell expansion and wall loosening (Schopfer 1990), sexual development (Pennell and Roberts 1990), cell proliferation (Serpe and Nothnagel 1994) and programmed cell death (Langan and Nothnagel 1997).

7. BIOSYNTHESIS

Some features of AGP biosynthesis are now known, although much detail remains to be described. The transfer of Gal from UDP-Gal by unfractionated membranes from suspension-cultured *Lolium multiflorum* cells into material bound by the (1→6)-β-oligogalactosyl-specific myeloma protein J539 was demonstrated by Mascara and Fincher (1982). Block co-polymerisation was suggested for the galactan framework of the AG portion of *Acacia* gum on the basis of its regular structural features (Churms and Stephen 1984). The assembly of AGs in the Golgi apparatus was demonstrated by Schibeci and colleagues (1984) and the hydroxylation of prolines in the AGP polypeptide backbone by Cohen *et al* (1983), using the procedures developed by Cooper and Varner (1983) for Hyp-rich glycoproteins. More recently, the presence of a signal sequence on the AGP polypeptide was recognized (Mau *et al* 1995), indicative of its export from the endoplasmic reticulum.

8. OVERVIEW

The foregoing discussion illustrates once again how dependent the understanding of biological roles of natural compounds, and in the case of the AGPs, their discovery, has followed the invention or adaptation of appropriate tools for their detection and analysis. For AGPs, the techniques of polysaccharide and protein structural analysis, immunological detection, and recombinant DNA have figured prominently. Although not considered here for reasons of conciseness, it is apparent that the development of the AGP story has been influenced in many ways by the parallel development of knowledge and methodologies used to elucidate the structures and biological roles of the Hyp-rich glycoproteins, and to a lesser extent the AGs from gymnosperms and the Hyp-rich solanaceous lectins.

The results of this historical analysis also permit reflection on those aspects of AGP chemistry and biology that are due for further study. Here, a definitive demonstration of the causal relation between AGP structure and the various biological phenomena in which they have been implicated will be the most challenging. An integral part of the answer will lie in the elucidation of the molecular genetic control of AGP biosynthesis leading to the structural differences in the polysaccharide and protein parts of the AGP molecules that presumably provides the basis of their various specific biological roles.

REFERENCES

Anderson, D. M. W., and Dea, I. C. M., 1969, Chemotaxonomic aspects of the chemistry of *Acacia* gum exudates, *Phytochemistry* **8**: 167–176.

Anderson, D. M. W., Hirst E., and Stoddart, J. F., 1966, Studies on uronic acid materials. Part XVII. *Acacia senegal* gum (gum arabic), *J. Chem. Soc.* **C**: 1959–1966.

Anderson, D. M. W., Hendrie, A., and Munro, A. C., 1972, The amino acid and amino sugar composition of plant gums, *Phytochemistry* **11**: 733–736.

Anderson, M. A., Sandrin, M. S., and Clarke, A. E., 1984, A high proportion of hybridomas raised to a plant extract secrete antibody to arabinose or galactose, *Plant Physiol.* **75**: 1013–1016.

Anderson, R. L., Clarke, A. E., Jermyn, M. A., Knox, R. B., and Stone, B. A., 1977, A carbohydrate-binding arabinogalactan-protein from liquid suspension cultures of endosperm from *Lolium multiflorum*, *Aust. J. Plant. Physiol.* **4**: 143–158.

Andrew, I. G., and Stone, B. A., 1983, Chromatography of arabinogalactan-proteins, *Carbohyd. Polym.* **3**: 227–238.

Aspinall, G. O., 1969, Gums and mucilages, *Adv. Carbohyd. Chem. Biochem.* **24**: 333–379.

Aspinall, G. O., Molloy, J. A., and Craig, J. W. T., 1969, Extracellular polysaccharides from suspension-cultured sycamore cells, *Can. J. Biochem.* **47**: 1063–1070.

Bacic, A., Churms, S. C., Stephen, A. M., Cohen, P. B., and Fincher, G. B., 1987, Fine structure of the arabinogalactan-protein from *Lolium multiflorum*, *Carbohyd. Res.* **162**: 85–93.

Bacic, A., Du, H., Stone, B. A., and Clarke, A. E., 1996, Arabinogalactan proteins: a family of cell-surface and extracellular matrix plant proteoglycans, *Essays Biochem.* **31**: 91–101.

Baldo, B. A., Neukom, H., Stone, B. A., and Uhlenbruck, G., 1978, Reaction of some invertebrate and plant agglutinins and a mouse myeloma anti-galactan protein with an arabinogalactan from wheat, *Aust. J. Biol. Sci.* **31**: 149–160.

Basile, D. V., 1979, Hydroxyproline-induced changes in form, apical development, and cell wall protein in the liverwort *Plagiochila arctica*, *Am. J. Bot.* **66**: 776–783.

Basile, D. V., 1980, A possible mode of action for the morphoregulatory hydroxyproline-proteins, *Bull. Torrey Bot. Club* **107**: 325–338.

Caius, J. F., and Radha, K. S., 1939, The gum arabic of the bazaars and shops of Bombay, *J. Bombay Nat. Hist. Soc.* **41**: 261–271.

Chen, C.-G., Mau, S.-L., and Clarke, A. E., 1993, Nucleotide sequence and style specific expression of a novel proline-rich protein gene from *Nicotiana alata*, *Plant Mol. Biol.* **21**: 391–395.

Chen, C-G., Pu, Z-Y., Moritz, R. L., Simpson, R. J., Bacic, A., Clarke, A. E., and Mau, S-L., 1994, Molecular cloning of a gene encoding an arabinogalactan-protein from pear (*Pyrus communis*) cell suspension culture, *Proc. Natl. Acad. Sci. USA*, **91**: 10305–10309.

Churms, S. C., and Stephen, A. M., 1984, Structural studies of an arabinogalactan-protein from the gum exudate of *Acacia robusta*, *Carbohyd. Res.* **133**: 105–123.

Clarke, A. E., Gleeson, P. A., Jermyn, M. A., and Knox, R. B., 1978, Characterisation and localization of β-lectins in lower and higher plants, *Aust. J. Plant Physiol.* **5**: 707–722.

Clarke, A. E., Anderson, R. L., and Stone, B. A., 1979, Form and function of arabinogalactans and arabinogalactan-proteins, *Phytochemistry* **18**: 521–540.

Cohen, P. B., Schibeci, A., and Fincher, G. B., 1983, Biosynthesis of arabinogalactan-proteins in *Lolium multiflorum* (ryegrass) endosperm cells. III. Subcellular distribution of prolyl hydroxylase, *Plant Physiol.* **72**: 754–758.

Cooper J. B., and Varner, J. E., 1983, Selective inhibition of proline hydroxylation by 3,4-dehydroproline. *Plant Physiol.* **73**: 324–328.

Dillon, T., O'Ceallachain, D. F., and O'Colla, P., 1954, The constitution of arabic acid. II. *Proc. R. Irish Acad.* **57B**: 31–38.

Du, H., Simpson, R. J., Moritz, R. L., Clarke, A. E., and Bacic, A., 1994, Isolation of the protein backbone of an arabinogalactan-protein from the styles of *Nicotiana alata* and characterization of a corresponding cDNA, *Plant Cell* **6**: 1643–1653.

Du, H., Clarke, A. E., and Bacic, A., 1996a, Arabinogalactan-proteins: a class of extracellular matrix proteoglycans involved in plant growth and development, *Trends Cell Biol.* **6**: 411–414.

Du, H., Simpson, R. J., Clarke, A. E., and Bacic, A., 1996b, Molecular characterisation of a stigma-specific gene encoding an arabinogalactan-protein AGP from *Nicotiana alata*, *Plant J.* **9**: 313–323.

Fincher, G. B., Sawyer, W. H., and Stone, B. A., 1974, Chemical and physical properties of an arabinogalactan-peptide from wheat endosperm, *Biochem. J.* **139**: 535–545.

Fincher, G. B., Stone, B. A., and Clarke, A. E., 1983, Arabinogalactan-proteins: structure, biosynthesis, and function, *Annu. Rev. Plant Physiol.* **34**: 47–70.

Garfield, E., 1979, Citation analysis as a method of historical research into science, *Citation Indexing — Its Theory and Application in Science, Technology and Humanities,* John Wiley and Sons, New York, pp. 81–97.

Gay-Lussac, L. J., and Thenard, L. J., 1810, D'un memoire sur l'analyse vegetale et animale, *Ann. Chim.* **74**: 47–64.

Glaudemans, C. P. J., Zissis, E., and Jolley, M. E., 1974, Binding studies on a mouse-myeloma immunoglobulin A having specificity for β-D-(1→6)-linked D-galactopyranosyl residues, *Carbohyd. Res.* **40**: 129–135.

Gleeson, P. A., and Clarke, A. E., 1979, Structural studies on the major component of *Gladiolus* style mucilage, an arabinogalactan-protein, *Biochem. J.* **181**: 607–621.

Gleeson, P. A., and Clarke, A. E., 1980, Arabinogalactans of sexual and somatic tissues of *Gladiolus* and *Lilium*, *Phytochemistry* **19**: 1777–1782.

Heidelberger, M., Adams, J., and Dische, Z., 1956, Fractionation of gum arabic by chemical and immunological procedures, *J. Am. Chem. Soc.,* **78**: 2853–2855.

Jermyn, M. A., and Yeow, Y. M., 1975, A class of lectins present in the tissues of seed plants, *Aust. J. Plant Physiol.* **2**: 501–531.

John, J. F., 1812, Analyse einiger Gummiarten welche aus den Fruchten und Stammen der Prunusarten quellen, Schweigger, *J. Chem. Physik,* **6**: 374–380.

Jones, J. K. N., 1953, Larch ε-galactan. II. The isolation of 3-β-L-arabinopyranosyl-L-arabinose, *J. Chem. Soc.* 1672–1675.

Kiliani, H., 1880, Ueber die Identitat von Arabinose und Lactose, *Ber. Deutsch. Chem. Ges.* 2304–2306.

Knox, J. P., 1995, Developmentally regulated proteoglycans and glycoproteins of the plant cell surface, *FASEB J.* **9**: 1004–1012.

Knox, J. P., 1996, Arabinogalactan-proteins: Developmentally regulated surface proteoglycans of the plant cell surface. In *Membranes: Specialized Functions in Plants* (M. Smallwood, J. P. Knox, and D. J. Bowles, eds.), Bios Scientific, Oxford, pp. 93–102.

Knox, J. P., Day, S., and Roberts, K., 1989, A set of cell surface glycoproteins forms an early marker of cell position, but not cell type, in the root apical meristem of *Daucus carota* L., *Development* **106**: 47–56.

Labarca, C., and Loewus, F., 1972, The nutritional role of the pistil exudate in pollen tube wall formation in *Lilium longiflorum*. I. Utilization of injected stigmatic exudate, *Plant Physiol.* **50**: 7–14.

Laemmli, U. K., 1970, Cleavage of structural proteins during the assembly of the head of bacteriophage T4, *Nature* **227**: 680–685.

Langan, K. J., and Nothnagel, E. A., 1997, Cell surface arabinogalactan-proteins and their relation to cell proliferation and viability, *Protoplasma* **196**: 87–98.

Larkin, P. J., 1977, Plant protoplast agglutination and membrane-bound β-lectins. *J. Cell Sci.* **26**, 31–46.

Larkin, P. J., 1978, Plant protoplast agglutination by artificial carbohydrate antigens, *J. Cell Sci.* **30**, 283–292.

Li, S.-X., and Showalter, A. M., 1996, Cloning and developmental/stress regulated expression of a gene encoding a tomato arabinogalactan protein, *Plant Mol. Biol.* **32**: 641–652.

Mascara, T., and Fincher, G. B., 1982, Biosynthesis of arabinogalactan-protein in *Lolium multiflorum* (ryegrass) endosperm cells II, *In vitro* incorporation of galactosyl residues from UDP-galactose into polymeric products, *Aust. J. Plant Physiol.* **9**: 31–45.

Mau, S-L., Chen, C-G., Pu, Z-Y., Moritz, R. L., Simpson, R. J., Bacic, A., and Clarke, A. E., 1995, Molecular cloning of cDNAs encoding the protein backbones of arabinogalactan-proteins from the filtrate of suspension-cultured cells of *Pyrus communis* and *Nicotiana alata*, *Plant J.* **8**: 269–281.

McNamara, M., and Stone, B. A., 1981, Isolation, characterization and chemical synthesis of a galactosyl-hydroxyproline linkage compound from wheat endosperm arabinogalactan-peptide, *Lebensm.-Wiss.-Technol.* **14**: 182–187.

Nakamura, K., Tsumuraya, Y., Hashimoto, Y., and Yamamoto, S., 1984, Arabinogalactan-proteins reacting with eel anti-H agglutinin from leaves of cruciferous plants, *Agric. Biol. Chem.* **48**: 753–760.

Neubauer, C., 1854, Ueber das Arabin, Erdm., *J. Prak. Chem.* **62**: 193–202.

Nothnagel, E. A., 1997, Proteoglycans and related components in plant cells, *Int. Rev. Cytol.* **174**: 195–291.

O'Sullivan, C., 1884, Researches on the gums of the arabin group. Part I. Arabic acid: its composition and the products of its decomposition, *J. Chem. Soc.* **45**: 41–57.

O'Sullivan, C., 1890, Arabinon, the saccharon of arabinose, *J. Chem. Soc.* **57**: 59–63.

Pennell, R. I., and Roberts, K., 1990, Sexual development in the pea is presaged by altered expression of arabinogalactan protein, *Nature* **344**: 547–549.

Qi, W., Fong, C., and Lamport, D. T. A., 1991, Gum arabic glycoprotein is a twisted hairy rope, a new model based on O-galactosylhydroxyproline as the polysaccharide attachment site, *Plant Physiol.* **96**: 848–855.

Randall, R. C., Phillips, G. O., and Williams, P. A., 1989, Fractionation and characterisation of gum from *Acacia senegal*, *Food Hydrocol.* **3**: 65–75.

Samson, M. R., Jongeneel, R., and Klis, F. M., 1984, Arabinogalactan protein in the extracellular space of *Phaseolus vulgaris* hypocotyls, *Phytochemistry* **23**: 493–496.

Schibeci, A., Pnjak, A., and Fincher, G. B., 1984, Biosynthesis of arabinogalactan-protein in *Lolium multiflorum* (Italian ryegrass) endosperm cells, subcellular distribution of galactosyltransferases, *Biochem. J.* **218**: 633–636.

Schopfer, P., 1990, Cytochemical identification of arabinogalactan protein in the outer epidermal wall of maize coleoptiles, *Planta* **183**: 139–142.

Serpe, M. D., and Nothnagel, E. A., 1994, Effects of Yariv phenylglycosides on *Rosa* cell suspensions: Evidence for the involvement of arabinogalactan-proteins in cell proliferation, *Planta* **193**: 542–550.

Serpe, M.D., and Nothnagel, E.A., 1995, Fractionation and structural characterization of arabinogalactan-proteins from the cell wall of rose cells, *Plant Physiol.* **109**: 1007–1016.

Showalter, A. M., 1993, Structure and function of plant cell wall proteins, *Plant Cell* **5**: 9–23.

Smith, F., 1939a, The constitution of arabic acid. Part I. The isolation of 3-d-galactosido-l-arabinose, *J. Chem. Soc.* 744–753.

Smith, F., 1939b, The constitution of arabic acid. Part II. Degraded arabic acid, *J. Chem. Soc.* 1724–1738.

Smith, F., 1940, The constitution of arabic acid. Part III. Methylated arabic acid. *J. Chem. Soc.* 1035–1051.

Smith, F., and Montgomery, R 1959, *The Chemistry of Plant Gums and Mucilages and Some Related Polysaccharides*, Reinhold Pub. Corp., New York.

Sommer-Knudsen, J., Bacic, A., and Clarke, A. E., 1998, Hydroxyproline-rich plant glycoproteins, *Phytochemistry* **47**: 483–497.

Stacey, N. J., Roberts, K., and Knox, J. P., 1990, Patterns of expression of the JIM4 arabinogalactan-protein epitope in cell cultures and during somatic embryogenesis in *Daucus carota* L., *Planta*, **180**: 285–292.

Stephen, A. M., 1951, *Acacia mollissima*. I. The component sugars and aldobiuronic acid of black wattle gum, *J. Chem. Soc.* 646–649.

Strahm, A., Amado, R., and Neukom, H., 1981, Hydroxyproline-galactoside as a protein-polysaccharide linkage in a water soluble arabinogalactan-peptide from wheat endosperm, *Phytochemistry* **20**: 1061–1063.

Tsumuraya, Y., Hashimoto, Y., Yamamoto, S., and Shibuya, N., 1984, Structure of L-arabino-D-galactan-containing glycoproteins from radish leaves, *Carbohyd. Res.* **134**: 215–228.

van Holst, G.-J., and Fincher, G. B., 1984, Polyproline II conformation in the protein component of arabinogalactan-protein from *Lolium multiflorum*, *Plant Physiol.* **75**: 1163–1164.

van Holst, G.-J., and Clarke, A. E., 1986, Organ-specific arabinogalactan-proteins of *Lycopersicon peruvianum* (Mill) demonstrated by crossed electrophoresis, *Plant Physiol.* **80**: 786–789.

Yariv, J., Rapport, M. M., and Graf, L., 1962, The interaction of glycosides and saccharides with antibody to the corresponding phenylazo glycosides, *Biochem. J.* **85**: 383–388.

Yariv, J., Kalb, A. J., and Katchalski, E., 1967a, Isolation of a L-fucose binding protein from *Lotus tetragonolobus* seed, *Nature* **215**: 890–891.

Yariv, J., Lis, H., and Katchalski, E., 1967b, Precipitation of arabic acid and some seed polysaccharides by glycosylphenylazo dyes, *Biochem. J.* **105**: 1C–2C.

Youl, J. J., Bacic, A., and Oxley, D., 1998, Arabinogalactan-proteins from *Nicotiana alata* and *Pyrus communis* contain glycosylphosphatidylinositol membrane anchors, *Proc. Natl. Acad Sci USA* **95**: 7921–7926.

Chapter 2

Structural Classes of Arabinogalactan-Proteins

A. Bacic,[1] G. Currie,[1] P. Gilson,[1] S-L. Mau,[2] D. Oxley,[1] C. Schultz,[2] J. Sommer-Knudsen,[2] and A. E. Clarke[1]

[1]Plant Cell Biology Research Centre and [2]Cooperative Research Centre for Industrial Plant Biopolymers, School of Botany, University of Melbourne, Parkville, Victoria 3052, Australia

1. INTRODUCTION

Any consideration of the structural classes of arabinogalactan-proteins (AGPs) also raises the question, "What is an AGP?" The AGPs belong to the Hyp-rich glycoprotein (HRGP) family of molecules that also includes the extensins, Pro/Hyp-rich glycoproteins (P/HRGPs) and the solanaceous lectins (Showalter 1993, Kieliszewski and Lamport 1994, Du *et al* 1996a, Sommer-Knudsen *et al* 1998). In general, three criteria have defined AGPs: the presence of arabinogalactan chains, a Hyp-rich protein backbone and the ability to bind to a class of synthetic phenylazo dyes, the β-glycosyl Yariv reagents (Yariv *et al* 1962, Clarke *et al* 1979, Fincher *et al* 1983). It may now be necessary to reconsider our definitions. Arabinogalactan chains are found on proteins that do not bind the Yariv reagent [e.g., AG-peptide from wheat (Fincher *et al* 1974) and two glycoproteins from styles (Lind *et al* 1994, Sommer-Knudsen *et al* 1996)]. Some AGPs are Hyp-deficient, and others have short oligoarabinosides previously thought to be characteristic of the extensins and solanaceous lectins (Qi *et al* 1991, Baldwin *et al* 1993). This degree of variability is similar to that observed in glycosylation and protein backbones of the animal extracellular matrix (ECM) proteoglycans (Hardingham and Fosang 1992, Verma and Davidson 1994). These general criteria easily distinguish AGPs from the extensins and solanaceous lectins, but the boundaries between AGPs and P/HRGPs are less clearly defined. This variability raises several issues: (1) Are the criteria outlined above sufficient? (2) The AGPs are a family of molecules with different protein backbones, each existing as multiple glycoforms. Is there a distinct boundary between AGPs and

P/HRGPs, or is the HRGP family a continuum of molecules? (3) Can the knowledge and experiences from the animal proteoglycan field provide additional criteria that would clarify our definitions?

We will consider the uncertainties in the definitions of AGPs and other HRGPs with common structural features by drawing on the work from our laboratory on the HRGPs from the pistils of *Nicotiana alata* (see review by Sommer-Knudsen *et al* 1997) and from suspension culture cells. For extensive reviews on AGPs the following references should be consulted: Clarke *et al* (1979), Fincher *et al* (1983), Knox (1995), Du *et al* (1996a), Nothnagel (1997), Schultz *et al* (1998).

2. THE CARBOHYDRATE COMPONENT OF AGPS

Two models for the molecular structure of AGPs have been proposed: the "wattle-blossom" (Fig 1) and the "twisted hairy rope" (Fig 2). They have been generated from analysis of extensive data on the carbohydrate moiety, the solution properties of AGPs, and visualization of AGPs by electron microscopy, but in the absence of information on the sequence of amino acids in the protein backbone. In both models, the carbohydrate is O-glycosidically linked to Hyp residues of the protein through Gal and/or Ara. A major distinction between the models is the predicted shape: the wattle-blossom model is envisaged to be spheroidal, whereas the twisted hairy rope model is rod-like.

The carbohydrate is usually in the form of polysaccharide chains, Type II arabinogalactans (AGs), that consist of a backbone of (1→3)-linked β-D-Gal*p* residues substituted at C(O)6 by short side chains of (1→6)-linked β-D-Gal*p*. The side chains usually terminate in α-L-Ara*f* (Fig 1). Some AGPs are rich in glucuronic acid (e.g., gum arabic AGP), resulting in a negatively charged polysaccharide moiety (Nothnagel 1997), and others have short oligosaccharides of Ara*f* residues (Fig 2). Chemical degradation of the AG chains of some AGPs suggests a repetitive structure in which the β(1→3)-galactan backbone is interrupted at regular intervals by periodate-sensitive residues, possibly (1→6)-Gal*p* or (1→5)-Ara*f* (Churms *et al* 1981, Bacic *et al* 1987).

The occurrence of Type II AG chains is not, however, restricted to AGPs (see Table 1). Thus, Type II AG chains are also found: (1) on P/HRGPs [e.g., GaRSGP (Sommer-Knudsen *et al* 1996) and the 120-kDa glycoprotein (Lind *et al* 1994)], (2) covalently linked to pectins (see Yamada 1994, Zhang *et al* 1996 and references therein), and (3) as polysaccharide chains without covalent linkage to proteins in the secretion from larch (*Larix*) (Prescott *et al* 1995, Ponder and Richards 1997). Type II AG chains are not present on either extensins or solanaceous lectins (Kieliszewski and Lamport 1994). In addition to being present on some AGPs, short arabino-oligosaccharides are also found on extensins, P/HRGPs and the solanaceous lectins (Table 1). Both Type II AG chains and short arabino-oligosaccharides are present on P/HRGPs. Some AGPs also contain single Gal residues linked by an alkali-labile linkage to Ser (Fincher *et al* 1983), but this linkage is also present on P/HRGPs, extensins and solanaceous lectins (Sommer-Knudsen *et al* 1998). No direct evidence supports the presence of N-linked glycans on AGPs; the predicted consensus sequence (Asn-Xaa-Ser/Thr, where Xaa is any amino acid except Pro) for N-glycosylation is usually absent, and GlcNAc is rarely detected in monosaccharide analyses of AGPs (Table 1).

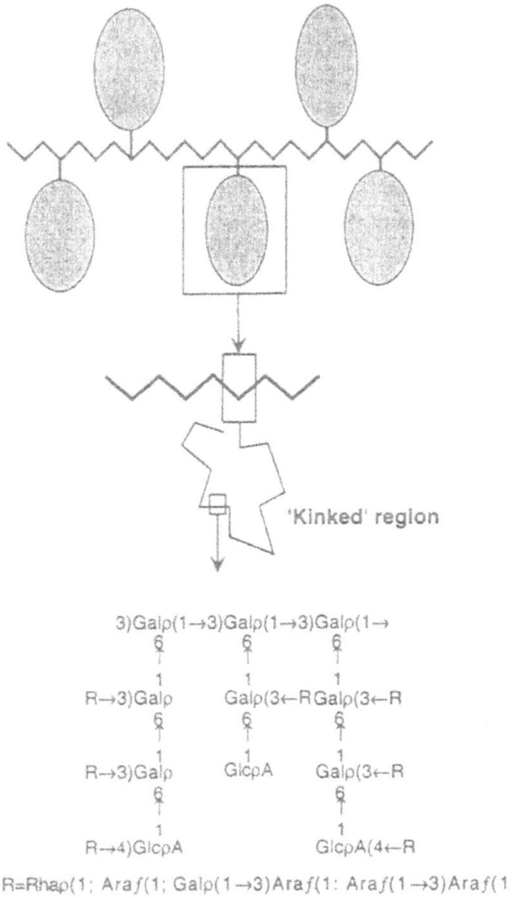

Figure 1. The "wattle-blossom" model of the structure of arabinogalactan-proteins (AGPs). In this model, there are 5 Hyp residues in an AGP, each of which bears an arabinogalactan chain. Each arabinogalactan chain contains 10 or more repeats of a β-(1–3)-linked galactose (Gal) oligosaccharide with a degree of polymerization (d.p.) of 12. The AGP as a whole is spheroidal. GlcA, glucuronic acid; Rha, rhamnose; Gal, galactose; Ara, arabinose; *p*, pyranose; *f*, furanose. Reproduced with permission from the *Annual Review of Plant Physiology*, Vol. 34, © 1983, by Annual Reviews, www.annualreviews.org

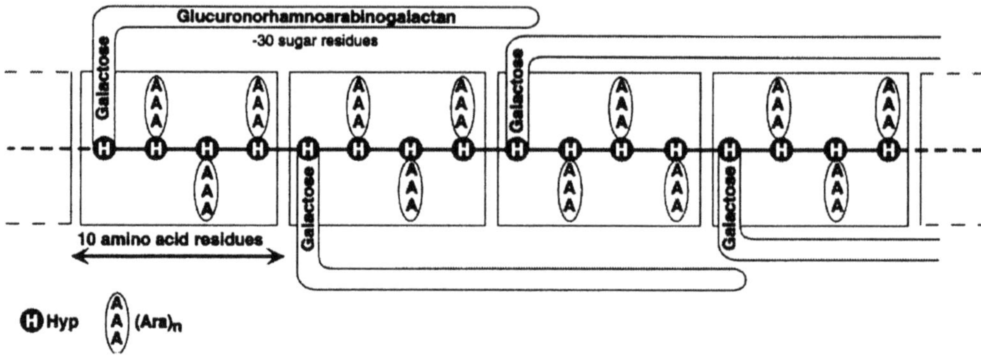

Figure 2. The "twisted hairy rope" model of the structure of arabinogalactan-proteins (AGPs). In this model, a hypothetical block size of 7 kDa contains 10 amino acid residues (1 kDa), 30 sugar residues (4.4 kDa) and 3 Hyp-triarabinosides (1.32 kDa). The glucuronorhamnoarabinogalactan probably has a galactan backbone with glucuronic acid (GlcA), rhamnose (Rha) and arabinose (Ara) side chains. Reproduced with permission from Qi, W., Fong, C., and Lamport, D. T. A., 1991, *Plant Physiol.* **96**: 848–855.

Table 1. The occurrence of O- and N-glycans on the family of HRGPs

	Linkage	AGPs	P/HRGPs	Extensins
O-glycans				
AG (Type II)	Hyp	+	+	–
$Ara_{(n=1\text{-}4)}$	Hyp	+	+	+
Gal	Ser/Thr	–(?)	+	+
N-glycans	Asn	–(?)	+	–

+, present –, absent; –(?) insufficient data to draw conclusion

Antibodies raised to AGPs have been used to study the spatial and temporal expression of AGPs and in some instances to infer function (for a review see Knox 1995). The antibodies raised to AGPs are directed to glycan epitopes, perhaps unsurprisingly, as they are so heavily glycosylated. Since the glycan classes on AGPs are also present on other families of glycoproteins and polysaccharides (see Table 1) the antibody data should be interpreted cautiously, as there is a real possibility that antibodies raised to AGPs will react with glycan chains on otherwise unrelated molecules. For example, J539 (Glaudmans 1975; Table 2) and MAC 207 (Knox 1995) bind a wide range of glycoconjugates in a range of tissue types. In some instances, however, a carbohydrate epitope on a Type II AG can be expressed in a tissue-specific but not class-specific manner. For example, the monoclonal antibody PCBC4 will detect AGPs and P/HRGPs (GaRSGP and 120 kDa) but only from stylar tissue of *Nicotiana alata* and not AGPs from suspension culture or gum arabic (see Table 2, Sommer-Knudsen 1995). The presence of tissue-specific oligosaccharide sequences on a particular family of proteoglycans is well documented in other eukaryotes.

Table 2. Examples of differing specificities of antibodies raised to glycan epitopes

		AGPs			P/HRGPs	
Antibody	Specificity	RT25/ RT35	Suspension culture	Gum arabic	GaRSGP	120 kDa GP
J539[a]	Galβ1-6Gal	+	+	+	+	+
PCBC4[b]	[CHO]	+	–	–	+	+
Anti-KLH[b]	[CHO]	–	–	?	+	+

+, binding; –, no binding; ?, not tested
[a]Glaudmans, 1975
[b]Sommer-Knudsen, 1995

3. THE PROTEIN COMPONENT OF AGPS

Most AGPs differ from the extensins and the P/HRGPs by having a neutral to acidic backbone and a protein content typically between 1 and 10% (w/w). The protein backbones of AGPs are normally rich in Hyp/Pro, Ser, Thr and Ala. The cloning of genes encoding the protein backbone of AGPs has revealed that they belong to a family of molecules that can be grouped into two broad classes designated "classical" and "non-classical" (Fig 3). Little sequence identity occurs at either the nucleotide or amino acid level within or between these two classes of backbones, and as a consequence, comparison to the P/HRGPs is not informative as a means of discriminating between the two classes of HRGPs. It is, however, possible to identify closely related ("homologues") genes in other plant species.

Nevertheless, analysis of the cDNA clones does reveal some shared domains/ regions. Both classes of cDNA clones have:

- An N-terminal hydrophobic secretion signal sequence, as expected for secreted molecules.
- A central Pro-rich domain/region, which contains most of the Hyp/Pro, Ala, Ser, Thr residues, suggesting that it is the likely region of O-glycosylation. In some classical AGPs (e.g., *Le*AGP1, Pogson and Davies 1995, Li and Showalter 1996) this is interrupted by a basic region of unknown function.
- A variable C-terminal domain ranging from a predicted hydrophobic GPI-anchor signal sequence (Fig 4) in the classical cDNAs to a hydrophilic (Asn-rich) domain for some of the non-classical AGPs (Fig 3).
- Generally no consensus sites (Asn-X-Ser/Thr) for N-glycosylation in the mature proteins.

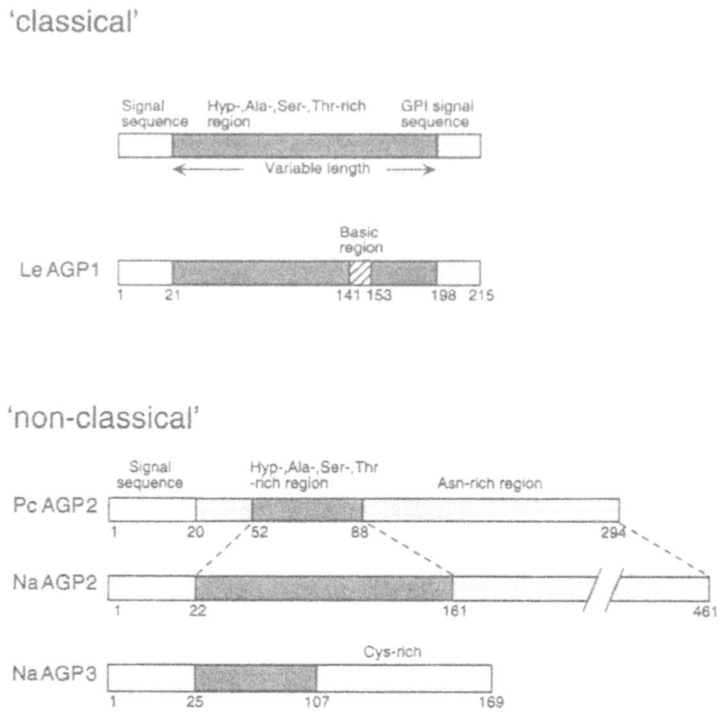

Figure 3. Schematic representation of the domain structure of "classical" and "non-classical" cDNA clones of AGPs. The clones were isolated from styles of *Nicotiana alata* (*Na*AGP3; Du *et al* 1994 1996b) and suspension-cultured cells of *Pyrus communis* (*Pc*AGP2; Mau *et al* 1995), *N. alata* (*Na*AGP2; Mau *et al* 1995) and *Lycopersicon* (*Le*AGP1; Pogson and Davies 1995, Li and Showalter 1996).

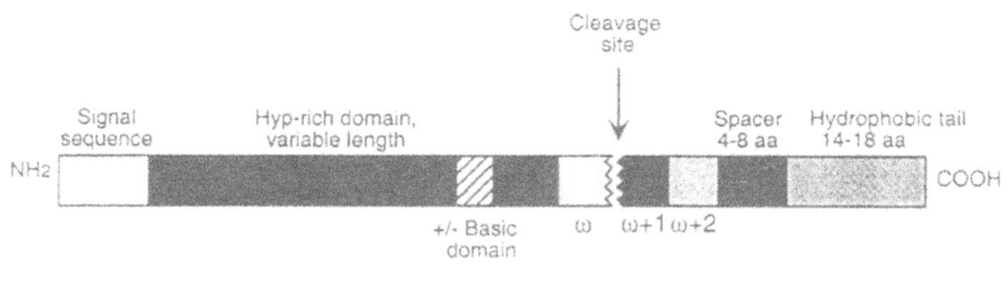

Figure 4. Classical AGP protein backbones include a GPI-anchor signal sequence. The cDNA clones encoding the protein backbones of AGPs display the following domain structure: an N-terminal signal sequence, a domain rich in Pro/Hyp, Ala, Ser and Thr of variable length, sometimes interrupted by a short basic region (e.g., *Le*AGP1), followed by a C-terminal GPI-anchor signal sequence. The C-terminal signal contains small aliphatic amino acids at the ω and ω+2 sites, followed by a short spacer, and terminates in a stretch of hydrophobic amino acids. Some plant, yeast and mammalian proteins possess a basic residue in the linker region between the ω+2 residue and the hydrophobic domain. The cleavage site for the addition of the GPI-anchor is indicated by an arrow. Diagram not drawn to scale. Reproduced from *Trends in Plant Sciences,* Vol. 3, Schultz, C., Gilson, P., Oxley, D., Youl, J., Bacic, A., GPI-anchors on arabinogalactan-proteins: implications for signalling in plants, pp. 426–431, 1998, with permission from Elsevier Science.

When cDNAs encoding the protein backbones of classical AGPs, *Na*AGP1 and *Pc*AGP1, were first isolated and sequenced, the C-terminal hydrophobic regions were predicted to form transmembrane helices (Du *et al* 1994, Chen *et al* 1994). However, the AGPs purified from the styles of *Nicotiana alata* (*Na*AGP1) and cell suspension cultures of *Pyrus communis* (*Pc*AGP1) were present in buffer-soluble extracts rather than membrane fractions. These observations prompted a thorough analysis to determine whether the C-terminal domains of *Na*AGP1 and *Pc*AGP1 were removed by post-translational processing. Mass-spectrometry of the deglycosylated protein backbones and C-terminal peptides showed that the hydrophobic C-terminal domains were absent from these AGPs (Youl *et al* 1998). Furthermore, the C-terminal amino acid carried an amide-linked ethanolamine residue. This feature is characteristic of a GPI-anchor, and this was confirmed by the detection of low levels of inositol, mannose and glucosamine, diagnostic components of GPI-anchors (Hooper 1997), in the native AGP (Youl *et al* 1998). Further evidence for the presence of GPI-anchors on AGPs comes from work on a GPI-anchored AGP from *Arabidopsis* callus culture (Sherrier *et al* 1999) and from a GPI-anchored AGP identified in the plasma membrane of *Rosa* cells in suspension cultures (Svetek *et al* 1999).

The GPI-anchor attachment site (designated ω) of approximately 20 animal and protozoan proteins has been determined (Udenfriend and Kodukula 1995). From these studies the following characteristics were found in the primary translation products of the GPI-anchored proteins. The GPI-anchored proteins include an N-terminal secretion signal and a C-terminal GPI-anchor signal sequence. The C-terminal signal contains small aliphatic amino acids at the ω and ω+2 sites, followed by a short spacer containing one or more basic amino acids, and the sequence terminates in a hydrophobic stretch of amino acids (Fig 4). A survey of the classical AGP protein backbones predicted from the domain structures indicates that a putative GPI-anchor signal sequence is present on all the known classical AGPs from *Nicotiana, Pyrus, Brassica, Pinus, Lycopersicon* and five putative classical AGPs from *Arabidopsis* (Schultz *et al* 1998). The non-classical AGPs do not possess the GPI-anchor signal.

A further level of complexity is observed within the non-classical cDNA clones. One member of this group, represented by *Na*AGP3 (Du *et al* 1996b) has an "atypical" amino acid composition for both the deduced protein and the isolated AGP protein, whereas the other members, represented by *Na*AGP2/*Pc*AGP2, have "atypical" amino acid compositions for the deduced protein, but the isolated AGP proteins have a "typical/classical" amino acid composition. In this latter group, it is suggested that the C-terminal Asn-rich domain of these protein backbones may be processed post-translationally by proteases, either intracellularly (in the vacuole) or extracellularly (Mau *et al* 1995). Extracellular processing, particularly of components of the extracellular matrix, is known for animal cells (Blei *et al* 1993). Some Asn/X junctions are common sites for proteolytic cleavage of plant proteins (Scott *et al* 1992), and cleavage at Asn residues in the domains flanking the Pro-rich domain would release a fragment that could be trimmed by exopeptidases. From these studies, it is clearly not possible to predict that a randomly obtained cDNA clone encodes a non-classical AGP without reference to corresponding protein sequence data and knowledge of the purification protocol.

Common motifs are also used to differentiate various classes of HRGPs (Showalter 1993, Kieliszewski and Lamport 1994, Sommer-Knudsen *et al* 1998). Thus, the extensins contain the motif Ser-$(Pro)_4$, and the P/HRGPs are characterized by either the Pro-Pro-Xaa-Yaa-Lys or Pro-Pro-Xaa-Lys motifs. The Ala-Pro, Pro-Ala sequences have been suggested to be motifs characteristic of classical AGPs (Showalter 1993). However, whether these "motifs" are non-random and/or significant remains to be resolved. The gene sequences from both classes of AGPs appear to lack motifs typically associated with proteins involved in protein–protein and protein–carbohydrate interactions and/or signal transduction pathways (Walker 1994, Iozzo and Murdoch 1996).

4. BINDING OF AGPS TO β-GLUCOSYL YARIV REAGENT

Yariv *et al* (1962) prepared a series of artificial antigens for detecting antibodies to carbohydrates by coupling diazotized 4-amino-phenyl glycosides with phloroglucinol. These reagents were later observed to precipitate AGPs in a range of plant extracts (Yariv *et al* 1967, Jermyn and Yeow 1975). Subsequently the β-glucosyl Yariv reagent (β-Glc Yariv reagent) has been widely used in the purification of AGPs (Gane *et al* 1995, Mau *et al* 1995) and, through its strong absorption in the visible spectrum (λ max 430 nm), in their quantification (Van Holst and Clarke 1985), detection in gels (Van Holst and Clarke 1986), and visualization in plant sections by light microscopy (Schopfer 1990). Notwithstanding the utility of the β-Glc Yariv reagent in these applications, the molecular basis for its specific interaction with certain AGPs is not understood.

The interaction depends both on the state of the Yariv reagent in solution and the chemical structure and organization of the AGP molecule. The reaction is usually performed in a 1% NaCl solution, and the precipitated complex can be dissociated by adding dimethylformamide or by adding saturated sodium dithionite. The saccharide moiety of the Yariv reagent is implicated in the binding reaction, since the glycopyranose residue must be in the β-D or α-L- configuration. In addition, the OH group at C(O)2 must be in the D-gluco configuration, and the diazo-group, which substitutes the phenylglycoside, must be at the C4 position of the phenyl ring (Jermyn and Yeow 1975, Jermyn 1978).

However, not all AGPs bind the β-Glc Yariv reagent. Arabinogalactans from larch (*Larix*) and the Hyp-containing AG-peptide from *Lolium multiflorum* (Anderson *et al* 1977) do not bind, but the AGP from carrot (*Daucus carota*), which lacks Hyp, binds the reagent (Baldwin *et al* 1993). Extensins and P/HRGPs do not bind the β-Glc Yariv reagent.

5. PARALLELS WITH ANIMAL PROTEOGLYCANS

The animal ECM and cell surface proteoglycans have been studied intensively and offer well-developed concepts for consideration of AGP structure and function. Like AGPs, the animal proteoglycans were originally thought to function only as gel-forming structural components of the ECM because of their dominant carbohydrate content. Subsequently, the animal proteoglycans were shown to have the capacity to interact with other proteins and to regulate the action of growth factors. In this way, some are able to regulate cell growth, others promote differentiation and neurite outgrowth, and others act as biological barriers and repellents (for a review see Iozzo and Murdoch 1996). The two major classes of animal proteoglycans are those with glycosaminoglycan chains (Hardingham and Fosang 1992) and the mucins (Verma and Davidson 1994). One common feature of the proteoglycans is the occurrence of structural/functional domains typically belonging to distinct protein families (see Fig 5A). These domains separate from the Ser/Thr-Gly repeats, which form the glycosylation domains. Such domains have yet to be defined on AGPs (Fig 5B). At present, the discovery of a GPI-anchor on classical AGPs provides the first clue of a potential mechanism by which AGPs may be involved in signal transduction pathways.

6. CONCLUSIONS

The last decade has seen a plethora of information accumulate on AGPs. We now know they are a family of molecules with different protein backbones, each existing as multiple glycoforms. Classical genetic approaches, including the isolation of mutants of *Arabidopsis*, offer a powerful new approach and will be important in ultimately determining the function of AGPs. The sequencing of the Type II AG chains, which largely define the multiple glycoforms, remains a significant challenge into the future. Our knowledge of the biosynthetic machinery responsible for the assembly of these glycan chains is also rudimentary. However, by drawing upon recent work on glycosyltransferases in animals and other eukaryotes, it should now be possible to identify the plant genes for the glycosyltransferases and thereby begin to unravel the mechanisms of tissue-specific glycan epitopes.

ACKNOWLEDGEMENTS

Our work is funded by a Special Research Centre grant from the Australian Research Council and a Cooperative Research Centre grant to the CRC for Industrial Plant Biopolymers from the Australian Government.

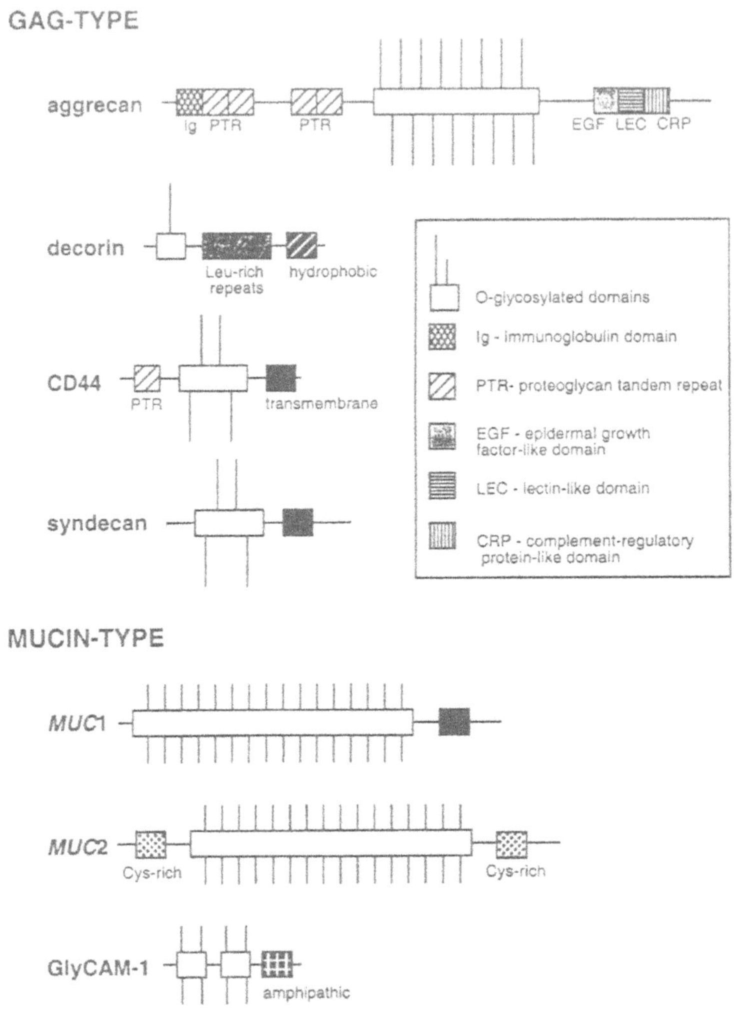

Figure 5. A. Schematic representation of animal proteoglycans [GAG (glycosaminoglycan)-type and mucin-type] showing different structural/functional protein domains.

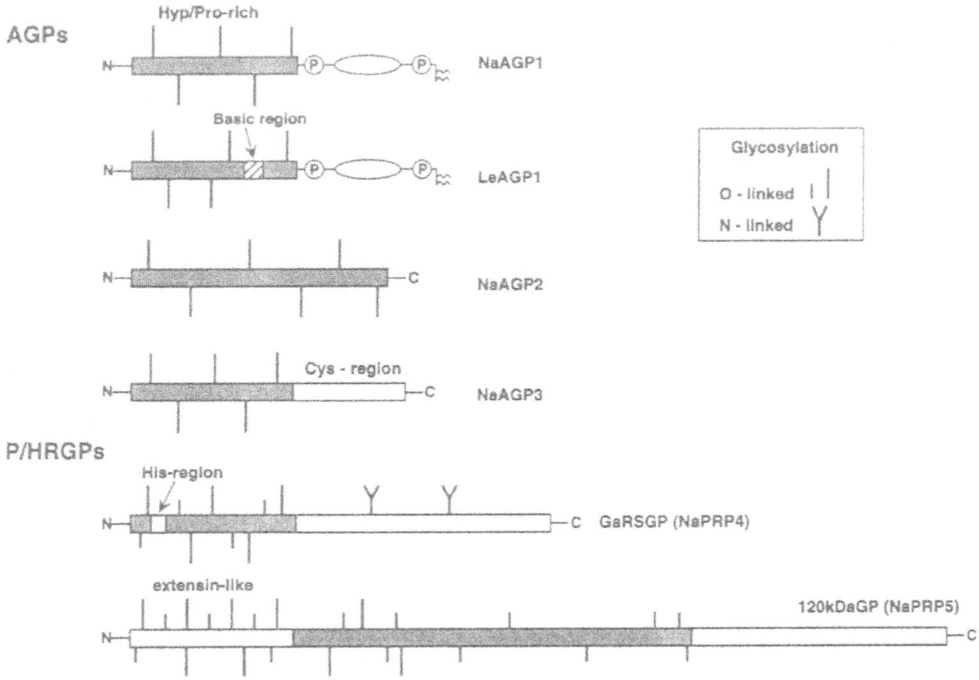

Figure 5. (continued) B. Schematic representation of representative plant AGPs and P/HRGPs showing different regions of the mature proteoglycans. The models are based on the structures predicated from Du *et al* (1994) and Youl *et al* (1998) for *Na*AGP1, Pogson and Davies (1995) and Li and Showalter (1996) for *Le*AGP1, Mau *et al* (1995) for *Na*AGP2, Du *et al* (1996b) for *Na*AGP3, Chen *et al* (1993) and Sommer-Knudsen *et al* (1996b) for *Na*PRP4 and Schultz *et al* (1997) for *Na*PRP5.

REFERENCES

Anderson, R. L., Clarke, A. E., Jermyn, M. A., Knox, R. B., and Stone, B. A., 1977, A carbohydrate-binding arabinogalactan-protein from liquid suspension cultures of endosperm from *Lolium multiflorum, Aust. J. Plant Physiol.* **4**: 143–158.

Bacic, A., Churms, S. C., Stephen, A. M., Cohen, P. B., and Fincher, G. B., 1987, Fine structure of the arabinogalactan protein from *Lolium multiflorum, Carbohydr. Res.* **162**: 85–93.

Baldwin, T. C., McCann, M. C., and Roberts, K., 1993, A novel hydroxyproline-deficient arabinogalactan-protein secreted by suspension cultured cells of *Daucus carota, Plant Physiol.* **103**: 115–123.

Blei, F., Wilson, E. L., Mignatti, P., Rifkin, D. P., 1993, Mechanism of action of angiostatic steroids: suppression of plasminogen activator activity via stimulation of plasminogen inhibitor, *J. Cell Physiol.* **155**: 568–578.

Chen, C-G., Mau, S-L., and Clarke, A. E., 1993, Nucleotide sequence and style-specific expression of a novel proline-rich protein gene from *Nicotiana alata*, *Plant Mol. Biol.* **21**: 391–395.

Chen, C-G., Pu, Z-Y., Moritz, R. L., Simpson, R. J., Bacic, A., Clarke, A. E., and Mau, S-L., 1994, Molecular cloning of a gene encoding an arabinogalactan-protein from pear (*Pyrus communis*) cell suspension culture, *Proc. Natl. Acad. Sci. USA* **91**: 10305–10309.

Churms, S. C., Stephen, A. M., and Siddiqui, I. R., 1981, Evidence for repeating sub-units in the molecular structure of the acidic arabinogalactan from rapeseed (*Brassica campestris*), *Carbohydr. Res.* **94**: 119–122.

Clarke, A. E., Anderson, R. L., and Stone, B. A., 1979, Form and function of arabinogalactans and arabinogalactan-proteins, *Phytochemistry* **18**: 521–540.

Du, H., Simpson, R. J., Moritz, R. L., Clarke, A. E., and Bacic, A., 1994, Isolation of the protein backbone of an arabinogalactan-protein from the styles of *Nicotiana alata* and characterization of a corresponding cDNA, *Plant Cell* **6**: 1643–1653.

Du, H., Clarke, A. E., and Bacic, A., 1996a, Arabinogalactan-proteins: a class of extracellular matrix proteoglycans involved in plant growth and development, *Trends Cell Biol.* **6:** 411-414.

Du, H., Simpson, R. J., Clarke, A. E., and Bacic, A., 1996b, Molecular characterization of a stigma-specific gene encoding an arabinogalactan-protein (AGP) from *Nicotiana alata*, *Plant J.* **9**: 313–323.

Fincher, G. B., Sawyer, W. H., and Stone, B. A., 1974, Chemical and physical properties of an arabinogalactan-peptide from wheat endosperm, *Biochem. J.* **139**: 535–545.

Fincher, G. B., Stone, B. A., and Clarke, A. E., 1983, Arabinogalactan-proteins: structure, biosynthesis and function, *Annu. Rev. Plant Physiol.* **34:** 47–70.

Gane, A. M., Craik, D., Munro, S. L. A., Howlett, G. J., Clarke, A. E., and Bacic, A. 1995, Structural analysis of the carbohydrate moiety of arabinogalactan-proteins from stigmas and styles of *Nicotiana alata*, *Carbohyd. Res.* **277**: 67–85.

Glaudmans, C. P. J., 1975, The interaction of homogenous murine myeloma immunoglobulins with polysaccharide antigens, *Adv. Carbohydr. Chem. Biochem.* **31**: 313–346.

Hardingham, T. E., and Fosang, A. J., 1992, Proteoglycans: many forms and functions, *FASEB J.* **6**: 861–870.

Hooper, N. M., 1997, Glycosyl-phosphatidylinositol anchored membrane enzymes, *Clin. Chim. Acta* **266**: 3–12.

Iozzo, R. V., and Murdoch, A. D., 1996, Proteoglycans of the extracellular environment: clues from the gene and protein side offer novel perspectives in molecular diversity and function, *FASEB J.* **10**: 598–614.

Jermyn, M. A., 1978, Comparative specificity of Concanavalin A and the β-lectins, *Aust. J. Plant Physiol.* **5**: 687–696.

Jermyn, M. A., and Yeow, Y. M., 1975, A class of lectins present in the tissues of seed plants, *Aust. J. Plant Physiol.* **2**: 501–531.

Kieliszewski, M. J., and Lamport, D. T. A., 1994, Extensin: repetitive motifs, functional sites, post-translational codes and phylogeny, *Plant J.* **5**: 157–172.

Knox, J. P., 1995, Developmentally regulated proteoglycans and glycoproteins of the plant cell surface, *FASEB J.* **9**: 1004–1012.

Li, S.-X., and Showalter, A. M., 1996, Cloning and developmental stress-regulated expression of a gene encoding a tomato arabinogalactan-protein, *Plant Mol. Biol.* **32**: 641–652.

Lind, J. L., Bacic, A., Clarke, A. E., and Anderson, M. A, 1994, A style-specific hydroxyproline-rich glycoprotein with properties of both extensins and arabinogalactan proteins, *Plant J.* **6**: 491–502.

Mau, S-L., Chen, C-G., Pu, Z-Y., Moritz, R. L., Simpson, R. J., Bacic, A., and Clarke, A. E., 1995, Molecular cloning of cDNAs encoding the protein backbones of arabinogalactan-proteins from the filtrate of suspension-cultured cells of *Pyrus communis* and *Nicotiana alata*, *Plant J.* **8**: 269–281.

Nothnagel, E. A., 1997, Proteoglycans and related components in plant cells, *Int. Rev. Cytol.* **174**: 195–291.

Pogson, B. J., and Davies, C., 1995, Characterisation of a cDNA clone encoding the protein motif of a putative arabinogalactan-protein from *Lycopersicon peruvianum*, *Plant Mol. Biol.* **28**: 347–352.

Ponder, G. R., and Richards, G. N., 1997, Arabinogalactan from Western larch, Part III: alkaline degradation revisited, with novel conclusions on molecular structure, *Carbohydr. Polymers* **34**: 251–261.

Prescott, J. H., Enriquez, P., Jung, C., Menz, E., and Groman E. V., 1995, Larch arabinogalactan for hepatic drug delivery: isolation and characterisation of a 9 kDa arabinogalactan fragment, *Carbohydr. Res.* **278**: 113–128.

Qi, W., Fong, C., and Lamport, D. T. A., 1991, Gum arabic glycoprotein is a twisted, hairy rope, *Plant Physiol.* **96**: 848–855.

Schopfer, P., 1990, Cytochemical identification of arabinogalactan-protein in the outer epidermal wall of maize coleoptiles, *Planta* **183**: 139–142.

Schultz, C., Hauser, K., Lind, J. L., Atkinson, A. H., Pu, Z-Y., Anderson, M. A., and Clarke, M. A., 1997, Molecular characterisation of a cDNA sequence encoding the backbone of a style-specific 120kDa glycoprotein which has features of both extensins and arabinogalactan proteins, *Plant Mol. Biol.* **35**: 833–845.

Schultz, C., Gilson, P., Oxley, D., Youl, J., and Bacic, A., 1998, GPI-anchors on arabinogalactan-proteins: implications for signalling in plants, *Trends Plant Sci.* **3**: 426–431.

Scott, M. P., Jung, R., Muntz, K., and Neilson, N. C., 1992, A protease responsible for post-translational cleavage of a conserved Asn-Gly linkage in glycinin, the major seed protein of soybean, *Proc. Natl. Acad. Sci. U.S.A.* **89**: 658–662.

Sherrier, D. J., Prime, T. A., and Dupree, P., 1999, Glycosylphosphatidylinositol-anchored cell surface proteins from Arabidopsis, *Electrophoresis* **20:** 2027-2035.

Showalter, A. M., 1993, Structure and function of plant cell wall proteins, *Plant Cell* **5**: 9–23.

Sommer-Knudsen, J., 1995, A proline and hydroxyproline-rich cell wall glycoprotein from styles of *Nicotiana alata*, Ph.D. Thesis, University of Melbourne, Australia.

Sommer-Knudsen, J., Clarke, A. E., and Bacic, A., 1996, A galactose-rich, cell-wall glycoprotein from styles of *Nicotiana alata*, *Plant J.* **9**: 71–83.

Sommer-Knudsen, J., Clarke, A. E., and Bacic, A., 1997, Proline- and hydroxyproline-rich gene products in the sexual tissues of flowers, *Sex. Plant Reprod.* **10**: 253–260.

Sommer-Knudsen, J., Bacic, A., and Clarke, A. E., 1998, Hydroxyproline-rich glycoproteins, *Phytochemistry* **47**: 483–497.

Svetek, J., Yadav, M. P., and Nothnagel, E. A., 1999, Presence of a glycosylphosphatidylinositol lipid anchor on rose arabinogalactan proteins, *J. Biol. Chem.* **174:** 14724-14733.

Udenfriend, S., and Kodukula, K., 1995, How glycosyl-phosphatidylinositol-anchored membrane proteins are made, *Annu. Rev. Biochem.* **64**: 563–591.

Van Holst, G.-J., and Clarke, A. E., 1985, Quantification of arabinogalactan-protein in plant extracts by single radial gel diffusion, *Anal. Biochem.* **148**: 446–450.

Van Holst, G.-J., and Clarke, A. E., 1986, Organ-specific arabinogalactan proteins of *Lycopersicon peruvianum* (Mill) demonstrated by crossed-electrophoresis, *Plant Physiol.* **80**: 786–789.

Verma, M., and Davidson, E. A., 1994, Mucin genes: structure, expression and regulation, *Glycoconjugate J.* **11**: 172–179.

Walker, J. C., 1994, Structure and function of the receptor-like protein kinases of higher plants, *Plant Mol. Biol.* **26**: 1599–1609.

Yamada, H., 1994, Pectic polysaccharides from Chinese herbs: structure and biological activity, *Carbohydr. Polymers* **251**: 269–276.

Yariv, J., Rapport, M. M., and Graf, L., 1962, The interaction of glycosides and saccharide with antibody to the corresponding phenylazo glycosides, *Biochem. J.* **85**: 383–388.

Yariv, J., Lis, H., and Katchalski, E., 1967, Precipitation of arabic acid and some seed polysaccharides by glycosyl-phenylazo dyes, *Biochem. J.* **105**: 1C–2C.

Youl, J., Bacic, A., and Oxley, D., 1998, Arabinogalactan-proteins from *Nicotiana alata* and *Pyrus communis* contain glycosylphosphatidylinositol membrane anchors, *Proc. Natl. Acad. Sci. USA* **95**: 7921–7926.

Zhang, Y., Kiyohara, H., Sakurai, M., and Yamada, H., 1996, Complement activating galactan chains in a pectic arabinogalactan (AGIIb-1) from the roots of *Angelica acutiloba* Kitagawa, *Carbohydr. Polymers* **31**: 149–156.

Chapter 3

Molecular Analysis of Genes Encoding Arabinogalactan-Proteins

Christophe Reuzeau, [1] Lars Snogerup, and Per Kjellbom
Department of Plant Biochemistry, Lund University, P.O. Box 117, 22100 Lund, Sweden;
[1]Present address: Crop Design France, INRA, Laboratoire de Biologie Cellulaire, Route de Saint Cyr, 78026 Versailles cedex, France

1. INTRODUCTION

Known arabinogalactan-proteins (AGPs) are acidic and hydrophilic glycoproteins associated with the cell wall and with the plasma membrane and contain 65–98% carbohydrate covalently linked to a protein backbone. The AGPs belong to the Hyp-rich glycoprotein (HRGP) family of proteins, which includes extensins, Pro-rich proteins, some nodulins and solanaceous lectins. The AGPs are present in all tissues and are likely components of cell–cell signaling pathways. The protein moiety is typically rich in Hyp, Pro, Ser, Thr and Ala residues (reviewed in Du *et al* 1996a, Nothnagel 1997). Most of the known AGP backbones are O-glycosylated. The polysaccharide is composed of mostly galactose and arabinose and consists of a backbone of (1-3)-linked β-D-Gal residues substituted at C(O)6 by short chains of (1-6)-linked β-D-Gal (Nothnagel 1997). It has been proposed that the HRGP family is a continuum of macromolecules substituted by mono-, oligo-, or polysaccharides (Du *et al* 1996a). The composition and the structure of recently characterized proteins (e.g., so-called non-classical AGPs and chimeric molecules) rather suggest that AGPs are distinct from the other HRGPs. The AGPs are defined by a high percentage of galactose and arabinose in the carbohydrate, a central Pro-rich domain in the protein backbone, and the ability to bind the β-glucosyl Yariv reagent (Yariv *et al* 1962). Classical AGPs have a three domain structure: a signal sequence, a central domain and a hydrophobic C-terminal domain. Most of the known AGPs are secreted into the extracellular matrix. The AGPs are also known to be associated with the plasma membrane and are expected

Cell and Developmental Biology of Arabinogalactan-Proteins
Edited by Nothnagel *et al.*, Kluwer Academic/Plenum Publishers, 2000

to be heterogeneous, since during Triton X-114 fractionation (Bordier 1981) they partition either to the detergent phase (Smallwood *et al* 1996) or to the water phase (Norman *et al* 1990, Pennell *et al* 1991, Stöhr *et al* 1996, Kjellbom *et al* 1997). Serpe and Nothnagel (1996) also obtained heterogeneous AGP fractions from plasma membranes of cultured rose cells. On the basis of the cDNA sequences of secreted AGPs of pear and tobacco cell cultures, Du *et al* (1996a) distinguished between "classical AGPs," with a predicted transmembrane helix in the C-terminal domain, and "non-classical" AGPs, with a hydrophilic Asn-rich domain in the C-terminal part of the molecule. This part of the protein is predicted to be cleaved and not to be present in the mature protein (Mau *et al* 1995).

The AGPs may act as glue, lubricant's and humectants (Showalter and Varner 1989, Nothnagel 1997), but nevertheless the exact function of the different AGPs is still unknown. The AGPs are involved in somatic embryogenesis (Kreuger and van Holst 1996), and studies using monoclonal antibodies show developmental regulation and transient cell position-specific expression of AGP carbohydrate epitopes (Pennell and Roberts 1990, Knox 1995). Thus it has been proposed that AGPs have major roles in cell–cell interactions and signaling (Knox 1995). The AGPs have also been implicated in stress responses (Pennell *et al* 1995, Kjellbom *et al* 1997). Biochemical and molecular data concerning AGPs, in comparison with that of structural HRGPs, are limited and derive from experiments carried out on a few plant species. The difficulties in understanding AGP function also stem from the fact that we do not know whether it is the protein or the carbohydrate part that is most important for the biological function. However it is generally considered that the carbohydrate possesses the signaling properties and that the protein backbone is merely a vehicle for delivering the carbohydrate at the right place. It has been recently shown that the same carbohydrate antigenic domain can be found on different protein backbones, indicating that AGP epitopes are involved in diverse signaling pathways.

To understand the relation between plasma membrane and secreted AGPs we are attempting to clone genes coding for the protein backbones of plasma membrane AGPs. To do so we have analyzed the molecular information available in the literature and in the sequence databases to discern specific regions of interest in the proteins that would allow the design of specific primers useful for cloning. In this chapter we are proposing a classification of AGPs based on amino acid sequence similarities. This classification will have to be refined when new sequences are discovered. The present study is based on sequence information and relates to putative biological functions of AGPs in the fields of glycoproteins, membranes, and development. In the first two sections we compare and analyze the sequences of known AGPs, translated expressed sequenced tags (ESTs) and related molecules. In the subsequent sections we correlate features of AGP sequences with their putative function.

2. ANALYSIS OF AGP SEQUENCES AND ESTS

Secreted AGPs from suspension-cultured cells of tobacco and pear have been purified and partially sequenced, and the corresponding cDNAs have been isolated (Chen *et al* 1994, Du *et al* 1994 1996b, Mau *et al* 1995, Schultz *et al* 1997 1998). The cDNA sequences did not show significant similarities at the amino acid or the nucleotide level, and there was no clear homology with other proteins. The AGPs were classified as "classical" or "non-classical" according to their overall structure (Mau *et al* 1995, Du *et al* 1996a). Table 1 lists all known AGPs and shows the common amino acid motifs. The alignment of all AGP sequences and the close examination (BLAST, Altschul *et al* 1990) of all sequences corresponding to identified AGPs and putative AGPs translated from cDNAs indicated that AGPs could be divided into at least four groups, each characterized by short conserved amino acid signature motifs (Table 2). A first group of sequences is characterized by the motif PAPSPA, whereas the motif QG-SDTR-LENGKYY is found in a second AGP group. A third AGP group, representing the majority of known AGPs, contains the motif SPPAP or SPPA. Most of these AGPs also contain the motif PAPAP described by Nothnagel (1997). An AGP from tobacco contains different motifs, shows weak homologies (SP, PP dipeptide repeats) to extensins and other Pro-rich proteins (BLAST) and may thus represent another group.

One can ask whether the presence of a conserved amino acid motif is related to a specific function or to a common ancestral origin, or to both parameters. An amino acid composition reflecting a protein rich in A, P and S, or motifs rich in A, P, S residues proposed to be characteristic for AGPs, may be found in other proteins as well (Table 2). A protein of unknown function, TED3 (Demura and Fukuda 1994), contains several motifs present in AGPs from *Nicotiana alata* and *Pyrus communis* (Table 2). Within its kinase domain a lectin-like receptor kinase (Hervé *et al* 1996) contains the RQGMS-FL motif present in group 2 AGPs. Sequences of known AGPs such as *Pinus taeda* AGP (U09556) also show similarities at the amino acid level with plant proteins such as a maize endosperm-specific protein (Carlson and Chourey 1997), an *Arabidopsis* cell surface protein (AC002387), a *Medicago truncatula* cell surface protein (U28149), a *Gossypium* Pro-rich protein (John and Keller 1995), and the algal cell adhesion molecule (CAM/chimeric extensin; Huber and Sumper 1994).

A closer examination of all of these sequences shows that the corresponding proteins might be included in one of the four major AGP groups (Table 2). A number of proteins with only partially known amino acid sequences show similarities with AGPs (Table 1; Kieliszewski *et al* 1992). Within the group 3 (Table 2), some polypeptides show similarities with AGPs in some portions of the sequence and with extensins and Pro-rich proteins in different parts of the sequence. If AGPs are considered to be a sub-group of HRGPs, it is not surprising to find proteins containing motifs characteristic for AGPs and a few motifs characteristic for extensins. This corroborates the suggestion by Kieliszewski and Lamport (1994) that the HRGPs are defined between the extremes corresponding to repeated motifs of SPPPX and less repeated motifs as in the AGPs.

Table 1. Conserved amino acid signature motifs in known AGPs. Sequences were aligned using Geneworks (Intelligenetics, Inc, Mountain View CA) and ATLAS (National Biomedical Research Foundation, Washington, DC).

Accession	Plant	Motif	Origin	Reference
–	*Daucus carota*	APAPAP, APAPSPA	protein	Jermyn and Guthrie 1985
U09556	*Pinus taeda*	APAPSPAS,	DNA	Loopstra *et al* 1995
U09555	*P. taeda*	PSPA, PPV, SSPPP		
–	*Rosa hybrida*	APAPSPA	protein	Komalavilas *et al* 1991
S79359	*Nicotiana alata*	G-SDTR-LENGKYY, SPTPAPATAP	cDNA	Mau *et al* 1995
S79358	*Pyrus communis*	G-SDTR-LENGKYY	cDNA	Mau *et al* 1995
AF082298	*Arabidopsis thaliana*	SPAPAP, APAPAP	cDNA	Schultz *et al* 1998
AF082299	*A. thaliana*	APAPAP	cDNA	
AF082300	*A. thaliana*	APAPAP, APAPAP	cDNA	
AF082301	*A. thaliana*	PAPA, PAPA, SPPAP	cDNA	
AF082302	*A. thaliana*	SPPAP	cDNA	
L47351	*Brassica napus*	VSPPAPTP,PSE	cDNA	Gerster *et al* 1996
L47352	*B. napus*		cDNA	
L17308	*Gossypium hirsutum*	SPPAP, PSPAPAPTP, PPTPAPAPA	cDNA	John and Keller 1995
–	*Lolium multiflorum*	SPPAP, PAPAP	Protein	Gleeson *et al* 1989
Z47980	*Lycopersicon esculentum*	MSPPAPPSE, TPAPA, TPAPATAP, PPAPAP	cDNA	Pogson and Davies 1995
X99148	*L. esculentum*	MSPPAPPSE, TPAPA, TPAPATAP, PPAPAP	DNA	Li and Showalter 1996
U14009	*P. communis*	PNSPPADA, ASPP	cDNA	Chen *et al* 1994
U88587	*N. alata*	SPPA, SPPAP	cDNA	Schultz *et al* 1997
U09554	*P. taeda*	TATPP, SSPPP, PPPV, PSPA	DNA	Loopstra *et al* 1995
U25628	*N. alata*	PSE, PPPPSS	cDNA	Du *et al* 1996b

Sequences of well-defined and putative AGPs (Table 1) were compared to collections of anonymous cDNAs (ESTs) of *Arabidopsis*, rice, soybean and maize and analyzed at both the nucleotide and the amino acid levels. About 20 non-redundant cDNA sequences available from *Arabidopsis*, rice, soybean and maize EST databases show similarities to AGP sequences. Also, several EST assemblies (TIGR database, Rockville, MD) indicating the presence of related sequences and sub-groups of AGPs have been detected. The analysis of amino acid sequences of several ESTs shows the presence of signal sequences; regions rich in A, P and S; hydrophobic C-terminal domains; and different conserved motifs. Some of these motifs are found in the amino acid sequences of the known AGPs, whereas other motifs appear to characterize sequences unique to specific ESTs. Again, four groups of AGPs could be defined on the basis of specific amino acid motifs (Table 2).

Alignment of ESTs from each group presented in Table 2 with sequences found in the databases indicates that all ESTs from group 1 show similarities with proteins associated with the plasma membrane, such as members of the CAM superfamily (Gumbiner 1993), the animal fasciclin (Zinn *et al* 1988) and a recently isolated algal CAM (Huber and Sumper 1994), for example. Translated ESTs from group 1 are shorter than known AGP sequences from groups 1, 2 and 3, which may reflect a post-transcriptional processing of the expressed RNAs. Sequences from group 1 are rich in A, P, T residues, contain hydrophobic amino acid stretches and hydrophobic C-terminal domains typically found among AGPs referred to as "classical," indicating putative membrane anchors. A soybean EST (T41457) shows high homology with the *Pinus* AGP within the TAILEK and the GFTLFAPTDNAF motifs found among the group 1 AGPs and also contains a PAPAP signature motif present in group 3 AGPs. It is thus likely that this group contains proteins with dual functions indicated by the presence of two very conserved motifs from two groups. This observation may also indicate the presence of sub-groups or processing mechanisms generating shorter proteins such as in group 3.

Table 2. The AGP groups according to common motifs found in known AGPs, in putative AGPs, and in AGP-like proteins. Results produced from the alignment of AGP amino acid sequences deduced from translation of ESTs and from translation of cloned cDNAs. At: *Arabidopsis thaliana*; Bn: *Brassica napus*; Gh: *Gossypium hirsutum*; Gb: *Gossypium barbadense*; Gm: *Glycine max*; Le: *Lycopersicon esculentum*; Mt: *Medicago truncatula*; Nt: *Nicotiana tabacum*; Na: *Nicotiana alata*; Os: *Oryza sativa*; Pc: *Pyrus communis*; Pt: *Pinus taeda*; Ps: *Pisum sativum*; Rc: *Ricinus communis*; Vc: *Volvox carteri*; Ze: *Zinnia elegans*; Zm: *Zea mays*. Sequences were analyzed using Geneworks (Intelligenetics, Inc, Mountain View, CA) and ATLAS (National Biomedical Research Foundation, Washington, DC). X is any amino acid

Group	Motifs	Accession	Type of protein or sequence	Plant	Similarity with AGP or AGP-like protein
1	APAPSPA, NLSGILDK, EGITIFAPSD, LKPGALN, HALPSYY, SNPVRTMA	U09556	AGP	Pt	-
	NSDNGITIFAP, LKAGTLN, PSYVS, SNPLRTQA	T76135	EST	At	U09556
	NSSDQGMTIFAP, LKPGTLN, PKYYS, SNPVRTQA	T88396	EST	At	U09556
	APAP, NITAILEK, NSSSEGMTVLAP, LKPGTLN,	Z25675	EST	At	U09556
	APAP, NITAILEK, NSSSEGMTVLAP, LKPGTLN,	T21870	EST	At	U09556
	APAP, NITAILEK, GMTVLAPTDNAF, LKPGTLN, SNPVRT	AB005239	Genomic sequence	At	U09556
	APAP, NITAILEK, NSSSEG	T44810	EST	At	U09556
	APAP, PINLTAILEK, NSSSEGMPVFAP, LKPGTLN	T41946	EST	At	U09556
	APAP, NITAILEK, NSSSEGMTVFAP, LKPGTLN,	T04258	EST	At	U09556
	APAP, NITAILEK, NSSSSNGLTVFAP, LKSGTLN	T04708	EST	At	U09556

Table 2 cont.

Group	Motifs	Accession	Type of protein or sequence	Plant	Similarity with AGP or AGP-like protein
	APAP, NIXAILEK, NSSSEGMTVFAQ, LKTPEPLN,	H36033	EST	At	U09556
	APAP, NITAILEA, NSSDQGMTIFAP, LKPGTLN	R64937	EST	At	U09556
	APAP, NITAILEA, NSSDQGMTILAP,	T44426	EST	At	U09556
	NVTAVLEK, GYTVFAXTD, LKPGTXN	D24393	EST	Os	U09556
	NVTAVLEK, GYTVFAPTD, LKPGTST	D48326	EST	Os	U09556
	RDGVLEK, GYTVFAXTD, LKPGTLN	D47867	EST	Os	U09556
	NVTAXLEK, GLTMFAPTD, SKPGTXN	D41853	EST	Os	U09556
	PLNLTEILTK, GLTVLAPTD, LRPGTLN	D48980	EST	Os	U09556
	PAPAP, TAILEK, GFTLFAPTD, LKPGALN	T41457	EST	Gm	U09556
	PAPAP, GITIFVPKDS, HALPRYY, PAPAP	U13760	Proline-rich protein	Gh	-
	SPPAP, PAPAPA, XITNFV,	T21172	EST	At	U13760
	SPPAP, PAPAPA, GITIFVPKDD, HALPHYYS	T41941	EST	At	U13760
	TIFVPKDD, HALPHYYS	T46424	EST	At	U13760
	HALPHYYS	T44632	EST	At	U13760
	APAP, NITGLLEK, GLTVFAPSD, PAPAP, SPPAP, VLLP	AC002387	Putative cell surface protein	At	-
	APAP, TGLLEK, GLTXFAPSD	Z34151	EST	At	AC002387
	GLTVFCPL	T15120	EST	Rc	AC002387
	GLTVFCPTDT,	AA661000	Genomic sequence	Mt	AC002387
	VLLP	H76512	EST	At	AC002387
	GIVNITVFPD, PAPA, GLTVFCPA	U95180	Endosperm specific protein	Zm	-
	GXTLFCPVDAAVD, HAVPDYYS	C19470	EST	Os	U95180
	SPPA, PYTIFVPTD	X80416	CAM	Vc	-
2	PAPA, ERQGLSDTRFLENGKYY	S79359	AGP	Na	-
	PAPA, QGMSDTRFLENGKYS	D30801	TED3	Ze	-
	QGMSDTRFLENGKYY,	S79358	AGP	Pc	-
	QGMSDTRYLENGKYY	U30506	E6	Gb	S39359
	QGMSDTRFLGNGKFYY	AB012854	Genomic sequence	Nt	-
	ERQRMSD, ERQGMSDTRFMANGKYY	L34288	Hypothetical protein	Bn	-

Table 2 cont.

Group	Motifs	Accession	Type of protein or sequence	Plant	Similarity with AGP or AGP-like protein
	ERQGMSDTRYMANGKYY	U78721	EST	At	L34288
	ERQGMXXTRFMEKGSYY	H37549	EST	At	L34288
	ERQGMSXTRFMEKGSYY	R90404	EST	At	L34288
3	APKASSP, EDDY, SPPAP, DGPS, TNVKL	L47351	AGP	Bn	-
	SPKASSP, EDDY, SPPXP, DGPS	R65535	EST	At	L47351
	PKASSP, EDDY, SPPAP, DGPS	AA394322	EST	At	L47351
	PKASSP, EDDY, SPPAP, DGPS, TNVKL	AA651118	EST	At	L47351
	PVSAP, SPPAP,	U13066	AGP	Na	-
	PAPA, PVSAP, SPPAP, PAPAP	X99147	AGP	Le	-
	SPPAP, PAPA	AC002391	Genomic sequence	At	X99147
	PAPAP, SPPAP	AA824752	EST	Le	X99147
	SPPA, SPPAP,	U88587	AGP	Na	-
	TATPP, SPPA, SPPAP,	L17308	Proline-rich protein	Gh	-
	TATPP, SASPPS, SPPA, LATSC	U14009	AGP	Pc	-
	AQSPAPAP, PAPAP, SASPPS	AF0828298	AGP	At	U14009
	LATSC, AQAPAPAP	AF0828299	AGP	At	U14009
	AQAPAPAP, PAPAP	AF0828300	AGP	At	U14009
	AQAPAP, PAPA, SPPAP	AF0828301	AGP	At	U14009
	AQAP, SPPAP	AF0028302	AGP	At	U14009
	TATPP	U09554	AGP	Pt	-
	PAPAP	S45139	NOD5	Ps	-
	PAPAP	Z16403	TTS1	Na	-
4	PSE, PPPPSS	U25628	AGP	Na	-

Sequences and ESTs from group 2 share similarities with "non-classical" AGPs (Du *et al* 1996a). It has been suggested that the C-terminal domain rich in G, N, S, and Y is processed (Du *et al* 1996a) to give a shorter polypeptide rich in A, P, and T residues. Processing of the C-terminal domain could lead to the release of a signaling peptide targeted to the extracellular matrix or recycled to a cytoplasmic target. In this group we also found TED3, a Gly-rich protein from *Zinnia elegans* preferentially expressed in cells that redifferentiate into tracheary elements (Demura and Fukuda 1994), and a Pro-rich protein from cotton cell wall (John 1995). Both proteins exhibit the very conserved amino acid motif QG-SDTR-LENGKY in the C-terminal domain. This signature motif might be considered as specific to the AGP group 2, since it is not found in any other proteins registered in the databases. Similar motifs are present in a hypothetical protein from *Brassica napus* (Robert *et al* 1994), in a recently isolated EST from *Nicotiana tabacum* (AB012854) shown to be up-regulated by cytokinins and in an *Arabidopsis* genomic clone. However, we do not know if these molecules exhibit the same carbohydrate epitopes. It is important to note that all AGP and EST sequences of group 2 end with a Pro residue. The significance is not evident but could reflect a site for a processing enzyme.

The ESTs and proteins from group 3 show similarities with animal-secreted proteins such as mucins (van Klinken *et al* 1995), with plant extensins and Pro-rich proteins, and with the "classical" AGPs (Du *et al* 1996a). The amino acid sequences are rich in P, A, S and T residues. The proteins have a C-terminal hydrophobic domain and internal stretches of hydrophobic amino acids. It has been shown that the C-terminal domain of the *Nicotiana* and *Pyrus* AGPs is processed (Schultz *et al* 1998, Youl *et al* 1998). The "classical" AGPs are thought to be C-terminally processed and attached to glycosylphosphatidylinositol (GPI) membrane anchors (Schultz *et al* 1998, Youl *et al* 1998). Sequences from group 3 show a higher content in A, P and S residues compared with sequences from groups 1 and 2. Two *Arabidopsis* ESTs and a *Brassica napus* AGP (Gester *et al* 1996), although very similar, show length variation in their sequences and, in addition, the insertion of several short amino acid stretches between the conserved motifs, as described for other members of group 1, possibly indicating post-transcriptional modifications.

3. SIGNIFICANCE OF AGP SUBFAMILIES

We report in this paper that, on the basis of the sequence homologies and the presence of specific conserved amino acid motifs, AGP genes and ESTs can be divided into at least four groups of AGPs and AGP-like molecules. The characteristics of these genes and the groups to which they belong are summarized in Tables 1 and 2.

The phylogenetic tree of AGPs and related sequences (Fig 1) also shows that the sequences divide into several major groups. The nodulin member (S45139) seems to be the most ancient molecule, whereas the *Volvox* chimeric extensin (X80416) does not seem to have given rise to any recent variants. The phylogenetic analysis indicates that the *Volvox* protein belongs to group 3 probably because of its extensin domain confirming the bimodular structure of the protein.

Since most of the ESTs have been isolated from libraries prepared with RNA from mixed tissues it is not possible to conclude anything about their location within the plant. Proteins belonging to different groups could have different expression patterns. In addition, they could have distinct functions, and their synthesis could be regulated by different mechanisms. Even if AGPs are ubiquitous in plants it does not appear that one group of AGPs is characteristic for specific families or species, although it has been shown that some of the carbohydrate epitopes are missing from the Solanaceae family (Snogerup 1997). In sugar beet leaves AGP isoforms of different electrical charge but the same molecular weight have different epitopes recognized by monoclonal antibodies, giving additional variability or specificity to the AGP molecule (Stöhr *et al* 1996, Kjellbom *et al* 1997). The related AGP sequences likely correspond to a common ancestor gene that is now divided into various gene subfamilies. This possibility was also suggested by Li and Showalter (1996) who found 15 putative AGP genomic clones

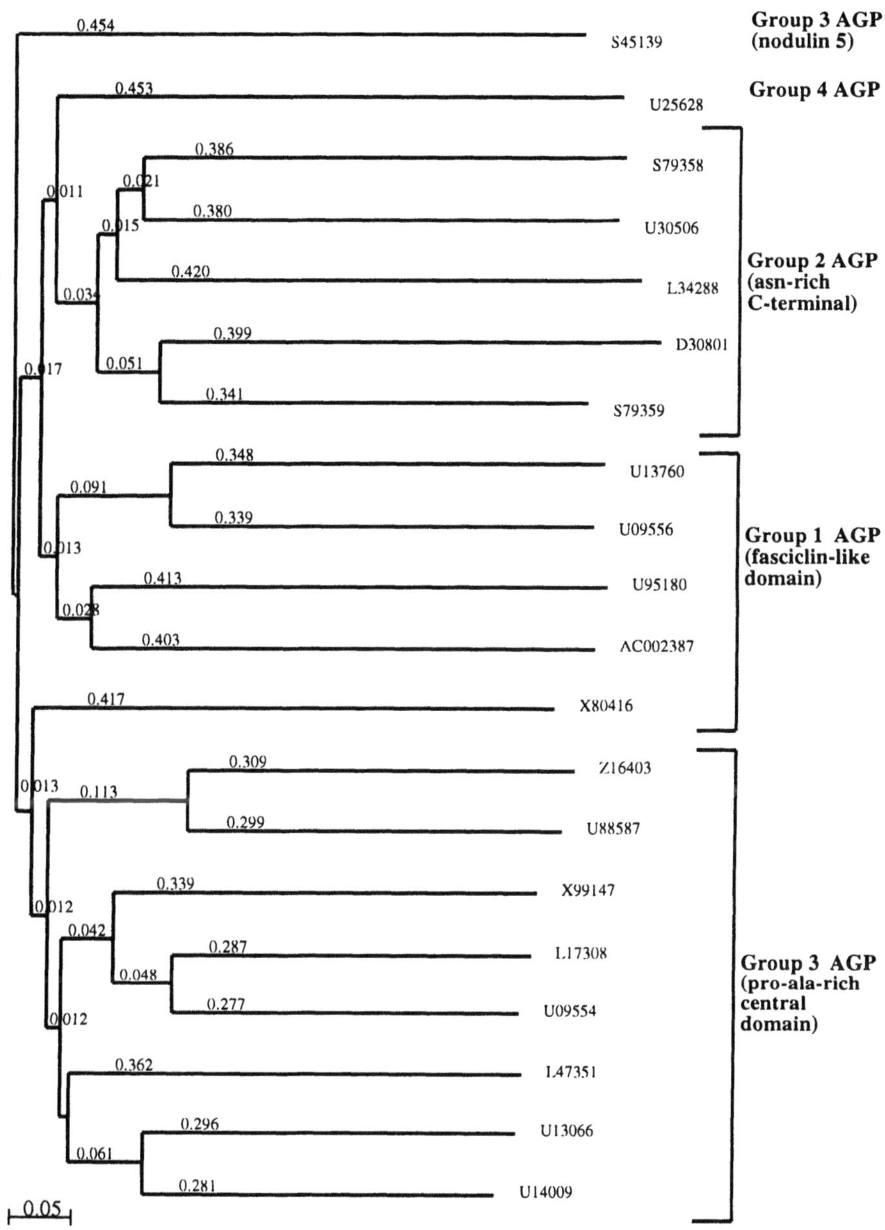

Figure 1. Representation of the phylogenetic relationships between AGPs and related proteins. Sequences were aligned using CLUSTALW (Higgins *et al* 1992) at the Washington University site (http://www.ibc.wustl.edu). The dendogram depicting the evolutionary relationships was constructed using the PHYLIP program package (http://evolution.genetics.washington.edu). The numbers above the branches refer to the corrected estimate of the genetic distance between the putative ancestor gene at the node and the gene at the end of the branch. The AGP sequences are indicated by their Genebank accession numbers.

during the screening with a nucleotide probe corresponding to the PAPAP domain representing related or unrelated genes.

The AGPs of group 1 show a high variability of the sequences but a strong conservation of specific motifs. They all show similarity with a highly conserved domain of fasciclin I (Hortsch and Goodman 1990), which is either localized at the plasma membrane or secreted. We observed a high variability in the C-terminal region of group 1 sequences as well as the presence of the PAPAP motif in different positions among the sequences. This variability may represent alternative splicing mechanisms similar to those that are well described in animal cells for molecules involved in cell–cell interactions and development such as integrins (Hynes 1992), fibronectins (Kornblihtt *et al* 1996) and neural CAMs (Wang *et al* 1998). It is thus possible that AGPs from group 1 are developmentally regulated (Pennell and Roberts 1990, Pennell *et al* 1991, Kjellbom *et al* 1997). Post-transcriptional processing events may offer the possibility of producing different isoforms of AGPs by exposing or retrieving protein domains that can be phosphorylated or glycosylated. This mechanism would produce molecules with different signaling activities that may give rise to different cell identities, which would be important for meristem development in which cell position and cell–cell interactions are governing cell fate (Meyerowitz 1997).

The AGPs of group 2 contain similar sequences regardless of plant species, all exhibiting a C-terminal Pro residue and a conserved stretch of 10 hydrophobic amino acids in their C-terminal parts, which are processed and not present in the mature proteins (Mau *et al* 1995). The TED3 has been identified in proto-xylem cells, which have been directed towards a cell death pathway and is consistent with a role of AGPs marking cells that are committed to cell death (Buckner *et al* 1998). Also, the TED3 promotor contains transcription factor binding elements responsible for the gene regulation in proto-xylem cells (Igarashi *et al* 1998). This observation is consistent with a time and cell type-specific expression pattern described for some AGPs. In addition the presence of an Asn-rich domain similar to the C-terminal of the group 2 AGPs from *Pyrus* and *Nicotiana* also occurs in a *Plasmodium falciparum* antigen (Wahlgren *et al* 1986) and in genes from *Dictyostelium discoideum* (Han and Firtel 1998) and *Synechocystis* sp. (Kaneko *et al* 1996) that encode, respectively, proteins related to transcription factors and a hypothetical AGP-like molecule.

The sequences of group 3 are dissimilar but include many identified AGP sequences in addition to ESTs. All the members of group 3 show homologies to extensins or Pro-rich proteins. Motifs such as PAPAP and SPPAP are present in addition to motifs such as SPPP, TAPTP and KASSP, which are characteristic of extensins and Pro-rich proteins. Pro-rich proteins could be involved in protein–protein interactions through their Pro-rich domains or could be linkers between the wall and the plasma membrane that enable dynamic and signaling connections (Pont- Lezica *et al* 1993) and signal transmission or provide a precise pattern for glycosylation. Like members of group 1, group 3 members have insertions of amino acid stretches between conserved motifs, possibly indicating RNA splicing events.

Such events might be involved in the developmental regulation of AGPs. The presence of introns in AGP genes has only been reported for the algal CAM gene (Huber and Sumper 1994) and for a genomic AGP clone of tomato (Li and Showalter 1996). The latter has an intron separating the long central domain from the putative membrane attachment domain. However no intron is present in the *Pinus* AGP genes nor in the gene encoding the *Arabidopsis* cell surface protein of group 1.

4. LOCALIZATION OF AGPS AT THE PLASMA MEMBRANE

To date there are no sequence data available for AGPs isolated from plasma membrane. Plasma membrane AGPs have been suggested to be receptors for extracellular signals and matrix molecules (Pennell *et al* 1989, Nothnagel 1997), and their localization at the cell surface (Norman *et al* 1990, Kjellbom *et al* 1997) is consistent with this hypothesis. Little is known about how plasma membrane AGPs are attached to the membrane (Norman *et al* 1990, Schultz *et al* 1998). Analysis of AGPs and putative AGP sequences shows that some AGPs contain one or more transmembrane domains, or a hydrophobic C-terminal segment (PSORT, Genome-Net, K. Nakai, Osaka University, Japan) that has been identified as a putative GPI membrane anchor domain (Schultz *et al* 1998, Youl *et al* 1998). The authors reported that several AGPs from *Nicotiana, Pyrus, Arabidopsis, Brassica, Gossypium* (all belonging to group 3) are likely to be attached to the plasma membrane by GPI anchors because they contain amino acids in positions typical for GPI anchor attachment sites.

Evidence for the presence of CAMs in plants is lacking. Some HRGPs and some of the plasma membrane AGPs could be true CAMs, as suggested by Huber and Sumper (1994). They reported the isolation of a *Volvox* gene coding for a membrane-associated chimeric extensin, which exhibited similarities to fasciclin I and spliced variants that were produced in a developmentally controlled fashion. The presence of spliced variants of CAMs in *Volvox* and their developmental regulation resemble the presence of different AGP molecules at different stages of plant development (Pennell *et al* 1991). In the case of *Drosophila* embryos, fasciclin I is found in a soluble form as well as in an insoluble membrane-bound form associated with the plasma membrane through a GPI-anchor (Hortsch and Goodman 1990). The occurrence of soluble and membrane-bound AGPs in plant cells would be consistent with a role of AGPs in cell–cell interactions during development and with the function of a molecule bearing positional information. The transient appearance of AGP carbohydrate epitopes at the cell surface during developmental processes, such as embryogenesis, is consistent with AGPs being exposed at the plasma membrane as GPI-anchored membrane proteins and thereafter released by lipases at appropriate developmental times (Kjellbom *et al* 1997, Schultz *et al* 1998,

Youl *et al* 1998). The AGPs from group 1, which contain the fasciclin-like signature motif, may represent developmentally regulated plasma membrane AGPs. Different sub-groups of AGPs could be active at different developmental stages or represent different cell fates. Bates *et al* (1994) suggested that the interactions between plasma membrane integrins and extracellular matrix molecules are necessary for maintaining the living state of the cell, and the disruption of these interactions activates an intrinsic suicide program leading to programmed cell death. Thus plasma membrane AGPs (or AGPs showing similarities with CAM) could be involved in the maintenance of these interactions and in the control of cell death, directly or by signaling mechanisms coupled to other membrane receptors. It is possible that only some AGPs undergo processing before being released from the plasma membrane. However in each group it seems that AGPs are likely to undergo processing mechanisms such as for GPI anchors (Schultz *et al* 1998, Youl *et al* 1998). It is also possible that some AGP genes give rise directly to shorter AGPs already "preprocessed" at the RNA level.

5. CHIMERIC AGPS AND RELATED MOLECULES

Some AGPs show similarities with chimeric cell wall proteins that contain at least two different structural modules involved in different functions. Chimeric proteins have recently been described in the plant HRGP superfamily (Kieliszewski and Lamport 1994). These peculiar HRGPs include the stigma-specific extensin (Goldman *et al* 1992), the *Chlamydomonas* HRGP (Waffenschmidt *et al* 1993), the *Volvox* embryo-specific extensin (Ertl *et al* 1992), the *Volvox* CAM (Huber and Sumper 1994) and the pollen extensin pex-1 (Rubinstein *et al* 1995a b). Also the lectin-receptor-like kinase (Hervé *et al* 1996), which presents two distinct functional domains, the recently described WAK proteins (He *et al* 1996) and the tangled gene product (Miller *et al* 1997) could well belong to this group of chimeric modular proteins.

The pex-1 protein shows sequence similarities with a cotton AGP (John and Crow 1992), various extensins, receptor-like kinases, Leu-rich repeat proteins (Jones and Jones 1997) and animal mucins, but not with fasciclin I. On the bases of its localization and the putative role of pex-1 in adhesion, Rubinstein *et al* (1995b) proposed that pex-1 is an AGP. This is not unlikely, since the protein contains signature motifs present in group 3 AGPs. Rubinstein *et al* (1995b) mentioned that the globular domains of the algal CAM and the pex-1 are structurally similar to *Drosophila* fasciclin I. The authors proposed that the extensin-like domain of this multidomain protein anchors the protein in the cell wall, whereas the globular domain has other functions (e.g., cell adhesion and cell signaling such as in the receptor-like kinases CLAVATA and ERECTA [Meyerowitz 1997]). This hypothesis is consistent with the suggested function of AGPs in cell–cell interactions and in developmental signaling (Pennell and Roberts 1990, Pennell *et al* 1991). Chimeric proteins might also represent ancestors that diverged differently by keeping several functional modules, and chimeric AGPs (group 1) would thus be more ancient than "mono-modular" AGPs (group 3).

Sequences of group 1 AGPs show a motif very similar to the *Drosophila* CAM fasciclin I and to the chimeric extensin/fasciclin CAM-like protein from *Volvox*. Three *Arabidopsis* genomic sequences that contain a domain similar to the fasciclin signature motif have been detected in the databases (Table 2). One of them has been identified and mapped to chromosome 2 and is predicted to encode a cell surface protein (AC002387). The amino acid sequence shows Pro-rich domains and strong homologies with other cell surface proteins (AA661000) and with most of the known AGPs of group 1. Interestingly, the C-terminal contains domains proposed to be typical for AGPs (PAPAP, SPPAP). The sequence also shows strong homology with the maize endosperm-specific protein and with the *Pinus* group 1 AGP. It should be noted that another *Arabidopsis* gene that also shows similarity with the fasciclin signature motif (AB005239) sequence has recently been identified on chromosome 5.

6. SIGNIFICANCE OF PLANT FASCICLIN MOTIFS AND PLANT CAMS

As already noted, information concerning CAMs in plants is lacking. No direct evidence has been found concerning a protein and its function in cell adhesion. Several studies concerning adhesion and migration of pollen grains have been performed on styles. Some AGP antigenic regions have been putatively involved in these processes. A few CAMs have been reported directly or indirectly in plants. Several proteins have been proposed as candidates for plant CAMs, and these include integrin homologs, various intrinsic membrane proteins such as WAK (He *et al* 1996), the tangled gene product, LRR proteins and CAM homologs (Huber and Sumper 1994). The discovery that plant cells may contain several CAMs with fasciclin motifs is an important step. Similar to their insect homologs, these proteins are likely to have a complex glycosylation pattern and be involved in cell adhesion. Fasciclins are present in insects and other animal cells, and all proteins containing the fasciclin I domain are involved in cell–cell contact and in developmental processes. In these organisms fasciclins are expressed in a very limited subset of tissues. The localization of fasciclin-related molecules in other organisms, in certain tissues or at certain developmental stages, will help to elucidate their function. For example, many AGPs have been found in flower tissues where numerous cell–cell interaction mechanisms occur. Such interactions are particularly important in styles where pollen tubes grow by following a track created or maintained by proteins such as AGPs.

The similarity between the algal CAM, the pine and the cotton AGPs, some *Arabidopsis* clones and the maize endosperm-specific protein is striking. Alignment of these protein sequences indicates that the H1 domain and the H-box from the *Drosophila* fasciclin I (Hortsch and Goodman 1990) are very conserved. It is suggested that these domains are involved in protein–protein interactions. We found fasciclin-like signature motifs in different sequences, originating from different plants, closely related to the pine AGP gene. It is thus likely that fasciclins and some AGPs (group 1) are related in structure and in function since both are involved in cell adhesion and in developmental processes. It also appears that animal fasciclins, animal CAMs, and plant AGPs are likely to share common regulatory mechanisms such as RNA splicing, complex glycosylation, and attachment to the plasma membrane through GPI anchors.

7. CONCLUSIONS

In this chapter we have identified conserved amino acid motifs that allowed the discrimination of different AGP groups (Figs 1, 2). In addition to genes coding for secreted AGPs, databases contain ESTs, genomic sequences and related sequences corresponding to a large number of additional AGP-like genes located on different chromosomes. This work led us to consider a classification of AGPs based on the structure and composition of the protein backbones. The AGPs and AGP-like molecules can be phylogenetically analyzed and divided into at least four groups. Each major group of sequences is defined by conserved and specific amino acid motifs. Some amino acid motifs are only found in AGPs, whereas others are also characteristic of other molecules such as insect fasciclins. By acknowledging these similarities it might be possible to discern AGP function in plant cell development. The presence of multi-domain proteins, such as group 1 AGPs, may reflect their complex evolution and indicate their multi-functional properties. The presence of different genes could explain the number of localizations and functions proposed for AGPs as well as their implication in different signaling pathways.

Post-transcriptional mechanisms such as RNA processing are likely to operate during plant cell development. These mechanisms might allow cells to produce and secrete proteins with certain carbohydrate domains exposed in the extracellular matrix. The presence or absence of amino acid motifs may define the pattern of glycosylation known to be correlated with cell–cell interactions and cell identity. Although very little is known about O-glycosylation in plants it is likely that carbohydrates confer specificity to AGPs and that the protein backbones provide the presence, the order and the spacing of these branches. Interestingly many of the sequences containing AGP signature motifs also contain putative membrane attachment consensus sequences as well as putative transmembrane helices. It seems that most of known AGPs contain features for processing at the gene or at the protein level. This is consistent with the fact that AGPs are cell surface molecules (Stöhr *et al* 1996, Kjellbom *et al* 1997). It would be interesting to know if there are differences between constitutively expressed basal AGPs and stress-induced AGPs, and between cell wall and plasma membrane-associated AGPs (Kjellbom *et al* 1997).

Kieliszewski and Lamport (1994) suggested that HRGPs belong to a common superfamily and that it is important to focus on the similarity of the different subfamilies. Our analysis shows that it is possible to describe, on the protein basis, an AGP family within the HRGP superfamily. With the increasing knowledge about AGPs and other HRGPs it might be possible to trace more precisely the origin of HRGPs and other cell wall proteins to a small number of protein modules (Woessner *et al* 1994). It is possible that the origin of AGP carbohydrate epitopes is more ancient than the origin of the protein backbones, and that the evolution of the two moieties has followed different constraints related to their respective functions.

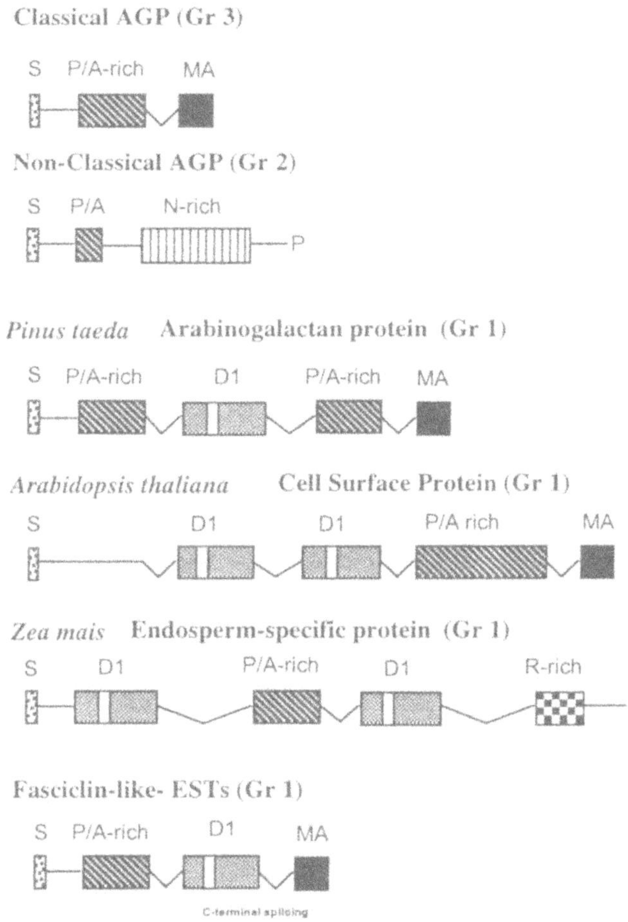

Figure 2. Structures of polypeptide sequences of AGPs and related molecules. Boxes indicate specific domains that are present in each group of sequences. A: Ala; D1: domain 1; MA: membrane attachment; N: Asn; P: Pro; R: Arg; S: secretion signal. Bars not to scale.

ACKNOWLEDGEMENTS

We thank Drs. Ian Max Möller, Department of Plant Physiology, and Maria Cordeiro and Marianne Sommarin, Department of Plant Biochemistry, Lund University, for many insightful discussions. CR is a recipient of a Marie Curie fellowship (European Union).

REFERENCES

Altschul, S. F., Gish, W., Miller W., Myers, E. W., and Lipman, D. J., 1990, Basic local alignment search tool, *J. Mol. Biol.* **215**: 403–410.

Bates, R. C., Buret, A., van Helden, D. F., Horton, M. A., and Burns, G. F., 1994, Apoptosis induced by inhibition of intercellular contact, *J. Cell Biol.* **125**: 403–415.

Bordier, C., 1981, Phase separation of integral membrane proteins in Triton X-114 solution, *J. Biol. Chem.* **256**: 1604–1607.

Buckner, B., Janick-Buckner, D., Gray, J., and Johal, G. S., 1998, Cell death mechanisms in maize, *Trends Plant Sci.* **3**: 218–223.

Carlson, S. J., and Chourey, P. S., 1997, Cloning of a novel maize endosperm-specific protein with partial sequence homology to a pollen surface protein, *Biochem. Biophys. Acta* **1341**: 101–104.

Chen, C.-G., Pu, Z.-Y., Moritz, R. L., Simpson, R. J., Bacic, A., Clarke, A. E., and Mau, S.-L., 1994, Molecular cloning of a gene encoding an arabinogalactan-protein from pear (*Pyrus communis*) cell suspension culture, *Proc. Natl. Acad. Sci. USA* **91**: 10305–10309.

Demura, T., and Fukuda, H., 1994, Novel vascular cell-specific genes whose expression is regulated temporally and spatially during vascular system development, *Plant Cell* **6**: 967–981.

Du, H., Simpson, R. J., Moritz, R. L., Clarke, A. E., and Bacic, A., 1994, Isolation of the protein backbone of an arabinogalactan-protein from styles of *Nicotiana alata* and characterization of a corresponding cDNA, *Plant Cell* **6**: 1643–1653.

Du, H., Clarke, A. E., and Bacic, A., 1996a, Arabinogalactan-proteins: a class of extracellular matrix proteoglycans involved in plant growth and development, *Trends Cell Biol.* **6**: 411–414.

Du, H., Simpson, R. J., Clarke, A. E., and Bacic, A., 1996b, Molecular characterization of a stigma specific gene encoding an arabinogalactan protein (AGP) from *Nicotiana alata*, *Plant J.* **9**: 313–323.

Ertl, H., Hallmann, A., Wenzl, S., and Sumper, M., 1992, A novel extensin that may organize extracellular matrix biogenesis in *Volvox carteri*, *EMBO J.* **11**: 2055–2062.

Gerster, J., Allard, S., and Robert, L. S., 1996, Molecular characterization of two *Brassica napus* pollen expressed genes encoding putative arabinogalactan proteins, *Plant Physiol.* **110**: 1231–1237.

Gleeson, P. A., McNamara, M., Wettenhall, R. E. H., Stone, B. A., and Fincher, G. B., 1989, Characterization of the hydroxyproline rich protein core of an arabinogalactan protein secreted from suspension cultures *Lolium multiflorum* (Italian ryegrass) endosperm cells, *Biochem J.* **264**: 857–862.

Goldman, M. H., De S., Pezzotti, M., Seurinck, J., and Mariani, C., 1992, Developmental expression of tobacco pistil-specific genes encoding novel extensin-like proteins, *Plant Cell* **4**: 1041–1051.

Gumbiner, B. M., 1993, Proteins associated with the cytoplasmic surface of adhesion molecules, *Neuron* **11**: 551–564.

Han, Z., and Firtel, R., 1998, The homeobox-containing gene *Wariai* regulates anterior-posterior patterning and cell type homeostasis in *Dictyostelium*, *Development* **125**: 313–325.

He, Z. H., Fujiki, M., and Kohorn, B. D., 1996, A cell wall associated, receptor like protein kinase, *J. Biol. Chem.* **271**: 19789–19793.

Hervé, C., Dabos, P., Galaud, J. P., Rougé, P., and Lescure, B., 1996, Characterization of an *Arabidopsis thaliana* gene that defines a new class of putative plant receptor kinases with an extracellular lectin-like domain, *J. Mol. Biol.* **258**: 778–788.

Higgins, D. G., Bleasby, A. J., and Fuchs, R., 1992, CLUSTAL V: improved software for multiple sequence alignment, *Comp. Appl. Biosci.* **8**: 189–191.

Hortsch, M., and Goodman, C. S., 1990, *Drosophila* fasciclin I, a neural cell adhesion molecule, has a phosphatidylinositol lipid membrane anchor that is developmentally regulated, *J. Biol. Chem.* **265**: 15104–15109.

Huber, O., and Sumper, M., 1994, Algal-CAM: Isoforms of a cell adhesion molecule in embryos of the alga *Volvox* with homology to *Drosophila* fasciclin I, *EMBO J.* **13**: 4212–4222.

Hynes, R. O., 1992, Integrins: versatility, modulation, and signalling in cell adhesion, *Cell* **69**: 11–25.

Igarashi, M., Demura, T., and Fukuda, H., 1998, Expression of the *Zinnia* TED3 promoter in developing tracheary elements of transgenic *Arabidopsis*, *Plant Mol. Biol.* **36**: 917–927.

Jermyn, M. A., and Guthrie, R., 1985, A final assault on the structure of carrot AGPs, *AGP News* **5**: 4–25.

John, M. E., 1995, Characterization of a cotton (*Gossypium hirsutum* L.) fiber mRNA (Fb-B6), *Plant Physiol.* **107**: 1477–1478.

John, M. E., and Crow, L. J., 1992, Gene expression in cotton (*Gossypium hirsutum* L.): fiber cloning of the mRNAs, *Proc. Natl. Acad. Sci. USA* **89**: 5769–5773.

John, M. E., and Keller, G., 1995, Characterization of mRNA for a proline rich protein of cotton fiber, *Plant Physiol.* **108**: 669–676.

Jones, D. A., and Jones, J. D. G., 1997, The role of leucine-rich repeat proteins in plant defences, *Adv. Bot. Res.* **24**: 89–167.

Kaneko, T., Sato, S., Kotani, H., Tanaka, A., Asamizu, E., Nakamura, Y., Miyajima, N., Hirosawa, M., Sugiura, M., Sasamoto, S., Kimura, T., Hosouchi, T., Matsuno, A., Muraki, A., Nakazaki, N., Naruo, K., Okumura, S., Shimpo, S., Takeuchi, C., Wada, T., Watanabe, A., Yamada, M., Yasuda, M., and Tabata, S., 1996, Sequence analysis of the genome of the unicellular cyanobacterium *Synechocystis* sp. strain PCC6803. II. Sequence determination of the entire genome and assignment of potential protein-coding regions, *DNA Res.* **3**: 109–136.

Kieliszewski, M. J., and Lamport, D. T. A., 1994, Extensin: repetitive motifs, functional sites, post-translational codes, and phylogeny, *Plant J.* **5**: 157–172.

Kieliszewski, M.. J., Kamyab, A., Leykam, J. F., and Lamport, D. T. A., 1992, A histidine-rich extensin from *Zea mays* is an arabinogalactan protein, *Plant Physiol.* **99**: 536–547.

Kjellbom, P., Snogerup, L., Stöhr, C., Reuzeau, C., McCabe, P., and Pennell, R. I., 1997, Variability and oxidative cross-linking of plasma membrane arabinogalactan-proteins, *Plant J.* **12**: 1189–1196.

Knox, P., 1995, Developmentally regulated proteoglycans and glycoproteins of the plant cell surface, *FASEB J.* **9**: 1004–1012.

Komalavilas, P., Zhu, J.-K., and Nothnagel, E. A., 1991, Arabinogalactan-proteins from the suspension culture medium and plasma membrane of rose cells, *J. Biol. Chem.* **266:** 15956–15965.

Kornblihtt, A. R., Pesce, C. G., Alonso, C. R., Cramer, P., Srebrow, A., Werbajh, S., and Muro, A. F., 1996, The fibronectin gene as a model for splicing and transcription studies, *FASEB J.* **10**: 248–257.

Kreuger, M., and van Holst, G. J., 1996, Arabinogalactan proteins and plant differentiation, *Plant Mol. Biol.* **30**: 1077–1086.

Li, S. X., and Showalter, A. M., 1996, Cloning and developmental/stress regulated expression of a gene encoding a tomato arabinogalactan protein, *Plant Mol. Biol.* **32**: 641–652.

Loopstra, C. A., and Sederoff, R. R., 1995, Xylem specific gene expression in loblolly pine, *Plant Mol. Biol.* **27**: 277–291.

Mau, S.-L., Chen, C.-G., Pu, Z.-Y., Moritz, R. L., Simpson, R. J., Bacic, A., and Clarke, A. E., 1995, Molecular cloning of cDNAs encoding the protein backbones of arabinogalactan-proteins from the filtrate of suspension-cultured cells of *Pyrus communis* and *Nicotiana alata*, *Plant J.* **8**: 269–281.

Meyerowitz, E. M., 1997, Genetic control of cell division patterns in developing plants, *Cell* **88**: 299–308.

Miller, D., Hable, W., Gottwald, J., Ellard-Ivey, M., Demura, T., Lomax, T., and Carpita, N., 1997, Connections: the hard wiring of the plant cell for perception, signaling and response, *Plant Cell* **9**: 2105–2117.

Nothnagel, E. A., 1997, Proteoglycans and related components in plant cells, *Int. Rev Cytol.* **174**: 195–291.

Norman, P. M., Kjellbom, P., Bradley, D. J., Hahn, M. G., and Lamb, C., 1990, Immunoaffinity purification and biochemical characterization of plasma membrane arabinogalactan rich glycoproteins of *Nicotiana glutinosa*, *Planta* **181**: 365–373.

Pennell, R. I., and Roberts, K., 1990, Sexual development in the pea is presaged by altered expression of arabinogalactan protein, *Nature* **344**: 547–549.

Pennell, R. I., Knox, J. P., Scofield, G., N., Selvendran, R. R., and Roberts, K., 1989, A family of abundant plasma membrane-associated glycoproteins related to the arabinogalactan proteins is unique to flowering plants, *J. Cell Biol.* **108**: 1967–1977.

Pennell, R. I., Janniche, L., Kjellbom, P., Scofield, G. N., Peart, J. M., and Roberts, K., 1991, Developmental regulation of a plasma membrane arabinogalactan protein in oilseed rape flowers, *Plant Cell* **3**: 1317–1326.

Pennell, R. I., Cronk, Q. C. B., Forsberg, L. S., Stöhr, C., Snogerup, L., Kjellbom, P., and McCabe, P. F., 1995, Cell-context signaling, *Phil. Trans. Roy. Soc. Lond. B* **350**: 87–93.

Pogson, B. J., and Davies, C., 1995, Characterization of a cDNA encoding the protein moiety of a putative arabinogalactan protein from *Lycopersicon esculentum*, *Plant Mol. Biol.* **28**: 347–352.

Pont-Lezica, R. F., McNally, J. G., and Pickard, B. G., 1993, Wall to membrane linkers in onion epidermis: Some hypotheses, *Plant Cell Environ.* **16**: 111–123.

Robert, L. S., Allard, S., Gerster, J. L., Cass, L., and Simmonds, J., 1994, Molecular analysis of two *Brassica napus* genes expressed in the stigma, *Plant Mol. Biol.* **26**: 1217–1222.

Rubinstein, A. L., Broadwater, A. H., Lowrey, K. B., and Bedinger, P. A., 1995a, *Pex-1*, a pollen specific gene with an extensin-like domain, *Proc. Natl. Acad. Sci. USA* **92**: 3086–3090.

Rubinstein, A. L., Marquez, J., Suarez-Cervera, M., and Bedinger, P. A., 1995b, Extensin-like glycoproteins in the maize tube wall, *Plant Cell* **7**: 2211–2225.

Schultz, C. J., Hauser, K., Lind, J. L., Atkinson, A. H., Pu, Z. Y., Anderson, M. A., and Clarke, A. E., 1997, Molecular characterisation of a cDNA sequence encoding the backbone of a style-specific 120 kDa glycoprotein which has features of both extensins and arabinogalactan proteins, *Plant Mol. Biol.* **35**: 833–845.

Schultz, C., Gilson, P., Oxley, D., Youl, J., and Bacic, A., 1998, GPI-anchors on arabinogalactan-proteins: implications for signalling in plants, *Trends Plant Sci.* **3**: 426–431.

Serpe, M. D., and Nothnagel, E. A., 1996, Heterogeneity of arabinogalactan proteins on the plasma membrane of rose cells, *Plant Physiol.* **112**: 1261–1271.

Showalter, A. M., and Varner, J. E., 1989, Plant hydroxyproline-rich glycoproteins. In *The Biochemistry of Plants* (A. Marcus, ed), Academic Press, New York, pp. 485–536.

Smallwood, M., Yates, E. A., Willats, W. G. T., Martin, H., and Knox, J. P., 1996, Immunochemical comparison of membrane associated and secreted arabinogalactan proteins in rice and carrot, *Planta* **198**: 452–459.

Snogerup, L., 1997, Plasma membrane arabinogalactan proteins, Ph.D. Thesis, Lund University, Sweden.

Stöhr, C., Snogerup, L., Pennell, R. I., and Kjellbom, P., 1996, An agarose gel electrophoretic method for the separation of arabinogalactan proteins, *Plant J.* **10**: 943–948.

van Klinken, B. J. W., Dekker, J., Büller, H. A., and Einerhand, A. W. C., 1995, Mucin gene structure and expression: protection vs. adhesion, *Am. J. Physiol.* **269**: G613–G627.

Waffenschmidt, S., Woessner J. P., Beer, K., and Goodenough, U. W., 1993, Isodityrosine cross-linking mediates cell wall insolubilization in *Chlamydomonas*, *Plant Cell* **5**: 809–820.

Wahlgren, M., Aslund, L., and Franzen, L., 1986, A *Plasmodium falciparum* antigen containing clusters of asparagine residues, *Proc. Natl. Acad. Sci. USA* **83**: 2677–2681.

Wang, B., Williams, H., Du, J. S., Terrett, J., and Kenwrick, S., 1998, Alternative splicing of human NrCAM in neural and nonneural tissues, *Mol. Cell. Neurosci.* **10**: 287–295.

Woessner, J., Molendick, A. J., van Egmond, P., Klis, F. M., Goodenough, U. W., and Haring, M. A., 1994, Domain conservation in several volvocalean cell wall proteins, *Plant Mol. Biol.* **26**: 947–960.

Yariv, J., Rapport, M. M., and Graf, L., 1962, The interaction of glycosides and saccharides with antibody of the corresponding phenylazo glycosides, *Biochem. J.* **85**: 383–3888.

Youl, J. J., Bacic, A., and Oxley, D., 1998, Arabinogalactan-proteins from *Nicotiana alata* and *Pyrus communis* contain glycosylphosphatidylinositol membrane anchors, *Proc. Natl. Acad. Sci. USA* **95**: 7921–7926.

Zinn, K., McAllister, L., and Goodman, C., 1988, Sequence analysis and neuronal expression of fasciclin I in grasshopper and *Drosophila*, *Cell* **53**: 577–587.

Chapter 4

The C-Terminal PAC Domain of a Secreted Arabinogalactan-Protein from Carrot Defines a Family of Basic Proline-Rich Proteins

Timothy C. Baldwin,[1,2] Arjon J. van Hengel[2] and Keith Roberts[2]
[1]*School of Biological Sciences, Universiti Sains Malaysia, Penang, Malaysia.*
[2]*Department of Cell Biology, John Innes Centre, Colney, Norwich, United Kingdom.*

1. INTRODUCTION

Since the first authenticated AGP core polypeptide was cloned by Chen *et al* (1994), a total of six cDNAs corresponding to authenticated AGPs have been reported (Nothnagel 1997, Gao *et al* 1999). In view of this recent molecular data, the criteria previously used to define an AGP (a Hyp/Pro, Ala and Ser rich core polypeptide; possession of an arabinogalactan Type II glycan moiety attached to the core protein; and binding to β-glucosyl Yariv reagent) needed to be amended to take account of this 'new' information. With this in mind, the protein sequences corresponding to the core polypeptides of the cloned AGPs have been divided into two types, "classical" and "non-classical" (Mau *et al* 1995, Du *et al* 1996). According to this convention if the predicted AGP sequence beyond the signal peptide has Pro/Hyp, Ala, and Thr/Ser as the most abundant residues, it is designated "classical". In general, "classical" AGP sequences encode a polypeptide with at least three distinct domains: an N-terminal signal sequence, a central Pro/Hyp-rich region and a C-terminal hydrophobic region that contains a glycosylphosphatidylinositol (GPI) anchor signal (Youl *et al* 1998). "Non-classical" AGP genes, however, encode polypeptides which contain an N-terminal secretion signal followed by a variety of domains, none of which seem to encode a C-terminal GPI membrane anchor signal sequence. Furthermore they may contain consensus sites for N-glycosylation, which do not occur in their "classical" counterparts.

Based upon the available cDNA sequences, Nothnagel (1997) divided the domain structures of the predicted AGP core polypeptides into four categories (A, B, C, D). Type (A) "classical" AGP structure is represented by AGPPc1 (Chen *et al* 1994) and AGPNa1 (Du *et al* 1994). Type (B) "non-classical" AGP structure contains a Pro and a Asn-rich domain as seen in AGPNa2 (Mau *et al* 1995). Type (C) "non-classical" structure contains a Pro-rich domain inserted in an Asn-rich domain as in AGPPc2 (Mau *et al* 1995), and type (D) AGP core protein structure contains a Pro-rich domain followed by a Cys-rich domain as exhibited by AGPNa3 (Du *et al* 1996). Since the publication of this landmark review, two further cDNAs corresponding to authenticated AGP core polypeptides have been cloned, which further expand the diversity of domain structures observed in this seemingly ubiquitous family of proteins. These additional cDNAs are *LeAGP1,* which encodes a "classical" AGP containing a highly basic domain and a potential N-glycosylation site (Gao *et al* 1999), and *DcAGP1,* which encodes a "non-classical" carrot AGP, which is defined by the presence of a C-terminal cysteine-rich domain and can be considered to have a type (D) AGP domain structure (Baldwin *et al* 2000).

All of the currently available sequence data on the AGP protein families seem to suggest an ever expanding complexity in the molecular structure of these proteins, which is undoubtedly linked to the ubiquity and diversity of functions attributed to AGPs and 'AGP-like' proteins.

2. THE PAC DOMAIN DEFINES A FAMILY OF BASIC PROLINE-RICH PROTEINS

Here we present a brief discussion of the C-terminal cysteine-rich domain that occurs in the recently cloned cDNA encoding the core polypeptide of DcAGP1, a previously isolated carrot AGP (Baldwin *et al* 1993 2000). This C-terminal cysteine-rich domain is found, well conserved, in a family of cell-wall associated proteins (Fig 1). We have designated this domain as the PAC domain, since it is found in Proline-rich proteins and Arabinogalactan-proteins, and contains conserved Cysteine residues (Baldwin *et al* 2000). From the results of database searches, it is clear that the DcAGP1 protein bears most 'similarity' to a basic, wound-regulated, proline-rich protein (PvPRP1) encoded by a cDNA isolated from *Phaseolus vulgaris* (Sheng *et al* 1991 1993), and the region that is best conserved is the PAC domain. Other members of this family of proteins containing a PAC domain include five cDNAs derived from *Nicotiana* species. The NtTTS and NaPRP4 glycoproteins all contain a PAC domain and were shown

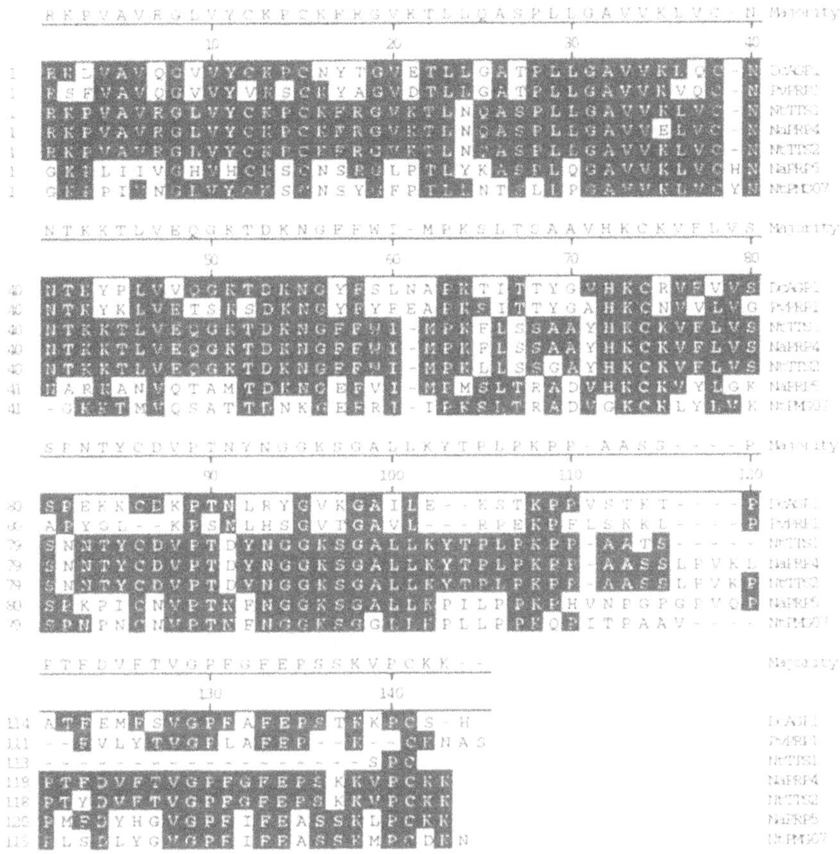

Figure 1. Pileup of the C-terminal PAC domains of several extracellular proteins showing the conserved positions and environment of the six cysteine residues. The alignment was created using the PILEUP program (Genetics Computer Group). The protein NaPRP4 is described in Chen *et al* (1993) and NaPRP5 in Schultz *et al* (1997).

to contain both O-glycans and N-glycans (Cheung *et al* 1993, Sommer-Knudsen *et al* 1996). The NaPRP5 and Pistil-Specific Extensin-Like Proteins (PELPs) such as NtPMG07 are 'AGP-like' and belong to the PAC domain family, but were also shown to contain extensin-like features (Goldman *et al* 1992, Schultz *et al* 1997). Common features of all these sequences are high homology in the cysteine-rich PAC domain (Fig 1) and the presence of N-glycosylation sites within this domain. Except for PvPRP1, the C-terminal domains all contain six conserved cysteine residues. Cysteines are likely to form disulfide bridges, and we suppose that the six cysteines form three pairs. The absence of the first and the fifth cysteine in

PvPRP1 suggests that these two Cys residues form such a pair in DcAGP1 and the *Nicotiana* AGP(-like) molecules. It is noteworthy that the six cysteines are separated by tracts of amino acids whose length is well conserved in the PAC domain family (Fig 2). The only exception to this rule is NtTTS1, where the distance between Cys residues 5 and 6 spans only 20 amino acids compared to 48-53 amino acids in the other PAC family members.

$$-C-[2]-C-\begin{bmatrix}20\\ to\\ 22\end{bmatrix}-C-\begin{bmatrix}33\\ to\\ 34\end{bmatrix}-C-[11]-C-\begin{bmatrix}48\\ to\\ 53\end{bmatrix}-C-$$

Figure 2. Diagram to illustrate the relatively constant number of amino acids between the six cysteines (C) in the PAC domains shown in Fig 1, excluding NtTTS1 and PvPRP1 (only four cysteines).

3. A FAMILY OF POLLEN PROTEINS SHARE FEATURES OF THE PAC DOMAIN

Using the Cys-rich PAC domain alone for database searches revealed that the presence and the spacing of the Cys residues as shown in Fig 2 also occur in another family of proteins. Figure 3 shows that several translated cDNAs encoding pollen-specific proteins contain the conserved cysteine positions of the PAC family, but without the upstream proline-rich domain. The spacing between the first five Cys residues is completely conserved, but the distance between the fifth and the sixth cysteines is considerably bigger in DcAGP1. Another difference between the two protein families is the conserved amino acid that follows the first and fourth Cys residue. The pollen-specific proteins show a switch from a basic to an acidic amino acid in these positions. Like the PAC domain family members, the pollen-specific proteins contain putative N-glycosylation sites. One of these proteins, the major allergen from olive tree pollen Ole e 1, has been purified, and using specific glycopeptide hydrolases the presence of N-glycans has been demonstrated (Villalba *et al* 1993). The presence of hydrophobic signal sequences suggests that these pollen specific proteins are secreted into the extracellular matrix. The absence of a proline-rich domain suggests that the pollen-specific proteins are only distantly related to AGPs. Divergence between these two classes is confirmed by the low molecular weight (20 kDa) of the glycosylated Ole e 1 and the fact that the carbohydrate component accounts for only 8% of the molecular weight. However, the conserved positions of the Cys residues suggest similar three dimensional structures for the PAC domain and the pollen specific cysteine-rich proteins.

```
                      10                  20                  30                  40
 1   - - - - - - - - - - - - - - - - - - - - - - - - - - - R K L V A V Q G V V Y C K  DcAGP1 C
 1   M E D I P Q P P V - - - - - - - - - - - - - - - - - - - S Q F H I Q G Q V Y C D  Ole e 1
 1   M A - - K A I V L L S - A L C I L A L A N F A H C - R P E V F D V E G K V Y C D  LAT52
 1   M A S L R T I P V I - F G I L F Y V L A S T A T A T D A P D Y V V Q G R V Y C D  Z16402
 1   M A S L R A L S V I A V A V V L F A L A D T A V A T K A P D Y V V Q G R V Y C D  U85499

                      50                  60                  70                  80
14   P C N Y T G V E T L L G A T P L L G A V V K L Q C N N - - - T K Y P L V V Q G K  DcAGP1 C
22   T C R - A G F I T E L S E F - I P G A S L R L Q C K D K E N G D V T F T E V G Y  Ole e 1
37   T C R - V Q F E T K L S E N - L E G A T V K L Q C R N I S T E A E T F S V E G V  LAT52
40   T C R - A E F E T N V T E Y - I K G A K V R L E C K H F G T D K V E R A I D G V  Z16402
41   R C R - A G F E T N V T E Y - I K G A K V R L E C R H F G T N V L E R A I D G V  U85499

                      90                 100                 110                 120
51   T D K N G Y F S L N A P K T I T T Y G V H K C R V F V V S S P E K K C D K P T N  DcAGP1 C
60   T R A E G L Y S M L V E R D - - - H K N E F C E I T L I S S G R K D C N E I P T  Ole e 1
75   T D K D G K Y K L T V N G D - - - H E N D I C E V T V V K S P R E D C K E S V S  LAT52
78   T D E T G T Y K I E L K D S - - - H E E D I C E V V L V H S P L A N C S E I E A  Z16402
79   T D E T G T Y K I E L R D S - - - H V E D I C E V V L V K S P L A D C H E I Q S  U85499

                     130                 140                 150                 160
91   L R Y G V K G A I L E K S T K P P V S T K T P A T F E M F S V G P F A - C - - -  DcAGP1 C
97   E G W A K P S L K F K L N T - V N G T T R T V N P L G F F K K E A L P K C A Q V  Ole e 1
112  G - Y E K A R I E C S D N V G I H N A V R F A N P L F F M K A E S V Q G C K E A  LAT52
115  E - R D R A R V L L T R N V G I C D N L R L A N P L G Y L K D Y H - - - C P S A  Z16402
116  L - R D R A P V L L T R N V G I S D N L R L A N P L G Y L K D V P L P V C G D L  U85499

                     170
126  - - - F E P S T K K P C S H                                                      DcAGP1 C
136  Y N K L G M Y P P N M                                                            Ole e 1
151  L D E L G L F P - - L - E F                                                      LAT52
151  A L L K Q F D L A D D D N - E                                                    Z16402
155  L K M F K L A D D D D - D Q                                                      U85499
```

Figure 3. Pileup of the complete sequences of four pollen-expressed cell wall proteins that contain signal sequences and a PAC domain, together with the PAC domain only of the carrot AGP gene *DcAGP1*. Although the sequences Ole e 1 (Villalba *et al* 1993), LAT52 (Twell *et al* 1989), *Oryza sativa* (Z16402) and *Phalaris coerulescens* (U85499) exhibit low overall amino acid homology to the PAC domains in Fig 1, the six cysteines are present in similar positions (boxed).

4. CONCLUSION

Since the highly conserved cysteine-rich sequence with its consensus spacing of the cysteine residues is clearly the defining feature of the PAC domain family of proteins and several pollen-specific proteins, this structure most likely reflects a common function for this protein domain. Unfortunately, the precise biological function of these highly conserved cysteine residues in *Daucus carota* , or in any of the other plant species listed above, is as yet unknown! However, recent studies of the animal blood serum glycoprotein lutropin have demonstrated that a cysteine-rich domain is located at the N-terminus of a mannose receptor for this protein and accounts for the binding of oligosaccharides with terminal GalNAc-4-SO_4 residues (Fiete *et al* 1998). This cysteine-rich domain thereby represents a novel carbohydrate-binding motif in animals. Similar cysteine-rich domains have also been described in three other members of the endocytic C-type family of lectin receptors (Fiete *et al* 1998) and thus may serve an equivalent carbohydrate-binding function in these receptors. These data suggest that the cysteine-rich PAC domain observed in this family of proteins may have a role in binding to carbohydrate or glycoprotein ligands in higher plants.

ACKNOWLEDGEMENTS

We had particularly useful conversations about AGP sequences with Antony Bacic, Thomas Schindler, Concha Domingo, Alice Cheung and Allan Showalter.

REFERENCES

Baldwin, T. C., McCann, M., and Roberts, K., 1993, A novel hydroxyproline-deficient arabinogalactan-protein secreted by suspension-cultured cells of *Daucus carota*: purification and partial characterisation, *Plant Physiol.* **103**: 115-123.

Baldwin, T. C., Domingo, C., Schindler, T., Seetharaman, G., Stacey, N., and Roberts, K., 2000, Molecular characterisation of DcAGP1, an atypical arabinogalactan-protein, related to a family of basic proline-rich proteins, (Submitted to *Plant Mol. Biol.*).

Chen, C-G., Mau, S-L., and Clarke, A. E., 1993, Nucleotide sequence and stylar-specific expression of a novel proline-rich protein gene from *Nicotina alata*, *Plant Mol. Biol.* **21**: 391-395.

Chen, C-G., Pu, Z. Y., Moritz, R. L., Simpson, R. J., Bacic, A., Clarke, A. E., and Mau, S. L., 1994, Molecular cloning of a gene encoding an arabinogalactan-protein from pear (*Pyrus communis*) cell suspension culture, *Proc. Natl. Acad. Sci. USA*, **92**: 10305-10309.

Cheung, A. Y., May, B., Kawata, E. E., Gu, Q., and Wu, H-M., 1993, Characterisation of cDNAs for stylar transmitting tissue-specific proline-rich proteins in tobacco, *Plant J.* **3**: 151-160.

Du, H., Simpson, R. J., Moritz, R. L., Clarke, A. E., and Bacic, A., 1994, Isolation of protein backbone of an arabinogalactan-protein from styles of *Nicotiana alata* and characterisation of a corresponding cDNA, *Plant Cell* **6**: 1643-1653.

Du, H., Simpson, R. J., Clarke, A. E., and Bacic, A., 1996, Molecular characterisation of a stigma-specific gene encoding an arabinogalactan-protein (AGP) from *Nicotiana alata*, *Plant J.* **9**: 313-323.

Fiete, D. J., Beranek, M. C., and Baeziger, J. U., 1998, A cysteine-rich domain of the 'mannose' receptor mediates GalNAc-4-$S0_4$ binding, *Proc. Natl. Acad. Sci. USA* **95**: 2089-2093.

Gao, M., Kieliswеski, M. J., Lamport, D. T. A., and Showalter, A. M., 1999, Isolation and characterisation of a novel, modular tomato arabinogalactan-protein corresponding to the *LeAGP-1* gene, *Plant J.* **18**: 43-55.

Goldman, M. H., Pezzotti, M., Seurinck, J., and Mariani, C., 1992, Developmental expression of tobacco pistil-specific genes encoding novel extensin-like proteins, *Plant Cell* **4**: 1041-1051

Mau, S. L., Chen, C. G., Pu, Z. Y., Moritz, R. L., Simpson, R. J., Bacic, A., and Clarke, A. E., 1995, Molecular cloning of cDNAs encoding the protein backbone of arabinogalactan-proteins from the filtrate of suspension-cultured cells of *Pyrus communis* and *Nicotiana alata*, *Plant J.* **8**: 269-281.

Nothnagel, E. A., 1997, Proteoglycans and related components in plant cells, *Int. Rev. Cytol.* **174**: 195-291.

Schultz, C. J., Hauser, K., Lind, J. L., Atkinson, A. H., Pu, Z-Y., Anderson, M. A., and Clarke, A. E., 1997, Molecular characterization of a cDNA sequence encoding the backbone of a style-specific 120 kDa glycoprotein which has features of both extensins and arabinogalactan proteins. *Plant Mol. Biol.* **35**: 833-845.

Schultz, C., Gilson, P., Youl, J., Oxley, D., and Bacic, A., 1998, GPI-anchors on arabinogalactan-proteins: implications for signalling in plants, *Trends Plant Sci.* **3**: 426-431.

Sheng, J., D'Olvido, O., and Mehdy, M. C., 1991, Negative and positive regulation of a novel proline-rich protein mRNA by fungal elicitor and wounding, *Plant J.* **1**: 345-354.

Sheng, J., Jeong, J., and Mehdy, M. C., 1993, Developmental regulation and phytochrome-mediated induction of mRNAs encoding proline-rich proteins, glycine-rich proteins and hydroxyproline-rich glycoproteins in *Phaseolus vulgaris* L., *Proc. Natl. Acad. Sci. USA* **90**: 828-832.

Sommer-Knudsen, J., Clarke, A. E., and Bacic, A., 1996, A galactose-rich, cell-wall glycoprotein from styles of *Nicotiana alata*. *Plant J.* **9**: 71-83.

Twell, D., Wing, R., Yamaguchi, J., and McCormick, S., 1989, Isolation and expression of an anther-specific gene from tomato. *Mol. Gen. Genet.* **217**: 240-245.

Villalba, M., Batanero, E., Lopes-Otin, C., Sanchez, L. M., Monsalve, R. I., Gonzales de la Pena, M. A., Lahoz, C., and Rodriguez, R., 1993, The amino acid sequence of Ole e 1, the major allergen from olive tree (*Olea europaea*) pollen. *Eur. J. Biochem.* **216**: 863-869.

Youl, J. J., Bacic, A., and Oxley, D.,1998, Arabinogalactan-proteins from *Nicotiana alata* and *Pyrus communis* contain glycosylphosphatidylinositol membrane anchors. *Proc. Natl. Acad. Sci. USA* **95**: 7921-7926.

Chapter 5

Structure and Biosynthesis of L-Fucosylated Arabinogalactan-Proteins in Cruciferous Plants

Yohichi Hashimoto
Saitama University, Urawa, Japan

1. INTRODUCTION

Yamamoto (1982), surveying blood group ABH-active substances in vegetable foods, found potent blood group H-like activity inhibiting hemagglutination of human type O erythrocytes by eel anti-H sera in the polysaccharides from radish and turnip leaves. This finding prompted us to isolate and characterize L-fucosylated (L-Fuc) AGPs (F-AGPs) from leaves of radish, turnip, and rape (Nakamura *et al* 1984). The F-AGPs were purified from hot phosphate-buffered saline (PBS) extracts of these cruciferous plants by precipitation with three volumes of 99% ethanol and successive chromatography on DEAE-cellulose and on Sepharose 6B or Sephadex G-100. Elution with a linear gradient of $NaHCO_3$ separated acidic AGPs from neutral and more acidic polysaccharides like starch and pectin. Treatment with α-amylase removed contaminating starch in the crude leaf polysaccharide fraction or unbound polysaccharide mixture. Rechromatography on anion exchange resin followed by gel filtration was effective for the preparation of higher purity AGPs differing in molecular masses. From radish, rape, and *Sisybrium officinale* leaves and radish primary roots, recoveries of total AGPs were 100–200 mg/kg fresh tissue weight. The purified AGPs were apparently homogenous upon ultracentrifugation and electrophoresis, and were subjected to serological and chemical analyses. Serological and immunological characterization of AGPs were conducted through precipitation reactions with eel anti-H sera, *Aleuria aurantia* L-Fuc-specific lectin (AAL), and rabbit antibodies raised against the β-1,6-galactotetraosyl group coupled to BSA, as well as with Yariv artificial antigen (Tsumuraya *et al* 1984a, Ogura *et al* 1985, Kikuchi *et al* 1993). Radish F-AGPs inhibited strong hemagglutination of human type O erythrocytes by eel anti-H sera and AAL, but their complete lack of such inhibitory activity on the hemagglutination by chicken anti-H sera and *Ulex europeus* lectin, as well as their antigenicity, clearly distinguished F-AGPs from human blood group O substances (Nakamura *et al* 1984). Application of these methods to the PBS extracts from growing organs of radish plants revealed the maximal expression of blood H-like activity in roots at the earlier stage (2–8 days after germination) of development, and its rapid disappearance toward maturation (Tsumuraya *et al* 1988). In a survey of blood group H-like activity in 14 species of cruciferous plants, leaf polysaccharides from seven species gave positive reactions, and Table 1 shows the distribution of blood group H-active F-AGPs in cruciferous plants. Leaf F-AGPs from four plants related to turnip were found to be positive in all the precipitation reactions tested. No blood group H-like activity could be detected in the PBS-extracts from leaves of cabbage and Brussels sprouts. Indeed, AGPs purified from a PBS-extract of cabbage leaves were totally devoid of L-Fuc. The F-AGPs were isolated from watercress leaves but failed to inhibit hemagglutination of human O erythrocytes by eel anti-H sera even at a concentration of 400 μg/ml, suggesting the possibility that L-Fuc residues were attached to α-L-Ara*f* at a location different from the α-(1→2)-L-fucosidic linkage in the active polysaccharides.

Table 1. Distribution of L-Fuc AGPs in leaves of cruciferous plants — hemagglutination inhibition test

Genus	Species	Fucosylated AGP[a]
Matttiola	*incana* cv. Purple heart (stock)	–
Capsella	*bursa-pastoris* (Shepherd's purse)	+
Raphanus	*sativus* L. (radish)	+[b]
Brassica	*campestris* L. (rape)	+[b]
	rapa var. *glabra* (turnip)	+[b]
	rapa var. *chinensis* (Chingensai)[d]	+[b]
	rapa var *amplexicaulis* (Chinese cabbage)	+[b]
	rapa var. *hakabura* (Nozawana)[d]	+[b]
	rapa var. *periviridis* (Komatsuna)[d]	+[b]
	rapa var. *narinosa* (Tahsai)[d]	–[c]
	juncea Czern et Coss (mustard)	+[b]
	oleracea var. *capitata* (cabbage)	–[b]
	oleracea var. *gemmifera* (Brussel sprouts)	–
Wasabia	*japonica* Matsumura (Wasabi)	+[b]
Nasturtium	*officinale* (watercress)	±[b]
Sisymbrium	*officinale* (Kakinegarashi)[d]	+[b]
Rorippa	*indica*	–

[a]Blood group H-like activity: activity inhibiting the hemagglutination of human type O erythrocytes by eel anti-H sera or *Aleuria aurantia* L-Fuc-specific lectin.

[b]Fucosylated AGPs were isolated from leaf tissues and characterized.

[c]Leaf AGPs of *B. rapa* var. *narinosa* contain an unidentified 6-deoxyhexose, possibly L-Rha, instead of L-Fuc.

[d]Japanese name.

2. STRUCTURE OF RADISH AGPS

Table 2 compares the compositions and properties of AGPs from radish leaves, seeds, and primary and mature roots (Tsumuraya *et al* 1984b 1987 1988). Variation among them is evident, indicating their organ-specific and developmentally regulated expression and the multiple nature of their biological functions. Radish primary root AGPs are fucosylated and exhibited blood group H-like activities comparable to those of radish leaf F-AGPs. The leaf and primary root F-AGPs formed single precipitin lines that fused to each other in the reactions with eel anti-H precipitin and AAL (Tsumuraya *et al* 1988). Radish mature root produces serologically inactive AGPs that lack L-Fuc and exhibit no precipitation reaction with L-Fuc-specific lectins. Reacting with Yariv artificial antigen, all purified radish AGPs formed a single precipitin line as is typical of AGP proteoglycans. Major AGPs from radish organs are acidic proteoglycans because of their contents of 4–14% (mol% of total sugars) of GlcA and its 4-O-methyl derivative. Table 3 shows the carbohydrate compositions and properties of F-AGPs from leaves of five species of cruciferous plants to verify a wide distribution of H-active proteoglycans among this plant family (Nakamura *et al* 1984). Some F-AGPs analagous to radish leaf AGPs have been isolated and characterized from leaves of cultivated species of turnip such as Chinese cabbage, *Brassica rapa* var. *amplexcaulis*, *B. rapa* var. *hakabura*, and *B. rapa* var. *perviridis*. Interestingly, AGPs from a Chinese vegetable, *B. rapa* var. *norinosa,* contain 2–4% of a 6-deoxyhexose, possibly Rha, instead of L-Fuc, suggesting mutation at the enzyme level.

Table 2. Properties and carbohydrate compositions (mol%) of AGPs in radish organs

	Seeds	Leaves		Primary Roots		Mature Roots	
	AGP	R-I	R-II	AGP-I	AGP-II	AGP-III	AGP-IV
Gal	41.6	58.0	56.9	33	61	40	62
Ara	44.8	27.8	33.0	24	27	55	24
L-Fuc	0.0	4.2	6.0	10	5	0	0
Xyl	3.2	0.0	0.0	0	0	0	0
Glc	0.2	0.0	0.0	16	0	0.1	0
Gal/Ara (mol/mol)	0.93	2.09	1.72	1.4	2.3	0.73	2.6
Uronic acids[a]	10.2	10.4	4.5	8	7	4	14
GlcA	6.0	1.8	0.8	n.d.	3.3	n.d.	0
4-O-Me-GlcA	4.2	8.6	3.7	n.d.	3.7	n.d.	14
Mol wt (kDa)	52	130	75	21	78	25,39	88
H-like activity (μg/ml)[b]	n.d.	10	5	5	5	n.d.	n.d.
Yariv[c]	+	+	+	+	+	+	+
AAL[d]	–	+	+	+	+	–	–
Anti-Gal$_4$[e]	+	+	+	+	+	+	+

[a]Determined as GlcA; n.d.: not determined; +: positive reaction; –: negative reaction.
[b]H-like activity (μg/ml): minimum amount required for complete inhibition of hemagglutination of human type O erythrocytes by eel anti-H agglutinin.
[c]Yariv: Precipitation reaction with Yariv artificial antigen
[d]AAL: Formation of a precipitin line with *Aleuria aurantia* L-Fuc-specific lectin.
[e]Anti-Gal$_4$: Precipitation reaction with rabbit antibodies against β-1,6-galactotetraosyl group coupled to BSA.

Table 3. Comparisons of carbohydrate compositions (mol%) and properties of fucosylated AGPs from leaves of five cruciferous plants

	Brassica campestris (rape)	*B. rapa* var. *glabra* (turnip)		*B. rapa* var. *perviridis* (Komatsuna)	*Wasabia japonica* (Wasabi)		*Sisybrium officinale* (Kakinegarashi)		
	B-N	B-I	B-II	BR-AGP	WJ-I	WJ-II	SOI-1	SOII-2	SOIII-1
Gal	52.6	22.4	57.0	57.2	48.5	40.3	48.5	55.9	60.9
Ara	38.2	72.0	38.1	41.8	45.3	50.8	37.5	36.2	33.0
Fuc	3.6	4.3	2.2	1.2	6.1	7.9	7.5	5.2	3.2
Xyl	0.0	1.3	0.0	0.0	0.0	0.0	<0.1	<0.1	<0.1
Glc	<0.1	0.0	0.0	0.0	0.0	0.0	0.0	0.0	0.0
Man	0.0	0.0	0.0	0.0	0.0	0.0	4.5	<1.0	<1.0
Gal/Ara (mol/mol)	1.37	0.31	1.50	1.37	1.07	0.8	1.28	1.47	1.85
Uronic acids[a]	5.5	2.7	5.5	6.0	5.0	4.0	<1.0	3.9	6.1
Mol wt (kDa)	50	40, 150	180	48	58	11	40	50	60
H-like activity (μg/ml)[b]	10	5	20	125	30	63	10	20	40
AAL[c]	+	+	+	+	+	+	+	+	+
Yariv[d]	+	+	+	+	+	+	+	+	+
Anti-Gal$_4$[e]	+	+	+	+	+	+	+	+	+

[a]Determined as GlcA; n.d.: not determined; +: positive reaction; –: negative reaction.
[b]H-like activity (μg/ml): minimum amount required for complete inhibition of hemagglutination of human type O erythrocytes by eel anti-H agglutinin.
[c]AAL: Formation of a precipitin line with *Aleuria aurantia* L-Fuc-specific lectin.
[d]Yariv: Precipitation reaction with Yariv artificial antigen.
[e]Anti-Gal$_4$: Precipitation reaction with rabbit antibodies against β-1,6-galactotetraosyl group coupled to BSA.

Radish F-AGPs and their modified products have been subjected to chemical and enzymatic analyses to elucidate the complex structures of their carbohydrate moieties (Tsumuraya *et al* 1984b 1988, Ogura *et al* 1985). *Aspergillus niger* β-glucuronidase, *Irpex lacteus* exo-β-1,3-galactanase, and *Trichoderma viride* endo-β-1,6-galactanase have been purified and characterized with regard to their specificity and mode of action. Highly purified preparations of these enzymes, together with other commercial glycosidases, have been effectively used for structural modification and analyses of the carbohydrate moieties of F-AGPs (Tsumuraya *et al* 1984b, Kikuchi *et al* 1993). Table 4 summarizes the sources of enzymes; substrates including synthetic glycosides, oligomers, and native and modified AGPs; and major sugars released from the substrates by the action of each enzyme. These enzymes, except for *Rhodotorula flava* α-L-arabinofuranosidase and *Fusarium oxysporum* α-L-fucosidase, have been purified from commercial enzyme preparations of fungi and a basidiomycete, which were mostly developed for the formation of plant protoplasts. Incubation of F-AGPs with *R. flava* α-L-Arafase released a large part of the L-Ara, thereby rendering the proteoglycan susceptible to the action of the other enzymes (Ogura *et al* 1985, Tsumuraya *et al* 1984b 1988). Smith degradation of radish and rape leaf F-AGPs resulted in a complete elimination of the peripheral sugar residues, leaving a core structure, and β-1,6-galactooligomers carrying D-glycerol at the reducing ends, which were generated from the side chains of the backbone structure (Tsumuraya *et al* 1984b 1988, Ogura *et al* 1985). The type of linkage and anomeric configuration of the β-D-Gal residues were established by methylation, action of *E. coli* β-galactosidase, and NMR (Tsumuraya *et al* 1984b). Methylation of radish leaf and primary root AGPs and their modified products corroborated that they differed unequivocally in the degrees of branching and side chain lengths, whereas L-Fuc, uronic acids and a large part of L-Ara*f* were equally located at the non-reducing ends (Tsumuraya *et al* 1984b 1988, Ogura *et al* 1985).

Table 4. Glycosidases useful in structural analyses of radish F-AGPs

Glycosidase	Source	Substrate	Product
α-L-Arabinofuranosidase	*Rhodotorula flava*	PNP α-L-Ara*f*	PNP + L-Ara
(α-L-Arafase)	(IFO-0407)	L-Ara*f*α(1→3)Galβ(1→6)Gal	L-Ara + Gal_2
		(AraGalGal)	
		Radish leaf and root AGPs	L-Ara
β-Glucuronidase	*Aspergillus niger*	PNP β-GlcA	PNP + GlcA
(β-GlcA-ase)	Pectinex Ultra SP-L	GlcAβ(1→6)Gal	GlcA + Gal
	(Novo Nordisk Ferment)	4-O-Me-GlcAβ(1→6)Gal α-L-Arafase-digested radish root AGP	4-O-Me-GlcA+Gal 4-O-Me-GlcA
α-L-Fucosidase	*Fusarium*	PNP α-L-Fuc	PNP + L-Fuc
(α-L-Fucase)	*oxysporum*	L-Fucα(1→2)L-Ara*f*α(1→3)Galβ(1→6)Gal (FucAraGalGal)	L-Fuc + AraGalGal
Exo-β-1,3-galactanase	*Irpex lacteus*	β-1,3-Galactan	Gal
(β-1,3-Galase)	(*Polyporus tulipiferae*) Driselase (Kyowa Co.)	α-L-Arafase-digested radish AGPs	Gal + neutral & acidic β-1,6-galactooligomers AraGalGal
Endo-β-1,6-galactanase	*Trichoderma viride*	α -L-Arafase-digested radish	β-1,6-Galactooligomers
(β-1,6-Galase)	Onozuka R-10	root AGP	(Gal_n: n=2-4)
	(Yakult Co.)		4-O-Me-GlcA-(Gal_n) n=1–3

PNP: *p*-nitrophenyl, *p*-nitrophenol; L-Ara: L-arabinose, L-Ara*f*: L-arabinofuranoside; Gal: D-galactose; GlcA: D-glucuronic acid, D-glucuronide; 4-O-Me-GlcA: 4-O-methyl D-glucuronic acid; L-Fuc: L-fucose, L-fucoside

An exo-β-1,3-galactanase (β-1,3-Galase) purified from Driselase, an enzyme product of *Irpex lacteus* (*Polyporus tulipiferae*), is a useful tool for structural analyses of radish F-AGPs. The enzyme is able to attack native and modified radish AGPs in an exowise manner, liberating D-Gal and oligomers carrying an additional Gal residue at the reducing ends, which are split off from the main chains. Thus, exhaustive β-1,3-galactanase digestion of native and α-L-Arafase-digested radish F-AGPs supplied neutral and acidic oligomers, which are a prerequisite for the determination of complex side chain structures, as well as for characterization of the strict acceptor specificities of glycosyltransferases. For instance, AraGalGal, and β-1,3- and β-1,6-galactooligomers were used as valuable substrates for the enzymes to determine their participation in the biosynthesis of F-AGPs (Tsumuraya *et al* 1984b 1990, Misawa *et al* 1996).

Methylation of galactooligomers containing equal amounts of L-Fuc and L-Ara demonstrated the attachment of L-Fuc to a terminal L-Ara*f* residue at O-2, which linked to O-3 of a β-D-Gal residue in the side chains, and the linkage of a small amount of L-Ara*f* to O-5 of an adjacent α-L-Ara*f* residue (Tsumuraya *et al* 1984b, Ogura *et al* 1985). Likewise, methylation of carboxyl-reduced AGPs and aldobiouronic acids revealed the attachment of GlcA and 4-O-Me-GlcA to the terminal β-D-Gal residue of the side chains through β-1,6-glycosidic linkage (Tsumuraya *et al* 1984b 1988, Ogura *et al* 1985). Anomeric configurations of the terminal sugars were determined by susceptibility of the glycosidic linkage to specific glycosidases and by NMR. For instance, the cleavage of L-Fuc-containing oligomers by the action of *Charonia lampas* or *F. oxysporum* α-L-fucosidases strongly supports α-L-fucosidic linkage (Tsumuraya *et al* 1984b, Misawa *et al* 1996). *Aspergillus niger* β-glucuronidase purified from Pectinex Ultra SP-L (Novo Nordisk Ferment), completely hydrolyzed aldobiouronic acids to uronic acids and D-Gal, indicative of β-glycosidic linkage of the terminal GlcA or 4-O-Me-GlcA.

Figure 1 shows a proposed structure of the carbohydrate moiety of a radish leaf AGP and the linkages susceptible to glycosidases used in structural analyses. The structure L-Fucα(1→2)L-Ara*f*α(1→3)D-Galβ(1→6)D-Galβ(1→, which is responsible for blood group H-like activity, closely resembles the H-antigenic structure L-Fucα(1→2)D-Gal(1→3)D-GlcNAcβ(1→ of human blood group O(H) substance but is completely inactive as the antigen (Tsumuraya *et al* 1984a). The biological significance of fucosylation is not feasible to surmise directly from the variable and highly complex structures of AGPs. Conceivably, the attachment of auxiliary sugars such as L-Ara*f*, L-Fuc and GlcA to the side chains of the β-3,6-galactan backbone and methylation at O-4 of the terminal GlcA might play a crucial role in enhancing the resistance of the carbohydrate moiety against the attacks of plant pathogens. It has been accepted that pathogenic fungi and bacteria are able to excrete various enzymes to destroy cell walls and matrix, thereby facilitating invasion of host plant tissues. It is possible that fragile tissues of primary roots and frequent infection through stomata in leaves induced cruciferous plants to develop a novel defense system for AGPs, which involved both a unique enzyme catalyzing fucosylation of α-L-Ara*f* residues attached to the side chains and methylation of terminal GlcA at O-4. Hence, organ-specific fucosylation of AGPs is likely to be an indispensable mechanism for the maintenance of the AGPs in biologically active forms, which are operating in developing roots and leaves.

Furthermore, it can be anticipated that chemically and enzymatically modified AGPs, and their degradation products such as neutral and acidic oligomers arising from the side chains, may exert biological effects on cell proliferation, development, growth and differentiation in organs. Exchange of the native AGPs in tissues or proliferating cells with those of different structures by mild infusion facilitates an extensive survey for biological effects of structurally modified AGPs on plants. These approaches are important toward finding a clue for understanding the biological role of the carbohydrate structures in AGPs.

3. BIOSYNTHESIS OF RADISH AGPS

From the structural data, it can be inferred that biosynthesis of the carbohydrate moiety of F-AGPs proceeds on the basis of concurrent elongation and branching of the β-1,3-galactan main chain, resulting in the formation of the β-3,6-galactan backbone of F-AGPs to which such auxiliary sugars such as L-Ara*f*, L-Fuc, and GlcA are attached.

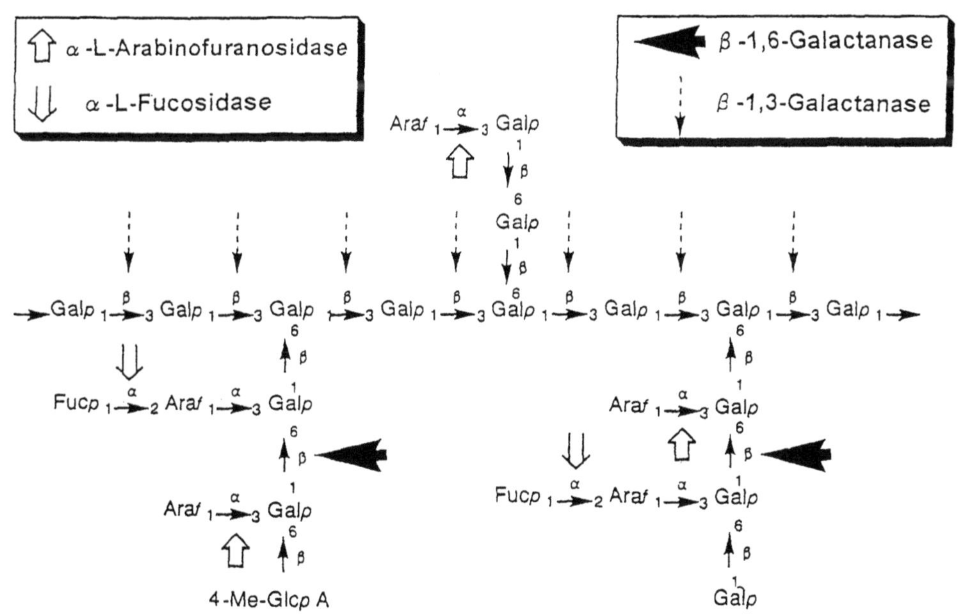

Figure 1. A proposed structure of the carbohydrate moiety of radish leaf AGP and the linkages (arrows) susceptible to glycosidases used in structural analyses.

The α-L-fucosyltransferase activities in microsomes from different organs of the growing radish plant after germination have been measured using L-Ara*f*α(1→3)

D-Galβ(1→6)D-Gal (AraGalGal) and tamarind xyloglucan (XG). As a result, separate enzymes for the two substrates were found to be expressed differently. Fucosylation of AraGalGal occurred in leaves and primary root, although in the latter the enzyme activity turned out to be undetectable toward maturation. In contrast, α-L-Fuc transfer to β-D-Gal in tamarind XG was found in microsomes prepared from leaves, hypocotyls, roots, and midribs in different stages of development, coincident with the ubiquitous distribution of fucosylated XG in tissues.

Regarding these findings, Golgi membranes were fractionated from fragile primary roots and used as the enzyme preparation because of the low contamination of fluorescent materials in such roots. The enzyme assay was carried out by fluorometric and radiolabeling methods using GDP-L-Fuc and the pyridylamino (PA)-derivative (AraGalGal-PA) as a fluorogenic substrate (Misawa *et al* 1996). Fluorescence (excitation at 320 nm, emission at 400) was recorded. A fluorogenic transfer product, L-FucAraGalGal-PA, resulted and was characterized by stepwise action of *F. oxysporum* α-L-fucosidase and *R. flava* α-L-arabinofuranosidase, followed by identification of the resulting products upon HPLC, as well as by methylation and liquid secondary ion mass spectrometric analyses, and measurement of blood group H-like activity. The enzyme exhibits maximal activity at pH 6.8, 25°C when L-Fuc transfer from GDP-L-Fuc to the standard acceptor AraGalGal was measured in the presence of 5 mM Mn^{2+}, 0.1% Zwittergent 3-16 or 0.1% Triton X-100, and 0.4 mM dithiothreitol. Relative activity (%) with various acceptors, taking that for AraGalGal-PA as unity (100), are shown in parentheses: 4-O-Me-GlcAβ (1→6)[L-Arafα(1→3)]D-Galβ(1→6)D-Gal(4-O-Me-GlcA(Ara)GalGal) (115), Galβ-(1→3)GlcNAc (7.5), D-Galβ(1→3)D-GlcNAcβ(1→3)D-Galβ(1→4)D-Glc (lacto-N-tetraose) (6.1), tamarind XG (66.3), XG-oligomer 8 ($Glc_4Xyl_3Gal_1$) (1), and XG-oligomer 9 ($Glc_4Xyl_3Gal_2$) (2.6). Radish root and seed AGPs were virtually inactive, owing to their relative activities of 2.2 and 3.6% even after prolonged incubation for 20 hours. However, nine fractions of neutral and acidic L-Ara-containing saccharides were resolved from an exo-β-1,3-galactanase-digest of a radish root AGP upon gel permeation chromatography and were shown to exhibit 6–80% of the acceptor activities as measured by radiolabeling. In particular, neutral and acidic galactooligomers containing 2 mol of L-Ara*f* residues acted as excellent acceptors, whereas higher molecular weight fragments were inactive. The K_M values (mM) for GDP-L-Fuc at fixed concentrations of AraGalGal-PA (10 mM), AraGalGal (33 mM), and 4-O-MeGlcA(Ara)GalGal (30 mM) were determined to be 0.17 (0.08, radiolabeling method), 0.08, and 0.04, respectively, on the basis of Hans-Woolf plots. Their respective K_M values (mM) at a fixed concentration of GDP-L-Fuc (1 mM) were 4.50 (3.72, radiolabeling method), 16.3, and 2.47. These data indicate that 4-O-Me-GlcA(Ara)GalGal is a good substrate.

The two enzymes involved in fucosylation at O-2 of α-L-Ara*f* in AGPs and β-D-Gal in tamarind XG can be clearly discriminated from each other in the following points: (1) the differential expression in organs and at the stages of development, (2) strict acceptor specificity, (3) response to detergents, (4) kinetic parameters, and (5) sensitivity to xylooligomers. Moreover, β-D-Gal(XG) :

2-α-L-fucosyltransferases in primary root and hypocotyl microsomes are essentially indiscernible from each other in their properties and kinetic parameters. Hence, it became obvious that a novel enzyme designated α-L-Ara*f* : 2-α-L-fucosyltransferase is specifically expressed in Golgi membranes of radish primary roots and leaves, resulting in the biosynthesis of organ-specific F-AGPs. Consistent with their contents of serologically active F-AGPs, appreciable activities (A, expressed as specific activity, pmol min^{-1} mg $protein^{-1}$) of this enzyme together with those (B, specific activity) of β-D-Gal(XG) : 2-α-L-fucosyltransferase could be detected as shown in the parentheses in primary root Golgi membranes of rape (A, 229; B, 471) *B. rapa* var. *hakubura* (A, 239; B, 289), *B. rapa* var. *perviridis* (A, 167; B, 454) and *B. rapa* var. *chinensis* (A, 586; B, 356).

β-Galactosyltransferases in Golgi membranes from radish primary roots have been investigated relative to the synthesis of the β-3,6-galactan backbone of F-AGPs. A β-1,3-galactan, a Smith degradation product of Acacia gum, was used as the standard substrate for β-D-Gal transfer from UDP-[^{14}C]-Gal. The enzyme is optimal at pH 6.0, 25°C, and requires Mn^{2+},Triton X-100, and dithiothreitol for full activity. β-1,3-Galactooligomers (DP = 2-5) are good substrates, and their relative activities (%) increase remarkably with increasing chain lengths in the following order; pentamer (160) > tetramer (151) > β-1,3-galactan (100) > trimer (73) > dimer (38). On the contrary, β-1,6-galactooligomers (DP = 2–6) were apparently inactive as the acceptor because their relative activities remained in the range of 8–9 (that for β-1,3-galactan was 100%), regardless of their chain lengths. Lacto-*N*-tetraose, lacto-*N*-neo-tetraose, and AraGalGal served as good acceptors with rates comparable to that of the dimer. β-1,6-Galactobiose was identified as a major labelled product in a β-1,3-Galase-digest of the transfer product of β-1,3-galactan. A radish mature root AGP is a poor acceptor but undergoes structural changes to be labelled after successive digestion with α-L-Arafase and *Trichoderma viride* endo-β-1,6-galactanase (β-1,6-Galase). The ^{14}C-Gal incorporated into the modified mature root AGP was recovered as ^{14}C-β-1,6-galactobiose and a small quantity of ^{14}C-Gal from the β-1,3-Galase-digest of the labelled root AGP. This indicates β-D-Gal transfer to O-6 or O-3 of the non-reducing terminal and the inner β-D-Gal residues, suggesting participation of 3-β-D- and 6-β-D-galactosyltransferase(s) in elongation and branching of the main and side chains of the substrates. β-1,4-Galactooligomers (DP = 2-5) and soybean β-1,4-galactan were labelled at similar rates to those of the β-1,3-isomers, and the transfer products formed were susceptible to *Penicillium citrinum* endo-β-1,4-galactanase giving rise to ^{14}C-β-galactobiose and -triose as the breakdown products. From these results, it is evident that two or three β-galactosyltransferases exist in radish primary root Golgi membranes and are distinguishable from each other by their specificities.

β-Glucuronosyl transfer from UDP-[^{14}C]-GlcA to O-6 of non-reducing terminal β-D-Gal residues by enzymes in radish primary root Golgi membranes has been demonstrated by identification of ^{14}C-aldobiouronic acid and the corresponding trisaccharide, which were isolated from β-1,3-Galase-digests of labelled products from β-1,3-galactan, and from an α-L-Arafase- and β-1,6-Galase-digested product of a radish root AGP. Radish β-glucuronosyltransferase is maximally active at pH 6.5, 30°C, in a reaction mixture containing β-1,3-galactan, Mn^{2+}, Triton X-100, and dithiothreitol. The reaction is inhibited by D-galacturonic acid, ATP, UDP and PP_i.

A specific transfer of α-L-Ara from UDP-L-Ara to β-1,6-galactobiose and -triose was shown to occur in the Golgi membranes, indicating the occurrence of the responsible α-L-arabinofuranosyltransferase for the peripheral structure of AGPs.

From the data, it was found that glycosyltransferases capable of catalyzing transfer of β-Gal, α-L-Ara, α-L-Fuc, and β-GlcA to the structurally related acceptors in AGPs are localized in Golgi membranes of primary roots and might be involved in biosynthesis of the carbohydrate moiety. These findings are only the first step to solving the biosynthetic pathway of highly branched carbohydrate structures of AGPs from a biochemical point of view. More elegant approaches and insights will be needed to break through the complex biochemical events.

REFERENCES

Kikuchi, S., Ohinata, A., Tsumuraya, Y., Hashimoto, Y., Kaneko, Y., and Matsushima, H., 1993, Production and characterization of antibodies to the β-(1→6)-galactotetraosyl group and the interaction with arabinogalactan-proteins, *Planta* **190**: 525–535.

Misawa, H., Tsumuraya, Y., Kaneko, Y., and Hashimoto, Y., 1996, α-L-Fucosyltransferase from radish primary roots, *Plant Physiol.* **110**: 665–673.

Nakamura, K., Tsumuraya, Y., Hashimoto, Y., and Yamamoto, S., 1984, Arabinogalactan-proteins reacting with eel anti-H agglutinin from leaves of cruciferous plants, *Agric. Biol. Chem.* **48**: 753–766.

Ogura, K., Tsumuraya, Y., Hashimoto, Y., and Yamamoto, S., 1985, An arabinogalactan-protein from rape leaves, *Agric. Biol. Chem.* **49**: 2851–2857.

Tsumuraya, Y., Nakamura, K., Hashimoto, Y., and Yamamoto, S., 1984a, Immunological properties of arabinogalactan-proteins from leaves of cruciferous plants, *Agric. Biol. Chem.* **48**: 2915–2917.

Tsumuraya, Y., Hashimoto, Y., Yamamoto, S., and Shibuya, N., 1984b, Structure of L-arabino-D-galactan-containing glycoproteins from radish leaves, *Carbohyd. Res.* **134**: 215-228.

Tsumuraya, Y., Hashimoto, Y., and Yamamoto, S., 1987, An arabinogalactan and an L-arabino-D-galactan containing proteoglycan from radish (*Raphanus sativus*) seeds, *Carbohyd. Res.* **161**: 113–126.

Tsumuraya, Y., Ogura, K., Hashimoto, Y., Mukoyama, H., and Yamamoto, S., 1988, Arabinogalactan-proteins from primary and mature roots of radish (*Raphanus sativus* L.), *Plant Physiol.* **86**: 155–160.

Tsumuraya, Y., Mochizuki, N., Hashimoto, Y., and Kovac, P., 1990, Purification of an exo-β-(1→3)-D-galactanase of *Irpex lacteus* (*Polyporus tulipiferae*) and its action on arabinogalactan-proteins, *J. Biol. Chem.* **265**: 7207–7215.

Yamamoto, S., 1982, Immunological comparison of carbohydrates with blood group ABH-activities contained in vegetable foods: a preliminary study, *Agric. Biol. Chem.* **46**: 1971–1974.

Chapter 6

Characterization and Localization of a Novel Tomato Arabinogalactan-Protein (LeAGP-1) and the Involvement of Arabinogalactan-Proteins in Programmed Cell Death

Allan M. Showalter,[1] Minggeng Gao,[1] Marcia J. Kieliszewski[2] and Derek T. A. Lamport[3]

[1]Departments of Environmental and Plant Biology and [2]Chemistry and Biochemistry, Molecular and Cellular Biology Program, Ohio University, Athens, OH 45701-2979; and [3]Emeritus Professor, MSU-DOE Plant Research Laboratory, Michigan State University, East Lansing, MI 48824-1312

1. INTRODUCTION

Arabinogalactan-proteins (AGPs) are a family of plant Hyp-rich glycoproteins containing Type II AG side chains (Fincher *et al* 1983, Nothnagel 1997). Several AGPs have been characterized over the years on the (glyco)protein level and, more recently, on the molecular level. Collectively, these studies indicate that AGPs are a diverse family that can be classified in one of two groups: the "classical" and the "non-classical" AGPs. Classical AGPs contain a polypeptide with at least three distinct domains: an N-terminal secretion signal sequence; a central domain, which contains most of the Pro/Hyp residues; and a C-terminal hydrophobic region, which is involved in the attachment of a glycosylphosphatidylinositol (GPI) membrane anchor (Youl *et al* 1998). The classical designation is given when Pro/Hyp, Ala, Thr, and Ser represent the most abundant residues in the encoded polypeptide, excluding the N-terminal signal sequence; otherwise the non-classical designation applies. Clones corresponding to non-classical AGPs specify N-terminal secretion signal sequences and various other domains, including Pro/Hyp-rich, Asn-rich, Cys-rich, and Hyp-poor domains, but, to date, do not specify any C-terminal hydrophobic domains. As additional structural data are obtained on AGPs, this working classification system will undoubtedly be refined. More importantly, additional structural data and the availability of purified AGPs will help elucidate the function of these extraordinary plant glycoproteins.

Currently, AGPs are believed to function in various aspects of plant growth and development (Nothnagel 1997). Much of this belief is based on studies performed using monoclonal antibodies directed against AGP carbohydrate epitopes, which indicate that AGPs are markers of cellular identity during development. Others have used certain AGPs to induce or inhibit plant embryogenesis. Several other groups have used Yariv reagent to perturb growth and development. For example, Serpe and Nothnagel (1994) treated rose cell cultures with Yariv reagent and observed inhibition of cell growth. Similar experiments using Yariv reagent to bind AGPs and perturb function resulted in inhibition of lily pollen tube growth *in vitro* (Jauh and Lord 1996, Roy *et al* 1998), in inhibition of *Arabidopsis* root growth (Willats and Knox 1996), in blocking elongation of suspension-cultured carrot cells (Willats and Knox 1996), and in the death of young rose cells in culture (Langan and Nothnagel 1997).

In this communication we focus on the structure and localization of LeAGP-1, a major AGP expressed in tomato cell cultures and plants. In addition, we present data dealing with the production of transgenic *LeAGP-1* antisense plantlets and the utilization of Yariv reagent to induce programmed cell death (PCD) in cultured cells. These later studies respectively represent our initial assault on elucidating the function of LeAGP-1 and AGPs.

2. RESULTS AND DISCUSSION

2.1 Molecular and Biochemical Characterization of LeAGP-1

The LeAGP-1 was first identified by DNA cloning in tomato (Fig 1) (Pogson and Davies 1995, Li and Showalter 1996). Verification that *LeAGP-1* cDNA and genomic clones encode an AGP core protein was accomplished by precipitating total AGPs from tomato cultured cells and purifying the corresponding core proteins. Yariv reagent was used to precipitate total AGPs from cultured tomato cells following homogenization or sonication. Subsequently, anhydrous hydrofluoric acid was used to deglycosylate these AGPs, and the resulting deglycosylated core proteins were separated by reverse phase HPLC on an PRP-1 column. Each of the resulting core protein fractions was subjected to amino acid composition analysis (Table 1). Although all four fractions had compositions consistent with classical AGPs, one fraction (HC-3) matched the composition predicted by the *LeAGP-1* cDNA clones the best. These fractions were also tested in Western blot and gel blots for their ability to react with an antibody produced from a synthetic peptide containing the Lys-rich subdomain, predicted by the *LeAGP-1* DNA clones, conjugated to keyhole lympet hemocyanin. Only fraction HC-3 reacted with this so-called PAP antibody named with respect to the first three residues in the synthetic peptide (see Fig 1). Fraction HC-3 was then subjected to trypsin digestion, and the resulting tryptic peptides were separated by reverse phase HPLC chromatography and sequenced. Three peptide sequences were obtained: 1) Hyp-Ala-Ala-Ala-Hyp-Thr-Lys-Pro-Lys, 2) Ala-Hyp-Ala-Ser-Ser-Hyp-Hyp-Val-Gln-Ser-Hyp-Hyp-Ala-Hyp-Ala-Hyp-Glu-Val-Ala-Thr-Hyp-Hyp-<u>Ala-</u>Val..., and 3) Ala-Hyp-Ala-Ser-Ser-Hyp-Hyp-Val-Gln-Ser-Hyp-Hyp-Ala-Hyp-Ala-Hyp-Glu-Val-Ala-Thr-Hyp-Hyp-Val-Ser-Thr-Hyp-Hyp-Ala-Ala-Ala-(Xaa)-Val-Ala-Ala.... These sequences corresponded precisely to the sequence predicted from the *LeAGP-1* clones and further indicated which prolyl residues were hydroxylated (Fig 1). It should be noted that two of these peptide sequences overlap and that a predicted Ala residue is absent in the longer overlapping sequence, indicating that LeAGP-1 isoforms exist or a minor amino acid sequencing error occurred. Taken together, these data indicate that fraction HC-3 represents an AGP core protein that corresponds to that predicted by the *LeAGP-1* clones.

(a)

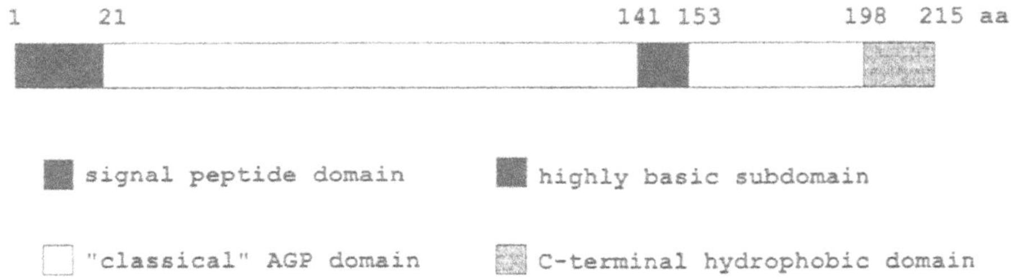

(b)

```
MetAspArgLysPheValPheLeuValSerIleLeuCysIleValValAlaSerValThr   20
GlyGlnThrProAlaAlaAlaProValGlyAlaLysAlaGlyThrThrProHypAlaAla   40
AlaHypThrLysProLysThrProAlaProAlaThrAlaProAlaSerAlaProProThr   60
AlaValProValAlaProValThrAlaProValThrAlaProThrThrProValValAla   80
AlaProValSerAlaProAlaSerSerProProLeuLysAlaHypAlaSerSerHypHyp  100
ValGlnSerHypHypAlaHypAlaHypGluValAlaThrHypHypAlaValSerThrHyp  120
HypAlaAlaAlaProValAlaAlaProValAlaSerGluThrThrProAlaProAlaPro  140
SerLysGlyLysValLysGlyLysLysGlyLysLysHisAsnAlaSerProAlaProSer  160
ProAspMetMetSerProProAlaProProSerGluAlaProGlyProSerMetAspSer  180
AspSerAlaProSerProSerLeuAsnAspGluSerGlyAlaGluLysLeuLysMetLeu  200
GlySerLeuValAlaGlyTrpAlaValMetSerTrpLeuLeuPhe                 215
```

Figure 1. (a) Domain organization for the predicted LeAGP-1 core protein consists of an N-terminal signal peptide, a central "classical" AGP domain sandwiching a highly basic subdomain, and a C-terminal hydrophobic domain, which may allow for GPI-mediated plasma membrane association. This modular organization was predicted by *LeAGP-1* cDNA and genomic clones. (b) Amino acid sequence of LeAGP-1 as predicted by *LeAGP-1* DNA clones and embellished with amino acid sequencing data. A potential N-glycosylation site (Asn-X-Ser/Thr) appears immediately following the basic domain and is italicized. A synthetic peptide encompassing the basic (Lys-rich) subdomain used here for production of a LeAGP-1 antibody is underlined in medium boldface; the peptide and its corresponding antibody are given a "PAP" designation in accordance with the first three residues of this synthetic sequence. Peptide sequences elucidated by amino acid sequencing are underlined; Pro residues found to be hydroxylated in these sequences are displayed as Hyp residues. Note that the Pro (or Hyp) residue at position 125 was undetermined by amino acid sequencing and hence is indicated with a dotted underline. The predicted site of C-terminal processing and GPI-anchor addition at Ser 192 is underlined in heavy boldface and discussed in the text.

Table 1. Amino acid compositions[a] of AGPs isolated from tomato suspension-cultured cells (i.e., HC-AGPs) compared with that predicted from *LeAGP-1* DNA clones

AA	LeAGP-1[b]	LeAGP-1[c]	HC-3	HC-1	HC-2	HC-4
Hyp	–	–	29	37	34	30
Pro	24.4	27.5	1	2	2	2
Asx	3.1	3.5	2	6	3	2
Thr	7.7	8.8	10	9	8	8
Ser	1.4	11.7	12	11	11	18
Glx	3.6	3.5	3	1	3	3
Gly	4.6	3.5	5	7	9	6
Ala	22.2	23.4	21	19	19	15
Val	8.2	8.2	9	3	4	7
Cys	0.0	0.0	0	0	0	0
Met	2.6	1.7	0	0	0	0
Ile	0.0	0.0	1	2	4	2
Leu	3.6	1.2	1	2	2	3
Tyr	0.0	0.0	0	0	0	0
Phe	0.5	0.0	1	0	0	1
His	0.5	0.6	0	0	0	0
Lys	6.7	6.4	5	2	2	2
Arg	0.0	0.0	0	0	0	1
Trp	1.0	0.0	nd	nd	nd	nd

[a] Expressed in mole %.
[b] LeAGP-1 excluding the N-terminal signal sequence.
[c] LeAGP-1 excluding the N-terminal signal sequence and the C-terminal hydrophobic domain presumably removed during GPI attachment.
nd = not determined.

Several interesting features emerge from the LeAGP-1 core protein sequence (Fig 1). First, its domain structure corresponds to other known classical AGPs; however, it is distinguished from them by the occurrence of the Lys-rich subdomain. This subdomain represents a prime site for intermolecular interactions with negatively charged cell surface molecules such as pectin. Second, this AGP is predicted to be modified by the addition of a GPI anchor near its C-terminus as demonstrated in two known classical AGPs (Youl *et al* 1998). The LeAGP-1 sequence contains the consensus sequence for such a modification. In addition, it is interesting to note that this consensus sequence (i.e., Ser-Gly-Ala) and C-terminal hydrophobic domain reside exclusively in exon 2 of the *LeAGP-1* gene and represent a likely functional exon domain (Fig 1) (Li and Showalter 1996). Third, a potential N-linked glycosylation site is encoded by the *LeAGP-1* clones and is located at the C-terminal side of the Lys-rich subdomain; such sites are not typical for AGPs. We predict that if such an N-linked glycan exists on LeAGP-1, then it will be in addition to numerous Type II AG side chains as well as other short oligoarabinosides as demonstrated for sycamore AGPs, gum arabic glycoprotein, and a maize AGP (Pope 1977, Qi *et al* 1991, Kieliszewski *et al* 1992). Clearly, we now hope to verify the existence of a GPI anchor and conduct a comprehensive carbohydrate analysis on LeAGP-1.

2.2 Localization of LeAGP-1

The PAP antibody was also used to immunolocalize LeAGP-1 in tomato cultured cells and various plant organs at the light and electron microscope levels (Fig 2). These data not only reveal where LeAGP-1 is located but also provide clues as to its functions. In tomato cultured cells, LeAGP-1 was detected at the cell surface (Fig 2a). Immunolocalization at the electron microscope level revealed that LeAGP-1 is located in the cell wall (Fig 2b). It should be noted that additional biochemical analysis indicated that LeAGP-1 exists as the major AGP in the media of cultured cells. The LeAGP-1 was also immunolocalized in tomato stems and floral styles. In stems, LeAGP-1 is prominently localized to developing metaxylem elements but is also found in outer phloem and inner phloem (Fig 2c). Ultrastructurally, LeAGP-1 is located primarily in the secondary cell wall of such developing metaxylem elements; however, some staining of the primary cell wall was also observed (Fig 2d). The most intense staining for LeAGP-1, however, was observed in the transmitting tissue of the style (Fig 2e). Immunolocalization at the electron microscope level showed LeAGP-1 to be copiously deposited in the intercellular spaces of transmitting tract cells as well as in their thin cell walls (Fig 2f).

Thus, LeAGP-1 is found in and outside tomato cell walls. The question of whether LeAGP-1 is also associated with the plasma membrane remains unanswered. Given that LeAGP-1 contains a putative site for GPI anchoring, it should be at least temporarily associated with the plasma membrane. We envision that LeAGP-1 moves from the endoplasmic reticulum to the Golgi and then to the plasma membrane, where processing of the GPI anchor occurs to release LeAGP-1 into the cell wall and space beyond. The inability of our antibody to immunolocalize LeAGP-1 to the plasma membrane may be due to a small pool of plasma membrane associated LeAGP-1 or to the antigenic epitope being blocked or inaccessible in this microenvironment. We also speculate that the amount of LeAGP-1 found beyond the wall is governed by the particular composition and amounts of cell wall components that may serve as a saturable affinity column, as opposed to a more elaborate trafficking scheme. In this regard, it will be extremely interesting to identify the cell surface components that interact with LeAGP-1.

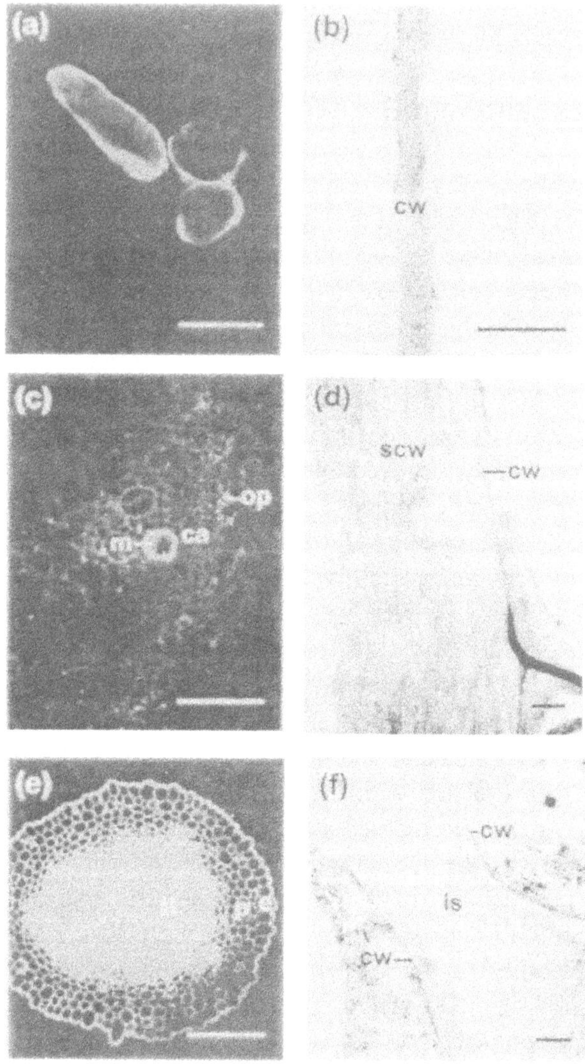

Figure 2. Immunolocalization of LeAGP-1 with the PAP antibody in tomato suspension-cultured cells, young stem, and style at the light and electron microscope levels. (a) Whole mount of suspension-cultured cells; (b) a portion of a suspension-cultured cell; (c) a portion of a cross-section of a young stem; (d) a portion of a maturing metaxylem element in young stem; (e) cross-section of a style; (f) neighboring transmitting tissue cells and their intercellular space; (a), (c), and (e) were immunolocalized at the light microscope level using FITC-conjugated secondary antibody, whereas (b), (d), and (f) were immunolocalized at the electron microscope level using 10 nm gold-conjugated secondary antibody. m, metaxylem element; ca, cambial zone; op, outer phloem; e, epidermis; p, stylar parenchyma; tt, stylar transmitting tissue. cw, (primary) cell wall; scw, secondary cell wall; is, intercellular space. Bars in (a), (c), and (e) = 100 μm. Bars in (b), (d), and (f) = 500 nm.

2.3 Production of Transgenic Anti-Sense *LeAGP-1* Tomato Plantlets

As one approach to elucidate the function of LeAGP-1, we have produced transgenic *LeAGP-1* antisense tomato plantlets. Here we used the 35S cauliflower mosaic virus promoter to drive expression of a nearly full-length *LeAGP-1* cDNA in an antisense orientation. The T-DNA transformed and untransformed tomato plantlets were used as controls. We obtained only 15 independent antisense transformed plantlets in contrast to hundreds of T-DNA transformed plantlets. Northern analysis and immunocytochemical analysis indicated that *LeAGP-1* mRNA and LeAGP-1 protein were severely down-regulated in comparison to the T-DNA transformed and untransformed control plantlets (data not shown). Visual analyses of the antisense plantlets showed them to be dramatically smaller than the control plantlets. Moreover, these antisense plantlets would not root, and all but six have now died. Clearly, this experiment needs to be repeated; however, if these initial results are correct, then the use of an inducible promoter to drive antisense expression warrants serious consideration.

2.4 AGPs and Programmed Cell Death

Previous experiments by Serpe and Nothnagel (1994) and Langan and Nothnagel (1997) demonstrated that Yariv reagent can cause growth inhibition and cell death in cultured cells. However, it was not known whether these cells died by necrosis or by PCD. Our initial studies in 1996 using tomato seedlings treated with 30 μM (β-D-Gal)$_3$ Yariv reagent showed ultrastructural changes characteristic of apoptosis in some cells at the root tip. Apoptosis, a type of PCD, was first identified in animals (Wyllie 1980). These changes included cytoplasmic shrinkage, nuclear membrane blebbing, and chromatin condensation; such changes were not observed in untreated control cells. Subsequently, we hypothesized that Yariv reagent causes cultured cells to die a PCD. A precisely controlled form of cellular suicide, PCD is characterized by internuclear DNA cleavage and changes in cell morphology. To test our hypothesis, we used cell suspension cultures of tomato and *Arabidopsis thaliana* and treated them for various times and with various concentrations (β-D-Gal)$_3$ and (α-D-Gal)$_3$ Yariv reagents before scoring them for PCD. Three criteria were used to determine PCD: 1) terminal deoxynucleotidyl transferase-mediated dUTP nick-end labeling (TUNEL) of DNA 3'-OH groups, 2) internucleosomal DNA cleavage, and 3) structural changes in cell morphology characteristic of PCD.

Although tomato cells were initially used in the TUNEL experiments and demonstrated appreciable TUNEL positive nuclei, *Arabidopsis* was rapidly adopted as our model system for this study, given the more robust nature of the TUNEL response in these cells (Table 2). For example, 100% of the *Arabidopsis* cells showed TUNEL positive nuclei when treated with 80 μM (β-D-Gal)$_3$ Yariv reagent for 72 hours in comparison with about 50% of the tomato cells. Note that untreated cells and cells treated with (α-D-Gal)$_3$ Yariv reagent [which does not react with

AGPs but is nearly identical to (β-D-Gal)$_3$ Yariv reagent] show very low percentages of TUNEL positive nuclei (Table 2). Moreover, with increasing incubation time, both the (α-D-Gal)$_3$ Yariv reagent-treated cells and the untreated cells demonstrated reduced percentages of TUNEL-positive nuclei. This observation provides support to the idea of social control of PCD as promulgated by Barres *et al* (1992) and Raff (1992) in animal cell cultures and by McCabe *et al* (1997) in plant cell cultures, whereby neighboring cells grown at high density provide signal molecules or growth factors that suppress PCD.

Table 2. Average percentage of apoptotic (i.e., TUNEL-positive) cells in *Arabidopsis* suspension-cultured cells treated with Yariv reagent

Time (hr)	Untreated	50 μM (α-D-Gal)$_3$	50 μM (β-D-Gal)$_3$	80 μM (α-D-Gal)$_3$	80 μM (β-D-Gal)$_3$
24	–	5.98	30.38	7.30	47.51
48	–	2.29	34.66	3.47	80.14
72	0.87	2.44	45.43	4.52	100.00

Internucleosomal DNA fragmentation was detected in (β-D-Gal)$_3$ Yariv reagent-treated *Arabidopsis* cells, but not in (α-D-Gal)$_3$ Yariv reagent-treated cells or in untreated cells. Such fragmentation was both time and concentration dependent and was detectable after 48 hours of treatment with 80 μM (β-D-Gal)$_3$ Yariv reagent or after 72 h of 60 μM (β-D-Gal)$_3$ Yariv reagent (data not shown). Such time-dependent internucleosomal DNA fragmentation is characteristic of PCD in contrast to necrosis, which is characterized by rapid cell death and non-specific DNA degradation.

Microscopic examination of *Arabidopsis* cells treated with (β-D-Gal)$_3$ Yariv reagent showed characteristic structural changes in cell morphology not observed in (α-D-Gal)$_3$ Yariv reagent-treated cells or in untreated cells (data not shown). At the light-microscope level, cells treated with the 80 μM (β-D-Gal)$_3$ Yariv reagent displayed cytoplasmic shrinkage and condensation; this was also observed at the electron-microscope level in addition to chromatin condensation and nuclear membrane blebbing. Notably, other organelles (e.g., mitochondria) were intact. Such microscopic changes are characteristic of animal apoptosis (Wyllie 1980). Some of these changes are also recorded for plant PCD (McCabe *et al* 1997). Plant PCD research is beginning to receive significant attention but currently comprises few studies, particularly at the ultrastructural level.

On the basis of these three criteria, (β-D-Gal)$_3$ Yariv reagent induces *Arabidopsis* suspension cultured cells to undergo PCD. The precise mechanism whereby Yariv reagent induces this process is unknown but clearly implicates AGP involvement. One idea is that Yariv reagent serves to disrupt plasma membrane–cell wall interactions involving AGPs, which initiates a PCD signal transduction cascade. Such a scenario is hypothesized and supported in the case of an animal extracellular matrix–plasma membrane interaction (Frisch and Francis 1994). Another related idea is that AGPs must be present or assemble as multimeric complexes at the cell surface to relay signal transduction information. Disruption of such complexes or their assembly may prevent growth signals from being relayed and thus trigger PCD.

3. CONCLUSIONS

1. LeAGP-1 is a novel classical AGP distinguished by its Lys-rich subdomain.
2. LeAGP-1 is the major AGP present in tomato cultured cells and media.
3. LeAGP-1 immunolocalizes in tomato to the cell walls of cultured cells, to the primary, and especially the secondary, cell walls of developing metaxylem elements of the stem and petiole, and to the cell walls and intercellular spaces present in stylar transmitting tissue.
4. Transgenic *LeAGP-1* antisense tomato plantlets were deficient in *LeAGP-1* mRNA and LeAGP-1 protein and showed severe overall growth inhibition and had difficulty rooting.
5. We speculate that LeAGP-1 is also associated with the plasma membrane via a GPI anchor where it could interact with cell wall components (including itself) and function as a cell adhesion molecule, possibly by regulating growth or PCD.
6. The ability of (β-D-Gal)$_3$ Yariv reagent to trigger PCD implicates AGPs in this process.
7. It is worth noting that xylem is produced as a result of PCD and that AGPs, including LeAGP-1, are often expressed in such cells.

ACKNOWLEDGEMENTS

This work was supported by NSF grant #IBN-9727757; by research funds from the Molecular and Cellular Biology Program, Ohio University, Athens, OH; and by an Ohio University Research Challenge Grant to AMS and MJK.

REFERENCES

Barres, B. A., Hart, I. K., Coles, H. S. R., Burne, J. F., Voyvodic, J. T., Richardson, W. D., and Raff, M. C., 1992, Cell death and control of cell survival in the oligodendrocyte lineage, *Cell* **70**: 31–46.

Fincher, G. B., Stone, B. A., and Clarke, A. E., 1983, Arabinogalactan-proteins: structure, biosynthesis, and function, *Ann. Rev. Plant. Physiol.* **34**: 47–70.

Frisch, S. M., and Francis, H., 1994, Disruption of epithelial cell-matrix interactions induces apoptosis, *J. Cell Biol.* **124**: 619–626.

Jauh, G. Y., and Lord, E. M., 1996, Localization of pectins and arabinogalactan-proteins in lily (*Lilium longiflorum* L.) pollen tube and style, and their possible roles in pollination, *Planta* **199**: 251–261.

Kieliszewski, M. J., Kamyab, A., Leykam, J. F., and Lamport, D. T. A., 1992, A histidine-rich extensin from *Zea mays* is an arabinogalactan protein, *Plant Physiol.* **99**: 538–547.

Langan, K. J., and Nothnagel, E. A., 1997, Cell surface arabinogalactan-proteins and their relation to cell proliferation and viability, *Protoplasma* **196**: 87–98.

Li, S., and Showalter, A. M., 1996, Cloning and developmental/stress-regulated expression of a gene encoding a tomato arabinogalactan protein, *Plant Mol. Biol.* **32**: 641–652.

McCabe, P. F., Levine, A., Meijer, P-J., Tapon, N. A., and Pennell, R. I., 1997, A programmed cell death pathway activated in carrot cells cultured at low cell density, *Plant J.* **12**: 267–280.

Nothnagel, E. A., 1997, Proteoglycans and related components in plant cells, *Int. Rev. Cytol.* **174**: 195–291.

Pogson, B., and Davies, C., 1995, Characterization of a cDNA encoding the protein moiety of a putative arabinogalactan protein from *Lycopersicon esculentum, Plant Mol. Biol.* **28**: 347–352.

Pope, D. G., 1977, Relationships between hydroxyproline-containing proteins secreted into the cell wall and medium by suspension-cultured *Acer pseudoplatanus* cells, *Plant Physiol.* **59**: 894–900.

Qi, W., Fong, C., and Lamport, D. T. A., 1991, Gum arabic glycoprotein is a twisted hairy rope, *Plant Physiol.* **96**: 848–855.

Raff, M. C., 1992, Social controls on cell survival and cell death, *Nature* **356**: 397–400.

Roy, S., Jauh, G. Y., Hepler, P. K., and Lord, E. M., 1998, Effects of Yariv phenylglycoside on cell wall assembly in the lily pollen tube, *Planta* **204**: 450–458.

Serpe, M. D., and Nothnagel, E. A., 1994, Effects of Yariv phenylglycosides on *Rosa* cell-suspensions: evidence for the involvement of arabinogalactan-proteins in cell proliferation, *Planta* **193**: 542–550.

Willats, W. G. T., and Knox, J. P., 1996, A role for arabinogalactan-proteins in plant cell expansion: evidence from studies on the interaction of β-glucosyl Yariv reagent with seedlings of *Arabidopsis thaliana, Plant J.* **9**: 919–925.

Wyllie, A. H., 1980, Cell death: The significance of apoptosis, *Int. Rev. Cytol.* **68**: 251–306.

Youl, J. J., Bacic, A., and Oxley, D., 1998, Arabinogalactan-proteins from *Nicotiana alata* and *Pyrus communis* contain glycosylphosphatidylinositol membrane anchors, *Proc. Natl. Acad. Sci. USA* **95**: 7921–7926.

Chapter 7

Cell Cycle Arrest by Perturbation of Arabinogalactan-Proteins with Yariv Phenylglycoside

Johnica A. Eyvazzadeh and Eugene A. Nothnagel
Department of Botany and Plant Sciences, University of California, Riverside, CA 92521-0124

1. INTRODUCTION

Arabinogalactan-proteins (AGPs) have been found in essentially all angiosperms, gymnosperms, and lower plants that have been tested for the presence of these proteoglycans (Jermyn and Yeow 1975, Clarke *et al* 1979, Basile and Basile 1990). Although the widespread occurrence of AGPs within the plant kingdom implies that these macromolecules may have important functions, the precise identities of these functions remain uncertain. Several lines of evidence suggest, however, that these functions may be in plant development. Developmentally regulated expression of AGPs in tissues of plants has been demonstrated at the level of the mature AGP by biochemical and immunological methods and at the level of the mRNA encoding the core polypeptide by hybridization techniques (Knox 1995, Du *et al* 1996, Nothnagel 1997, Serpe and Nothnagel 1999). Roles of AGPs in development of somatic embryos in culture has been suggested by immunolocalization studies that correlate particular AGP epitopes with embryogenic capacity and by experiments showing that success in embryogenesis can be influenced by the addition of AGPs from exogenous sources (Kreuger and van Holst 1996, Toonen *et al* 1996, McCabe *et al* 1997).

Other studies have shown that AGPs may act at the level of fundamental cellular processes that underlie plant development. Depending upon the experimental system, roles of AGPs in control of cell division, cell expansion, and/or programmed cell death have been suggested (Basile and Basile 1993, Zhu *et al* 1993, Dolan *et al* 1995, Schindler *et al* 1995). A particularly facile experimental approach to investigating AGP function in fundamental cellular processes has involved the use of Yariv phenylglycosides. Certain members of this class of synthetic, chromophoric molecules bind and aggregate AGPs. The most commonly used Yariv phenylglycoside is 1,3,5-tri-(*p*-β-D-glucosyloxyphenylazo)-2,4,6-trihydroxybenzene, or (β-D-Glc)$_3$ (Yariv *et al* 1962). Applied to living cells or plants, (β-D-Glc)$_3$ is a convenient probe for aggregating and presumably inactivating AGPs (Serpe and Nothnagel 1994, Jauh and Lord 1996, Willats and Knox 1996, Ding and Zhu 1997, Langan and Nothnagel 1997, Roy *et al* 1998, Thompson and Knox 1998). The specificity of the interaction and the attribution of the observed effects to AGP inactivation can be checked with other Yariv phenylglycosides, such as (α-D-Gal)$_3$ and (β-D-Man)$_3$, which do not bind AGPs and are thus powerful negative controls (Nothnagel 1997).

The present work is focused on the Rosa 57 line of suspension-cultured "Paul's Scarlet" rose cells from which results of experiments with Yariv phenylglycosides have been interpreted in terms of a relationship between AGPs and regulation of cell proliferation. Serpe and Nothnagel (1994) observed that growth of Rosa 57 cell cultures was completely inhibited by 50 μM (β-D-Glc)$_3$. Growth was not inhibited by (α-D-Gal)$_3$ or (β-D-Man)$_3$ Yariv phenylglycosides. The Rosa 57 cells remained viable for at least 7 days in the presence of (β-D-Glc)$_3$ and resumed growth at a near-normal rate upon subsequent transfer to control medium without (β-D-Glc)$_3$. Analysis of cell size in cultures treated with (β-D-Glc)$_3$ and halted in growth for 7 days showed that the cells were of normal or slightly larger-than-normal size, indicating that cell division had been blocked. Langan and Nothnagel (1997) performed a cell cycle analysis on unsynchronized cultures of Rosa 57 cells treated with (β-D-Glc)$_3$ and blocked in cell division. The resulting histograms of fluorescence from Hoechst 33258-stained nuclei showed that most of the (β-D-Glc)$_3$-treated cells were arrested in G1 phase. Because even randomly dividing control cultures always had at least 80% of the population in G1 phase, however, the data were inadequate to statistically distinguish whether (β-D-Glc)$_3$ blocked only in G1 or at other points in the cell cycle as well.

The first goal of the present investigation was to synchronize cell divisions in Rosa 57 cell cultures. Synchronization was expected to enable generation of populations having much fewer than 80% of the cells in G1 phase. The second goal was then to use synchronized cultures to determine whether (β-D-Glc)$_3$ arrests Rosa 57 cells only in G1 or at multiple points in the cell cycle.

2. MATERIALS AND METHODS

2.1 Growth of Cell Cultures

The Rosa 57 line of suspension-cultured cells of "Paul's Scarlet" rose (*Rosa* sp.) was grown in the dark in minimal organic medium as described by Nothnagel and

Lyon (1986). For the treatment flasks, the minimal organic medium was supplemented by the addition of (β-D-Glc)$_3$ Yariv phenylglycoside (Yariv *et al* 1962) to 50 μM, a concentration shown to inhibit division of Rosa 57 cells completely but reversibly (Serpe and Nothnagel 1994, Langan and Nothnagel 1997). Growth of the cell cultures was monitored as increase in settled cell volume (Langan and Nothnagel 1997).

2.2 Synchronization of Cell Cultures

The usual culture passage cycle period of 7 days was shortened to 3 or 4 days in preparation for synchronization. At the time of passage, 1 ml of settled cells was added to 19 ml of minimal organic medium in a 50-ml Erlenmeyer flask. Following Sala *et al* (1986), synchronization of cell divisions was accomplished through progressive additions of aphidicolin (Sigma Chemical Co., St. Louis, MO), a toxin from the fungus *Nigrospora sphaerica* that specifically and reversibly inhibits DNA polymerases α and δ. At 0, 22, and 46 hours after passage, aliquots of 10, 5, and 5 μl, respectively, from an aphidicolin stock solution of 40 mg/ml in dimethylsulfoxide were added to produce a total added aphidicolin concentration of approximately 40 μg/ml in the 20-ml cell culture. At 70 hours after passage, the cells were thoroughly washed four times in minimal organic medium without aphidicolin and then resuspended in 19 ml of minimal organic medium. This washing defined the 0 hour of time in the subsequent experimental protocols and data collections. Aliquots were withdrawn and fixed for nuclear DNA staining at 3-hour intervals thereafter until 24 hours.

2.3 (β-D-Glc)$_3$-treatment of Synchronized Cells

Cells were transferred to 12 replicate flasks and treated with aphidicolin as described in the previous section. At 0 hours (70 hours after the start of aphidicolin treatment), the cells were washed four times and resuspended in 17.5 ml of minimal organic medium. Each flask was then treated once with (β-D-Glc)$_3$ during the next 22.5 hours. The treatment times occurred at 4.5-hour intervals from 0 to 22.5 hours, with two flasks treated at each time point. The (β-D-Glc)$_3$ was added as 1.5 ml from a 667 μM stock solution in minimal organic medium. With 17.5 ml of medium, 1 ml of settled cells, and 1.5 ml of (β-D-Glc)$_3$ stock solution, the 50-ml Erlenmeyer flasks each contained a suspension culture totaling 20 ml with 50 μM (β-D-Glc)$_3$ concentration. One of the two flasks treated with (β-D-Glc)$_3$ at each time point was harvested by fixation at 22.5 hours after addition of (β-D-Glc)$_3$. The second flask was harvested by fixation at a later time, at 72 hours after washing from aphidicolin.

2.4 Fixation, Staining, and Fluorescence Microscopy of Cells

Cell fixation, softening with pectinase, and application of Hoechst 33258 dye for fluorescence staining of nuclear DNA were as described by Laloue *et al* (1980). Mitotic indices were estimated by microscopically examining at least 625 cells for each treatment and determining the proportion with partially or fully condensed chromosomes in any spatial distribution in the cell. Microfluorimetry of stained nuclei was performed as described (Langan 1996, Langan and Nothnagel 1997), and each histogram was generated from measurements of 300 nuclei. Statistical analyses of the results were performed with the Instat computer program (GraphPad Software, San Diego, CA).

3. RESULTS

3.1 Effect of Aphidicolin on Synchronization of Cell Divisions

Upon transfer of randomly dividing Rosa 57 cells to fresh minimal organic medium, the first doubling of settled cell volume occurred within 73 ± 22 hours (mean ± standard deviation). On the basis of this observation, a 70-hour treatment with aphidicolin was used with the expectation that this period was sufficiently long that most of the cells in the culture would have progressed to and halted at the G1/S interface. Figure 1 shows that after washing aphidicolin from the medium, the mitotic index was initially 0% and first became statistically greater than 0% at 18 hours. Most of the interphase cells in the 24-hour sample seemed to be in late-G2 phase. These cells had large, bright nuclei, although little chromosomal condensation was evident.

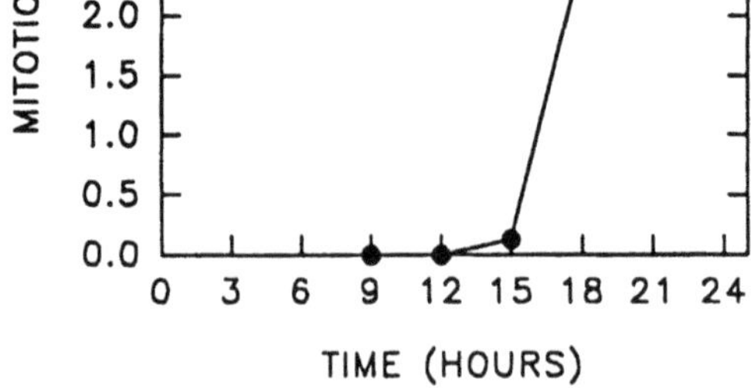

Figure 1. Onset of cell divisions as indicated by mitotic index in Rosa 57 cells after release from arrest with aphidicolin. At time 0 hours on the abscissa, the cells were washed from aphidicolin-containing medium into normal medium.

3.2 Effect of (β-D-Glc)$_3$ on Progression of the Cell Cycle

Growth of Rosa 57 cell cultures was strongly but reversibly inhibited by (β-D-Glc)$_3$, in agreement with previously reported results (Serpe and Nothnagel 1994, Langan and Nothnagel 1997). When cells were exposed to 50 μM (β-D-Glc)$_3$ for 7 days and then washed to minimal organic medium without (β-D-Glc)$_3$, the first doubling of settled cell volume occurred within 124 ± 14 hours after washing. Thus, the Rosa 57 cells remained viable during prolonged exposure to (β-D-Glc)$_3$, although this first doubling time was somewhat longer than the 73 ± 22 hours observed with control cells.

The effect of (β-D-Glc)$_3$ was studied during progression of the first cell cycle after cells were released from arrest by aphidicolin. In these experiments, (β-D-Glc)$_3$ was applied at various times after washing aphidicolin from the medium. For each time of (β-D-Glc)$_3$ application, one fixation was made 22.5 hours later. This interval was selected since Fig 1 showed that 22.5 hours was long enough for significant numbers of cells to progress through S and G2 phases up to M phase. To test whether the cells observed at the first fixation were actually arrested, rather than simply caught in a "snapshot" of synchronous progression, a second fixation for each (β-D-Glc)$_3$ application was made at 72 hours after washing from aphidicolin. Unarrested cells would have been able to complete a cell division cycle within about 72 hours, as judged from the 73 ± 22 hours first doubling time observed after passage of control cells. Thus, if (β-D-Glc)$_3$ arrested the cell cycle only in G1, then the fixations at 72 hours after washing from aphidicolin would have been expected to reveal low mitotic indices, indicating that most of the cells had completed the cell cycle and returned to G1 phase, irrespective of (β-D-Glc)$_3$ treatment.

Table 1 shows the mitotic indices resulting from these (β-D-Glc)$_3$ treatments after washing aphidicolin from the medium. Mitotic indices of approximately 5% were observed when (β-D-Glc)$_3$ was applied at or shortly after the washing from aphidicolin (protocols 1, 2, 7, 8). Mitotic indices gradually increased as the application of (β-D-Glc)$_3$ was delayed up to 22.5 hours after washing from aphidicolin (protocol 6). The two different fixations for each (β-D-Glc)$_3$ treatment seemed to reveal slightly different mitotic indices, although in most cases these differences were not significant, as indicated by overlap of the 95% confidence intervals for mitotic indices.

With all treatment protocols 1–12, most of the cells contributing to the mitotic index (Table 1) appeared to be in early prophase. The chromosomes were usually incompletely condensed, and arrangement of chromosomes in a pattern indicative of metaphase or anaphase was rarely observed.

Table 1. Mitotic indices of synchronized cells treated with (β-D-Glc)$_3$ Yariv phenylglycoside at various times after removal of aphidicolin. Cultures were treated with aphidicolin for the first 70 hours after passage and then, at time defined as t = 0 hours, washed to fresh medium free of aphidicolin. The washed cultures were treated with 50 μM (β-D-Glc)$_3$ at later times and subsequently fixed for microscopy at the indicated times.

Protocol Number	Time of (β-D-Glc)$_3$ Application (hr)	Time of Fixation (hr)	Mitotic Index (%)	95% Confidence Interval (%)
1	0	22.5	4.6	3.3–6.3
7	0	72.0	6.5	5.1–8.1
2	4.5	27.0	4.5	3.4–5.8
8	4.5	72.0	6.8	5.3–8.6
3	9.0	31.5	8.2	6.1–10.6
9	9.0	72.0	7.4	5.9–9.1
4	13.5	36.0	8.5	6.8–10.6
10	13.5	72.0	9.1	7.7–10.7
5	18.0	40.5	10.3	8.1–12.9
11	18.0	72.0	13.9	11.5–16.4
6	22.5	45.0	22.3	19.9–25.0
12	22.5	72.0	16.4	13.7–19.3

To facilitate interpretation of the mitotic index data, histograms of fluorescence intensity from stained nuclear DNA were generated for cells processed through some of the protocols. Figure 2 shows that cells processed through protocol 1 (Table 1) yielded a histogram containing a nearly symmetrical predominant peak. Cells processed through protocol 6, however, yielded an asymmetrical distribution with a higher frequency of more fluorescent nuclei.

4. DISCUSSION

A previous study examined the effects of (β-D-Glc)$_3$ on progression of the cell cycle in randomly dividing Rosa 57 cells (Langan and Nothnagel 1997). In that work, cells were grown in a 7-day culture passage cycle, and histograms of fluorescence from Hoechst 33258-stained nuclear DNA were presented for cells at 3, 5, and 7 days after transfer into control medium or into medium containing 50 μM (β-D-Glc)$_3$. Table 2 presents a quantitative summary of those six histograms. For the control cells, the proportion of the population in S/G2/M was 17.7% at 3 days after passage and then gradually decreased to 11.0% at 7 days after passage. These results seemed to indicate that, as expected, cell division was most frequent at 3 days after subculture and then gradually became less frequent later in the 7-day cell passage cycle. At 3 days after passage, fewer (β-D-Glc)$_3$-treated cells (13.0%) than control cells (17.7%) were in S/G2/M, indicating some extent of arrest by (β-D-Glc)$_3$ in G1. The proportion of (β-D-Glc)$_3$-treated cells in S/G2/M increased at 5 days to 17.0% and held there at 7 days, however, indicating that perhaps some of the cells gradually slipped through the G1 checkpoint, only to be subsequently blocked at the G2/M checkpoint. Although this interpretation of blockage by (β-D-Glc)$_3$ at both the G1 and G2/M checkpoints seemed consistent with the data (Langan 1996), the conclusion regarding G2/M was statistically weak because of overlaps among 95% confidence intervals on S/G2/M proportions in comparisons of the various treatments (Table 2).

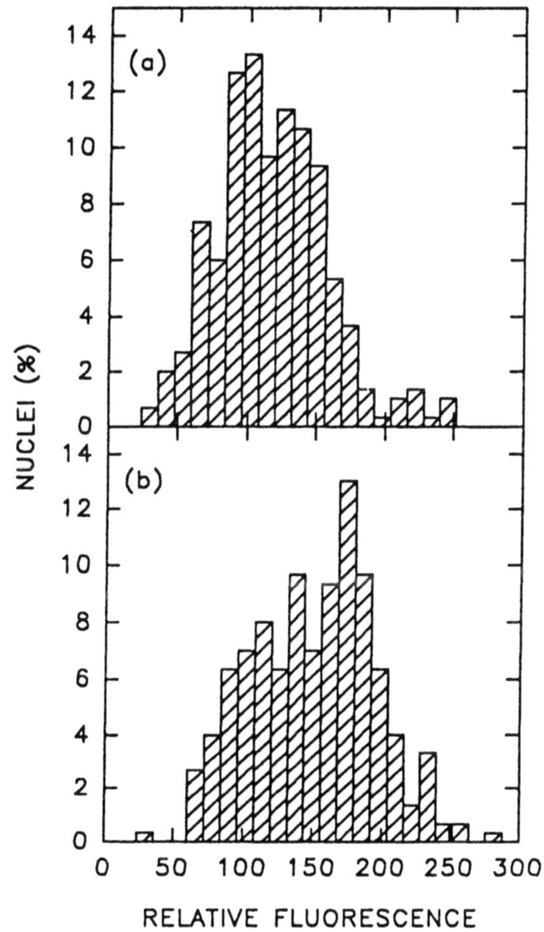

Figure 2. Histograms of fluorescence intensity from Hoechst 33258-stained nuclear DNA in Rosa 57 cells after washing from aphidicolin and treatment with (β-D-Glc)$_3$ according to the protocols described in Table 1. (a) Protocol 1. (b) Protocol 6. The G1 peak in (a) was centered at 113 relative fluorescence units.

Table 2. Proportion of population in S/G2/M phases in Rosa 57 cell cultures in various conditions. Histograms of fluorescence from Hoechst 33258-stained nuclear DNA were divided at the midpoint between the G1 and G2/M intensities. Cells with fluorescence less than this midpoint were classified as G1/S, and cells with fluorescence greater than this midpoint were classified as S/G2/M.

Treatment	Fixation Time or Protocol	Proportion in S/G2/M (%)	95% Confidence Interval (%)
Control[a]	3 days after passage	17.7	13.5–22.5
	5 days after passage	14.0	10.3–18.5
	7 days after passage	11.0	7.7–15.1
(β-D-Glc)$_3$[b]	3 days after passage	13.0	9.4–17.3
	5 days after passage	17.0	12.9–21.7
	7 days after passage	17.0	12.9–21.7
Aphidicolin/(β-D-Glc)$_3$[c]	Protocol 1	9.0	6.0–12.8
	Protocol 6	39.3	33.8–45.1

[a]Randomly dividing cells transferred into minimal organic medium and cultured for 7 days. Cell population results calculated (Langan 1996) from histograms of Langan and Nothnagel (1997).

[b]Randomly dividing cells transferred into minimal organic medium containing 50 μM (β-D-Glc)$_3$ and cultured for 7 days. Cell population results calculated (Langan 1996) from histograms of Langan and Nothnagel (1997).

[c]Experimental protocols as in Table 1. Cell population results calculated from histograms of Fig 2.

The first goal of the present investigation was to synchronize Rosa 57 cell cutures to enable examination of populations with larger S/G2/M proportions, thereby leading to data with greater statistical rigor. As a reversible inhibitor of DNA polymerases α and δ, aphidicolin has proven to be widely effective in arresting plant cell cycles at the G1/S interface (Sala *et al* 1986). Figure 1 shows that aphidicolin was similarly effective on Rosa 57 cells. Although the extent of synchronization cannot be accurately determined from Fig 1, the data subsequently collected (Table 1, Fig 2) show that the extent of synchronization was adequate to enable examination of populations with reasonably large S/G2/M proportions.

Synchronization of Rosa 57 cells with aphidicolin facilitated the study of effects of (β-D-Glc)$_3$ on the cell division cycle. The 4.6% mitotic index observed with protocol 1 cells (Table 1) was very similar to the mitotic index observed at the same 22.5-hour time in cells released from aphidicolin but not treated with (β-D-Glc)$_3$ (Fig 1). Although this coincidence might be interpreted as indicating that (β-D-Glc)$_3$ had no effect when applied at 0 hours, the histogram of Fig 2a showed otherwise. In the experiment of Fig 1, only about 4.5% of the cells were mitotic at 24 hours, but most of the interphase cells appeared to be in late G2 phase (see Section 3.1). In contrast, only a few protocol 1 cells were in G2 phase (Fig 2a, Table 2). Thus, when applied immediately upon washing aphidicolin from the medium, (β-D-Glc)$_3$ was quite effective in maintaining the arrest of Rosa 57 cells in G1. The 4.6% mitotic index observed with protocol 1 might have arisen from cells that were already in S phase when the aphidicolin was applied. These cells would have been unable to complete S phase until the aphidicolin was removed, at which time they finished S phase and progressed through G2 to M phase. The mitotic index observed at the later fixation (protocol 7) seemed slightly larger than that at the earlier fixation (protocol 1), which might indicate slippage of a few cells through the G1 arrest by (β-D-Glc)$_3$. The increase in mitotic index from protocol 1 to 7 was not statistically significant, however, since the 95% confidence intervals overlapped (Table 1).

Delaying the application of (β-D-Glc)$_3$ until 4.5 hours after washing from aphidicolin yielded the same results as applying (β-D-Glc)$_3$ immediately after washing from aphidicolin (Table 1, protocols 2, 8). As the delay between washing from aphidicolin and applying (β-D-Glc)$_3$ was increased beyond 4.5 hours, however, a gradual increase in mitotic index resulted. We interpret this observation as indicating that upon washing aphidicolin from the medium, a recovery period of somewhat more than 4.5 hours was required before cells began to progress from G1 into S phase. By 9.0 hours after washing from aphidicolin, some of the cells were progressing into S phase. Application of (β-D-Glc)$_3$ at this time blocked the departure of additional cells from G1. When the application of (β-D-Glc)$_3$ was further delayed, more cells departed G1 and contributed to the mitotic index at fixation time. For most of the (β-D-Glc)$_3$ application times from 4.5 to 18.0 hours, the mitotic indices at the second fixation times (protocols 8, 9, 10, 11) were slightly larger than at the first fixation times (protocols 2, 3, 4, 5), although none of these differences were statistically significant (Table 1). The higher, or at least equal, mitotic indices at the second fixation times were consistent with blockage by (β-D-Glc)$_3$ at another site besides G1. Since most of the cells blocked at this second site had chromosomes that were only partially condensed and were rarely in metaphase arrangement (see Section 3.2), this site of arrest was early in M phase, probably at the G2/M checkpoint.

Although the progression of cells from G1 into S phase seemed to be gradual in the time range between 4.5 and 18.0 hours after washing from aphidicolin, a much larger portion of the population moved into S phase in the time range between 18.0 and 22.5 hours after washing from aphidicolin. This conclusion was based on the high mitotic index of 22.3% observed with protocol 6 (Table 1) and was confirmed by Fig 2b, the corresponding histogram of nuclear fluorescence intensity. The S/G2/M proportion of 39.3% (Table 2) and the shape of the histogram were consistent with the presence of some S- and G2-phase cells in addition to those contributing to the 22.3% mitotic index. The statistically significant decrease in this mitotic index between the first fixation (protocol 6) and the second fixation (protocol 12) might indicate slippage of some cells through the G2/M arrest by (β-D-Glc)$_3$. Alternatively, this decrease in mitotic index might indicate that some of the mitotic cells observed in protocol 6 had already passed the G2/M checkpoint and were able to complete mitosis and return to G1 before the fixation for protocol 12.

The results of this study with synchronized Rosa 57 cells support the earlier, but at that time statistically weak, conclusion that (β-D-Glc)$_3$ causes a primary arrest of the cell cycle at the G1 checkpoint and a secondary arrest at the G2/M checkpoint (Langan 1996, Langan and Nothnagel 1997). The mechanisms through which perturbation of AGPs with (β-D-Glc)$_3$ causes these blockages remain to be elucidated. The cell division cycle can be halted at these classical checkpoints by various general problems, such as lack of sufficient cell size, lack of a favorable environment or, in the case of G2/M, lack of completely replicated DNA (Jacobs 1995). Perturbation of AGPs with (β-D-Glc)$_3$ may be interpreted by the cell as one of these general problems, such as lack of favorable environment. On the other hand, perturbation of AGPs might result in generation of a highly specific signal. Recent work has shown that AGPs in Rosa 57 cells are synthesized with glycosylphosphatidylinositol lipid anchors (Svetek *et al* 1999). Schultz *et al* (1998) have suggested that such anchors have important implications for signaling through interactions with other plant cell membrane proteins or through phospholipase cleavage of the anchor to release intracellular messengers such as phosphatidylinositol. Some phosphatidylinositols are cleaved to release inositolphosphates and diacylglycerol, both of which function in signal transduction pathways in some organisms. Release of diacylglycerol from rose AGPs seems unlikely, since the glycosylphosphatidylinositol anchors on rose AGPs have been shown to contain ceramide-class lipids principally composed of phytosphingosine and tetracosanoic acid (Svetek *et al* 1999). In yeast and animal cells, however, ceramides have been shown to have a role in mediating responses to extracellular stimuli leading to various events, including growth arrest, programmed cell death, and differentiation (Saba *et al* 1996).

ACKNOWLEDGEMENTS

This work was supported by an Intramural Research Grant from the Academic Senate of the University of California, Riverside.

REFERENCES

Basile, D. V., and Basile, M. R., 1990, Hydroxyproline metabolism and hydroxyproline-containing glycoproteins in leafy liverworts. In *Bryophytes, Their Chemistry and Chemical Taxonomy* (H. D. Zinsmeister and R. Mues, eds.), Clarendon Press, Oxford, pp. 275–288.

Basile, D. V., and Basile, M. R., 1993, The role and control of the place-dependent suppression of cell division in plant morphogenesis and phylogeny, *Memoirs Torrey Bot. Club* **25**: 63–83.

Clarke, A. E., Anderson, R. L., and Stone, B. A., 1979, Form and function of arabinogalactans and arabinogalactan-proteins, *Phytochemistry* **18**: 521–540.

Ding, L., and Zhu, J.-K., 1997, A role for arabinogalactan-proteins in root epidermal cell expansion, *Planta* **203**: 289–294.

Dolan, L., Linstead, P., and Roberts, K., 1995, An AGP epitope distinguishes a central metaxylem initial from other vascular initials in the *Arabidopsis* root, *Protoplasma* **189**: 149–155.

Du, H., Clarke, A. E., and Bacic, A., 1996, Arabinogalactan-proteins: a class of extracellular matrix proteoglycans involved in plant growth and development, *Trends Cell Biol.* **6**: 411–414.

Jacobs, T. W., 1995, Cell cycle control, *Annu. Rev. Plant Physiol. Plant Mol. Biol.* **46**: 317–339.

Jauh, G. Y., and Lord, E. M., 1996, Localization of pectins and arabinogalactan-proteins in lily (*Lilium longiflorum* L.) pollen tube and style, and their possible roles in pollination, *Planta* **199**: 251–261.

Jermyn, M. A., and Yeow, Y. M., 1975, A class of lectins present in the tissues of seed plants, *Aust. J. Plant Physiol.* **2**: 501–531.

Knox, J. P., 1995, Developmentally regulated proteoglycans and glycoproteins of the plant cell surface, *FASEB J.* **9**: 1004–1012.

Kreuger, M., and van Holst, G.-J., 1996, Arabinogalactan proteins and plant differentiation, *Plant Mol. Biol.* **30**: 1077–1086.

Laloue, M., Courtois, D., and Manigault, P., 1980, Convenient and rapid fluorescent staining of plant cell nuclei with "33258" Hoechst. *Plant Sci. Lett.* **17**: 174–179.

Langan, K. J., 1996, Cell surface arabinogalactan-proteins and their relation to plant cell proliferation, M.S. thesis, University of California, Riverside.

Langan, K. J., and Nothnagel, E. A., 1997, Cell surface arabinogalactan-proteins and their relation to cell proliferation and viability, *Protoplasma* **196**: 87–98.

McCabe, P. F., Valentine, T. A., Forsberg, L. S., and Pennell, R. I., 1997, Soluble signals from cells identified at the cell wall establish a developmental pathway in carrot, *Plant Cell* **9**: 2225–2241.

Nothnagel, E. A., 1997, Proteoglycans and related components in plant cells, *Int. Rev. Cytol.* **174**: 195–291.

Nothnagel, E. A., and Lyon, J. L., 1986, Structural requirements for the binding of phenylglycosides to the surface of protoplasts, *Plant Physiol.* **80**: 91–98.

Roy, S., Jauh, G. Y., Hepler, P. K., and Lord, E. M., 1998, Effects of Yariv phenylglycoside on cell wall assembly in the lily pollen tube, *Planta* **204**: 450–458.

Saba, J. D., Obeid, L. M., and Hannun, Y. A., 1996, Ceramide: an intracellular mediator of apoptosis and growth suppression, *Phil. Trans. Royal Soc. Lond. B Biol. Sci.* **351**: 233–241.

Sala, F., Galli, M. G., Pedrali-Noy, G., and Spadari, S., 1986, Synchronization of plant cells in culture and in meristems by aphidicolin, *Methods Enzymol.* **118**: 87–96.

Schindler, T., Bergfeld, R., and Schopfer, P., 1995, Arabinogalactan proteins in maize coleoptiles: developmental relationship to cell death during xylem differentiation but not to extension growth, *Plant J.* **7**: 25–36.

Schultz, C., Gilson, P., Oxley, D., Youl, J., and Bacic, A., 1998, GPI-anchors on arabinogalactan-proteins: implications for signalling in plants, *Trends Plant Sci.* **3**: 426–431.

Serpe, M. D., and Nothnagel, E. A., 1994, Effects of Yariv phenylglycosides on *Rosa* cell suspensions: evidence for the involvement of arabinogalactan-proteins in cell proliferation. *Planta* **193**: 542–550.

Serpe, M. D., and Nothnagel, E. A., 1999, Arabinogalactan-proteins in the multiple domains of the plant cell surface, *Adv. Bot. Res.* **30**: 207-289.

Svetek, J., Yadav, M. P., and Nothnagel, E. A., 1999, Presence of a glycosylphosphatidylinositol lipid anchor on rose arabinogalactan proteins, *J. Biol. Chem.* **274**: 14724–14733.

Thompson, H. J. M., and Knox, J. P., 1998, Stage-specific responses of embryogenic carrot cell suspension cultures to arabinogalactan protein-binding β-glucosyl Yariv reagent, *Planta* **205**: 32–38.

Toonen, M. A. J., Schmidt, E. D. L., Hendriks, T., Verhoeven, H. A., van Kammen, A., and de Vries, S. C., 1996, Expression of the JIM8 cell wall epitope in carrot somatic embryogenesis, *Planta* **200**: 167–173.

Willats, W. G. T., and Knox, J. P., 1996, A role for arabinogalactan-proteins in plant cell expansion: evidence from studies on the interaction of β-glucosyl Yariv reagent with seedlings of *Arabidopsis thaliana*. *Plant J.* **9**: 919–925.

Yariv, J., Rapport, M. M., and Graf, L., 1962, The interaction of glycosides and saccharides with antibody to the corresponding phenylazoglycosides, *Biochem. J.* **85**: 383–388.

Zhu, J.-K., Bressan, R. A., and Hasegawa, P. M., 1993, Loss of arabinogalactan-proteins from the plasma membrane of NaCl-adapted tobacco cells, *Planta* **190**: 221–226.

Chapter 8

A Major Antimicrobial Hybrid Chitin-Binding Protein from French Bean with Features Common to Arabinogalactan-Proteins and Hydroxyproline-Rich Glycoproteins

G. Paul Bolwell, Jonathan B. Trethowan and Przemyslaw Wojtaszek[1]

Division of Biochemistry, School of Biological Sciences, Royal Holloway, University of London, Egham, Surrey, TW20 0EX, UK; [1]Permanent Address: *Institute of Bioorganic Chemistry, Polish Academy of Sciences, Noskowskiego 12/14, 61-704 Poznan, Poland*

1. INTRODUCTION

Arabinogalactan-proteins (AGPs) were initially defined as Hyp-containing, highly glycosylated glycoproteins or proteoglycans. They contain arabinogalactan chains, and the linkage between the carbohydrate and protein is normally through Gal-Hyp. The protein is generally neutral to acidic and frequently contains Hyp/Pro-Ala motifs. However, the discovery of a number of developmentally important proteins showing hybrid features has broadened the definition (Sommer-Knudsen *et al* 1998). Some examples are those that contain Cys-rich domains at the C-terminus (Du *et al* 1996, Schultz *et al* 1997). Similarly, hybrid solanaceous lectins such as that from potato have an extensin-like domain at the N-terminus and a Cys-rich C-terminus, which contains chitin-binding domains analogous to other chitin-binding lectins (Allen *et al* 1996). A major chitin-binding extracellular protein of French bean has been identified and shows similar hybrid features with a Cys-rich C-terminal domain, but the N-terminus is that of a Pro-rich protein in its repeat sequence (Bolwell 1986 1987, Millar *et al* 1992). This French bean protein has been classified as a proline/hydroxyproline-rich glycoprotein (P/HRGP) (Sommer-Knudsen *et al* 1998), but it also shows some features common to AGPs. Further properties of this protein suggest an important role in defense against pathogen attack. In terms of this work, the protein has been defined as a chitin-binding Pro-rich protein (CBPRP; Brown *et al* 1998).

2. IDENTIFICATION AND CHARACTERIZATION OF M_r 42,000 CHITIN-BINDING GLYCOPROTEIN FROM FRENCH BEAN

The M_r 42,000 glycoprotein from French bean was first identified as the major target for arabinosylation in microsomal membranes from suspension-cultured cells of French bean incubated with UDP-[^{3}H]-arabinose (Bolwell 1986 1987). It could be purified from microsomal membranes by affinity chromatography on immobilized tri-N-acetylchitotriose and thus has lectin-like properties. It was subsequently found to be the major secreted glycoprotein in suspension-cultured cells and was present in the cell walls and medium (Millar *et al* 1992). It was purified to homogeneity and analyzed (Millar *et al* 1992). The M_r of the glycosylated protein was estimated by SDS-PAGE as 42,000 and the deglycosylated protein as 30,000. Amino acid analysis showed the presence of Hyp as 6.6%. Analysis of the derivatized deglycosylated protein showed the presence of 6.8% Cys. Estimating the core protein to be 300 amino acids implies the presence of approximately 70 arabinose and 60 galactose residues per mature glycosylated protein. These structural features provide one rationale for possibly classifying this protein as a P/HRGP (Sommer-Knudsen *et al* 1998). In addition to the expected O-glycosylations of Hyp, probably 3-5 N-glycosylation sites are also present. The N-terminus of the deglycosylated protein was determined as NYDKPOVEKPOVYKPOVEKPOVYKPOVEKPOVYK, where O represents Hyp (Millar *et al* 1992). This repeat sequence would classify it as a Pro-rich protein (PRP). However, there must be a second domain, which is Cys-rich, and which probably accounts for its chitin-binding properties. Thus, the French bean 42-kDa CBPRP resembles solanaceous lectins in its domain structure but differs from potato lectin, which has an extensin-like N-terminus (Allen *et al* 1996). It also differs in its location from potato lectin. Association of the CBPRP with the plasmalemma-wall interface in vascular cells (Millar *et al* 1992) is more consistent with an AGP. It also cross-reacts with monoclonal antibodies MCA 203 (D. J. Millar and G. P. Bolwell, unpublished data) and 265 (Millar *et al* 1992), which recognize other AGPs. The CBPRP is therefore at the limits of possible classification as an AGP but at the very least is an interesting variant reflecting the evolution of these hybrid-type glycoproteins, of which a number have now been described (Schultz *et al* 1997, Sommer-Knudsen *et al* 1998).

Evidence from biochemical and ultrastructural localization studies suggests that prolyl hydroxylation is completed before glycosylation of the 42-kDa CBPRP in the Golgi stack. Prolyl 4-hydroxylase was immunolocalized to pre-Golgi vesicles (Wojtaszek et al. 1999), whereas arabinosylated 42-kDa CBPRP can be first detected traversing the Golgi stack (Bolwell 1993). If ZM 226681 is used to inhibit prolyl hydroxylation and thus the subsequent arabinosylation and presumably galactosylation, then the 42-kDa CBPRP is retained in the microsomal fraction (Wojtaszek *et al* 1999).

3. MOLECULAR CLONING OF 42-kDa CBPRP

Several properties of the 42-kDa CBPRP have made it a difficult candidate for molecular cloning. The AGPs and other Hyp-rich glycoproteins have previously proved difficult to clone (Schultz *et al* 1997). Underlying reasons include excision of sequences by the bacterial host and recombinational events during cloning. We successfully isolated cDNAs using an oligonucleotide to the N-terminal sequence NYDKP, and using these cDNAs for subsequent re-screening of a cDNA library, we detected elicitor-induced mRNAs. It was not possible to improve the match between the amino acid sequences determined directly from the protein and predicted by the cDNAs, however, since it proved impossible to acquire internal protein sequence (Trethowan 1997). The 42-kDa CBPRP was refractory to trypsin, V8 protease, and lysC protease, even when the Cys residues were derivatized to prevent refolding. The protein seems remarkably stable.

Figure 1 shows the sequence of the longest clone obtained that had the N-terminal sequence of the native protein. This cDNA (D31) was apparently nearly full length for the open reading frame since it contained a leader sequence similar to that found for soybean PRP, but it lacked a 5' untranslated region. Expression of the mRNA corresponding to this cDNA was rapidly up-regulated by elicitation of suspension-cultured cells (Fig 2). This transcript was 1.4 kb in length. Hybridization to the identical band was observed using three separate probes with identical sequence to the repeat region (ZC12, 350bp; A52, 600bp; D31, 958 bp). Multiple transcript sizes were not observed, which might have been expected if there was a PRP equivalent to that in soybean, in addition to the 42-kDa protein, and which cross-hybridized. The absence of multiple transcripts suggests that the longest clone, D31, codes for the N-terminus of the 42-kDa CBPRP. The question remains as to whether D31 has been subjected to recombination and lost some of the C-terminal Cys-rich domain, the remains of which do not contain an open reading frame. Excision of sequence has been a common feature in efforts to clone these cDNAs. The A52 and D31 clones both diverge in a similar position around the sequence PSTN. A possibility remains that the Asp has some significance in that there are examples of proteins splicing at this residue in some lectins and storage proteins in legumes. A remote possibility exists that both domains are translated separately and spliced.

We can speculate on domain structure of the native protein, taking into account possible artefacts in the cDNA cloning or a possible protein-splicing mechanism. In this model the arabinose and galactose would be attached to the Hyp residues near the N-terminus. Such glycosylation would be reportedly unusual for PRP domains, since those for the soybean PRP are not thought to be glycosylated (Sommer-Knudsen *et al* 1998). The PRP domain would be 190 amino acids. The N-glycosylation sites would be in the C-terminal domain, which would also contain the Cys-rich chitin-binding sites. The Cys-rich domain would be ~110 amino acids for a 30-kDa core protein. This length would be sufficient to carry chitin–binding regions.

```
1    GGAATTCCGGCTTCCTTAAGCTCCCTAGTTCTGCTTCTTGCAGCTCTGCTTCTCTCCTCA  60
1     G  I  P  A  S  L  S  S  L  V  L  L  L  A  A  L  L  L  S  S   20

61   CAAAGGCTTGCTAACTATGACAAGCCCCCAGTAGAAAAACCTCCAGTTTACAAGCCCCCA 120
21    Q  R  L  A  N  Y  D  K  P  P  V  E  K  P  P  V  Y  K  P  P   40

121  GTTGAGAAACCACCAGTGTACAAGCCCCCAGTAGAGAAGCCACCAGTTTACAAGCCCCCA 180
41    V  E  K  P  P  V  Y  K  P  P  V  E  K  P  P  V  Y  K  P  P   60

181  GTTGAAAAACCACCAGTGTACAAGCCCCCAGTAGAGAAGCCACCAGTGTACAAGCCCCCA 240
61    V  E  K  P  P  V  Y  K  P  P  V  E  K  P  P  V  Y  K  P  P   80

241  GTTGAAAAACCACCAGTTTACAAACCCCCAGTAGAGAAGCCACCAGTGTACAAGCCCCCA 300
81    V  E  K  P  P  V  Y  K  P  P  V  E  K  P  P  V  Y  K  P  P  100

301  GTTGAGAAACCACCAGTTTACAAGCCCCCAGTTGAGAAGCCACCAGTTTACAAGCCCCCA 360
101   V  E  K  P  P  V  Y  K  P  P  V  E  K  P  P  V  Y  K  P  P  120

361  GTAAAGAAGCCACCAGTGTACAAGCCCCCAGTTGAGAAACCACCAGTTTACAAGCCCCCA 420
121   V  K  K  P  P  V  Y  K  P  P  V  E  K  P  P  V  Y  K  P  P  140

421  GTAGAGAAGCCACCAGTTTACAAGCCCCCAGTTGAGAAACCACCAGTTTACAAGCCACCA 480
141   V  E  K  P  P  V  Y  K  P  P  V  E  K  P  P  V  Y  K  P  P  160

481  GTGGAGAAGCCACCAGTGTACAAGCCCCCAGTTGAGAAGCCTCCGGTGTACAAGCCCCCA 540
161   V  E  K  P  P  V  Y  K  P  P  V  E  K  P  P  V  Y  K  P  P  180

541  GTTGAGAAGCCTCCGGTGTACAAGCCACCAGTAGAGAAGCCACCGGTTTACCAGCCACCA 600
181   V  E  K  P  P  V  Y  K  P  P  V  E  K  P  P  V  Y  Q  P  P  200

601  TACGGGAAGCCACCTCACCCAAAGTACCCTCCAAGCACCAACTGAGTATACACGCCATTC 660
201   Y  G  K  P  P  H  P  K  Y  P  P  S  T  N  *                 220

661  TGAAGTATGTTAAGAGTTGTATCTGAAGTATATGATAAATAAAAGTTGTAGGTACTTTAC 720

721  CATTGTCATAGTCAGAGATGTAGTTTACTTACAAATCATTTGTTTACAGTGTTGTGGCCG 780

781  AGATAAGAGTGTAATGGATGCATGTAAAACGACAAACAGAGGTATGGTTTTTTTTTTGAA 840

841  AGCTATGGATAATAGAAGTTTATGATTTTGATATTAAAAAAAAAAAAAAAAAAAAAAAAA 900

901  AAAAAAAAAAAAAAAAAAAAAAAAAAAAAAAAA                            950
```

Figure 1. Sequence analysis of a cDNA that codes for the N-terminus of the 42-kDa CBPRP. The nucleotide sequence of clone D31 is shown. A second clone, A52, corresponded to nucleotides 370–820 and was 98% homologous. A third clone, ZC12, corresponded to nucleotides 190–540. The translational open reading frame is shown. Both D31 and A52 have a stop codon after N214 in D31. However, there is no open reading frame that is Cys-rich. If sequence has been deleted, it is probably in this region of N214. The leader sequence (amino acids 1–24) has high homology to the leader sequence of soybean PRP. The mature protein starts at N25.

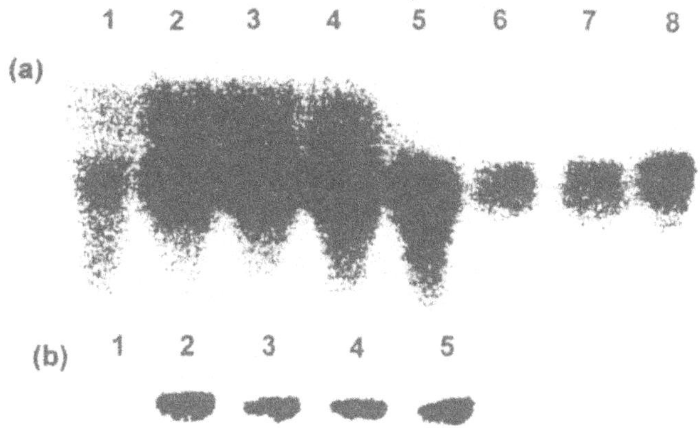

Figure 2. Northern blot analysis of the expression of the mRNA corresponding to the cDNA coding for the N-terminus of the 42-kDa CBPRP. Total RNA was isolated from unelicited cell culture (1) and from cultures following the addition of elicitor and harvested at 1.5 hours (2), 3 hours (3), 4.5 hours (4), 6 hours (5), 9 hours (6), 12 hours (7) and 24 hours (8). Blot (a) was probed with ^{32}P-labelled-D31, blot (b) with ^{32}P-labelled-A52.

4. CHITIN-BINDING PROPERTIES OF 42-kDa CBPRP

The 42-kDa CBPRP may function in two possible aspects of the response to pathogens. Firstly, it could be responsible for the perception of pathogen-derived chitin oligosaccharides. Such oligosaccharides have been shown to elicit the oxidative burst but not Phe ammonia-lyase in suspension-cultured cells of French bean (Wojtaszek *et al* 1995; Trethowan 1997). Figure 3 shows the binding properties of cells and purified 42-kDa CBPRP using equilibrium dialysis with radiolabeled chitin oligosaccharides. Both the cells and the 42-kDa CBPRP bound N-acetyl chitooligosaccharides rapidly with high affinity, but the binding was more reversible with the cells than the CBPRP. Scatchard analysis of this data gave a similar K_L of ~100 pM. This affinity would be in the range of a receptor, and the similar constant value suggests that the 42-kDa CBPRP could be responsible for much of the chitin-binding exhibited by the plant cells. However, it is difficult on the basis of its structure and its abundance to consider that the 42-kDa CBPRP is the chitin-binding receptor, as there is little to suggest that there are signal transduction domains in CBPRP. This paradox may change if full-length clones are ever established with certainty.

A second function could be related to the potential ability of 42-kDa CBPRP to bind to pathogens of French bean. It has been shown to bind to growing hyphal tips of the pathogen *Colletotrichum lindemuthianum* in culture (Millar *et al* 1992). In principle, CBPRP could be the first host molecule to function in this way and form a potential nucleation center for the deposition of other host-derived molecules, especially if it does prove to be esterified with ferulic acid (see below), a well-explored cross-linking agent in the primary wall. This ability was explored in elicitor-treated suspension cultures and in infection sites in the plant.

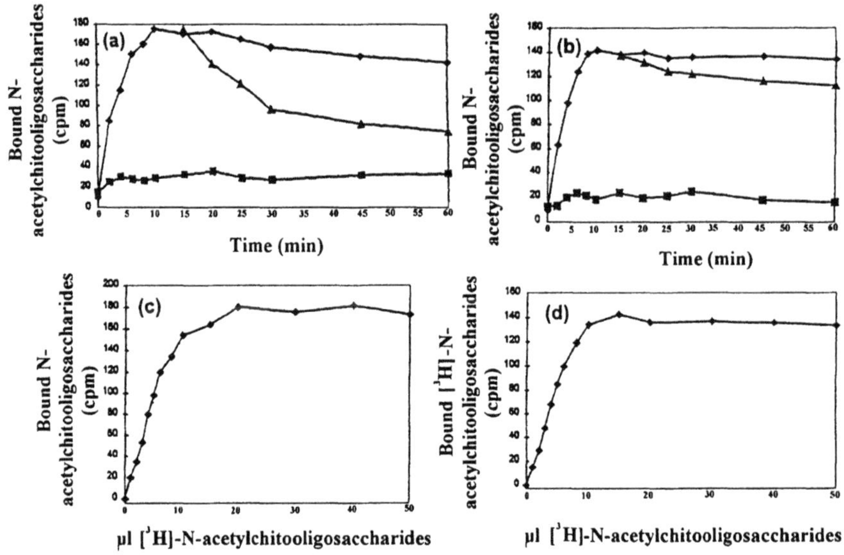

Figure 3. Chitin-binding properties of suspension-cultured French bean cells and the 42-kDa CBPRP. Chitin-binding properties were determined by equilibrium dialysis with [^{3}H]-N-acetylchito-oligosaccharides and unlabelled N-acetyl chitooligosaccharides of mean degree of polymerization = 7. (a) Time course of binding and reversibility of binding of radiolabeled chitooligosaccharides to 20 ml of suspension-cultured cells in the absence (♦) or presence of unlabelled oligosaccharides added in excess at zero time (■ non-specific binding) or 15 min (▲ competitive binding). (b) Time course of binding and reversibility of binding of radiolabeled oligosaccharides to 10 mg CBPRP in the absence (♦) or presence of unlabelled oligosaccharides added in excess at zero time (■ non-specific binding) or 15 min (▲ competitive binding). (c) Saturation curve of binding of radiolabeled oligosaccharides (cpm) to 20 ml of cells. (d) Saturation curve of binding of radiolabeled oligosaccharides (cpm) to 10-mg CBPRP.

5. ELICITOR-INDUCED OXIDATIVE CROSS-LINKING *IN VITRO* AND *IN PLANTA*

The 42-kDa CBPRP was found to be the major immobilized protein in walls of elicitor-treated suspension-cultured cells of French bean, the other abundant immobilized proteins being two extensins and a PR1-like extracellular protein (Wojtaszek *et al* 1995). Subsequently, conditions for the cross-linking of the 42-kDa CBPRP were determined *in vitro* (Wojtaszek *et al* 1997). Cross-linking requires extracellular alkalinization, the presence of a 46-kDa cell wall peroxidase (FBP1) and production of a substrate/reductant to be used for the generation of hydrogen peroxide by the peroxidase (Bolwell 1996). Hydrolysis of the product of cross-linking of the 42-kDa CBPRP followed by amino acid analysis is shown in Fig 4. We were unable to conclusively demonstrate the presence of iso-dityrosine, which has often been

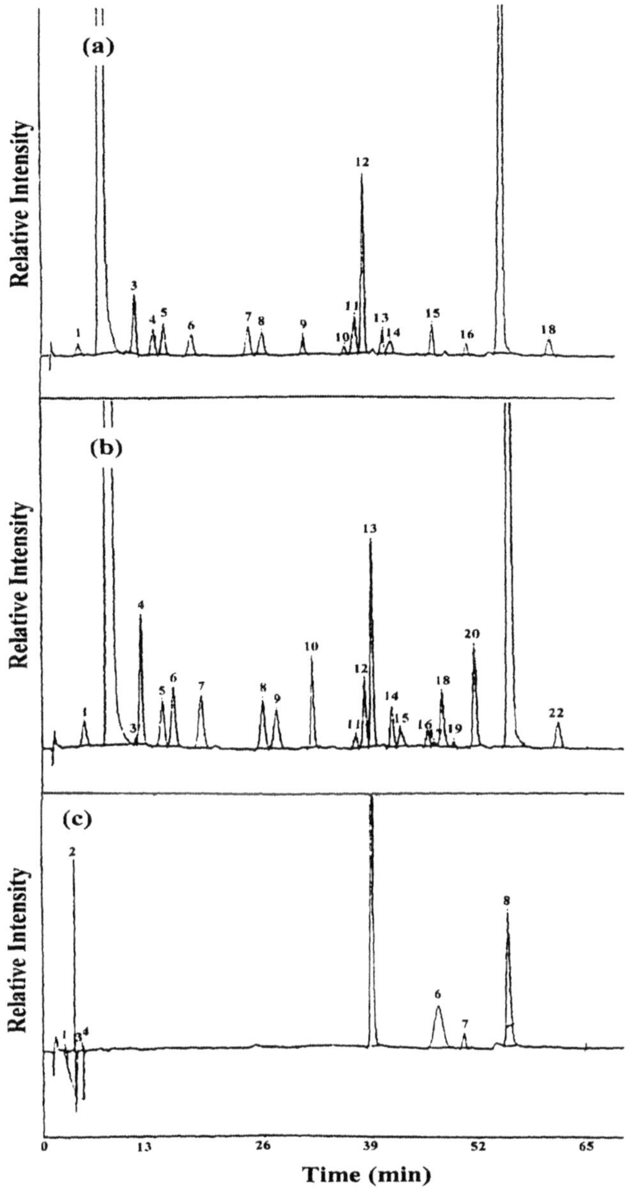

Figure 4. Amino acid analysis of monomeric and cross-linked CBPRP. Monomeric (a) and cross-linked (b) CBPRP were hydrolyzed and subjected to amino acid analysis on an Alpha Plus Analyzer (LKB, Sweden) and compared with authentic iso-dityrosine (a kind gift of Professor S. C. Fry, Edinburgh University, UK). No peak in either (a) or (b) corresponds exactly to the retention time of iso-dityrosine (peak 6 in c). The unusual peaks at position 1 in (a) and (b) are cysteic acid and those at 3 and 4 in (a) and (b), respectively, are Hyp. These amino acids are characteristic of the CBPRP domains.

postulated as the residue responsible for the cross-linking of HRGPs. Hydrolysis of the 42-kDa CBPRP, however, does show the presence of a blue fluorescent compound when analyzed by TLC (Bolwell 1986). We are currently investigating whether this compound is ferulic acid esterified to the arabinose residues. The FBP1 peroxidase shows high specificity for the formation of ferulic acid dimers (Zimmerlin *et al* 1994).

Evidence was obtained for the cross-linking of the 42-kDa CBPRP *in planta* in response to elicitation with chitooligosaccharide fragments (Fig 5). Tissue prints showed that following elicitor action the 42-kDa CBPRP could not be efficiently transferred to the nitrocellulose paper, suggesting it was immobilized. Furthermore, extracts of elicitation sites probed with anti-(bean 42-kDa CBPRP) serum on western blots showed that the protein was not extractable after elicitation. In extracts of unelicited hypocotyls the protein was detected as its monomer. However, in intercellular washing fluid the protein appears as a dimer. This propensity for dimerization was described previously (Bolwell 1987). Cross-linking could also be demonstrated in intercellular washing fluid when mixed with apoplastic fluid from elicited suspension cultured cells. The apoplastic fluid contains the substrate required for the generation of hydrogen peroxide (Wojtaszek *et al* 1997), and this substrate is presumably used by the peroxidase in the intercellular washing fluid to cross-link the CBPRP.

6. FUNCTION OF 42-kDa CBPRP IN THE DEFENSE RESPONSE

The function of the 42-kDa CBPRP in the defense response *in vivo* has been further investigated by immunolocalization. This technique has demonstrated that wounding of hypocotyls leads to increased abundance of CBPRP in the cortex, epidermis and vascular tissue immediately below the cut site (Millar *et al* 1992). The CBPRP was also identified as one of the components of papilla formation in bacterial infection sites (Brown *et al* 1998). Co-localization with the 46-kDa peroxidase, callose, and callose synthase was demonstrated in the cell wall and paramural deposits (Brown *et al* 1998). Figure 6 shows staining for the production of hydrogen peroxide in bacterial infection sites. The initial burst of peroxide surrounds the bacteria, and a later stage shows much more extensive staining along the wall. The peroxidase was co-localized with the areas of H_2O_2 production, consistent with a dual role in both generation and use of peroxide in cross-linking of proteins and phenolics during the construction of papillae. The CBPRP appears to be secreted into this region and is an important contribution to papilla structure. In the case of fungi species with chitinaceous cell walls, this role could be elevated in importance, since the CBPRP would actually bind to the penetrating hyphal tips. Cross-linking to host wall components would be very effective in immobilizing the pathogen so it could be attacked by active defense components such as reactive oxygen species, enzymes and low-molecular-weight compounds such as phytoalexins and other phenolics. Owing to the difficulty in "capturing" such fungal infection sites during sectioning, we have not yet been able to produce images of the quality that we have achieved with bacterial infection sites using immunogold labelling. However in preliminary studies of fungal penetration by *Colletotrichum lindemuthianum*, the CBPRP can be observed binding to hyphal tips precisely in a way that would predict a major role in anchoring fungal pathogens to the initial site of the interaction (P. Wojtaszek and G. P. Bolwell, unpublished data).

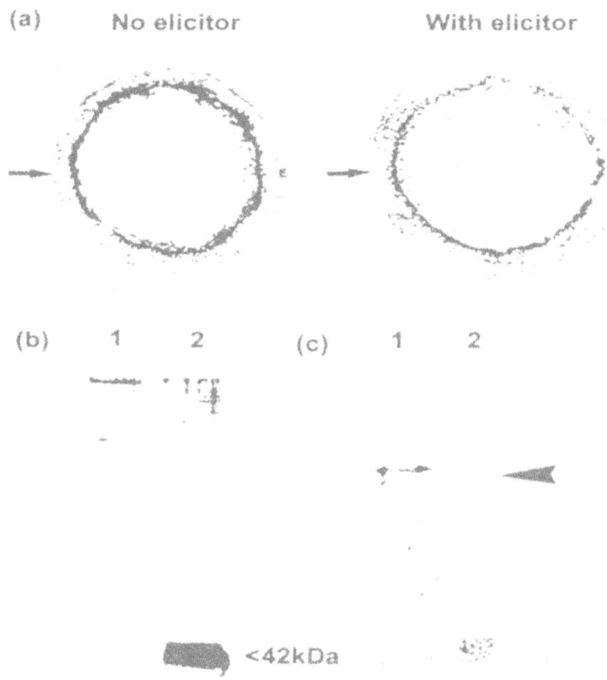

Figure 5. Immobilization of 42-kDa CBPRP *in planta.* (a) Tissue prints of hypocotyls of French bean mock treated with water droplets or with droplets containing chitooligosaccharides (1 μg/ml) at 1-cm intervals and incubated for 48 hours. Prints were developed with anti-(CBPRP) serum, which localizes to epidermis, cortex and vascular tissue (Millar *et al* 1992). Note inoculation site (arrowed) and that less CBPRP is transferred after elicitor treatment, showing it is immobilized. (X10) (b) Western blot of whole extracts of treated sites as in (a). Extracts were made in SDS-PAGE loading buffer, separated by SDS-PAGE and blotted onto nitrocellulose membrane. (1) Extracts from elicitor-treated hypocotyls, which show high-molecular-weight complexes only. (2) Extracts from water-treated sites showing predominantly monomeric CBPRP. (c) Cross-linking of CBPRP in intercellular washing fluid (IWF). Batches of untreated hypocotyls were vacuum infiltrated and gently centrifuged to isolate IWF. One batch of IWF was left untreated; a second was mixed with apoplastic fluid from elicitor-treated suspension-cultured cells, which supports oxidative cross-linking *in vitro* of the 42-kDa CBPRP (Wojtaszek *et al* 1997). Proteins were separated by SDS-PAGE, blotted and analyzed by Western blotting with anti-(42-kDa CBPRP) serum. (1) Untreated IWF, showing that the CBPRP is mainly dimer in this compartment. (2)Treated IWF, showing that the treatment with apoplastic fluid leads to high-molecular-weight complexes, which fail to enter the gel. The CBPRP has been cross-linked by the peroxidase also present in the IWF.

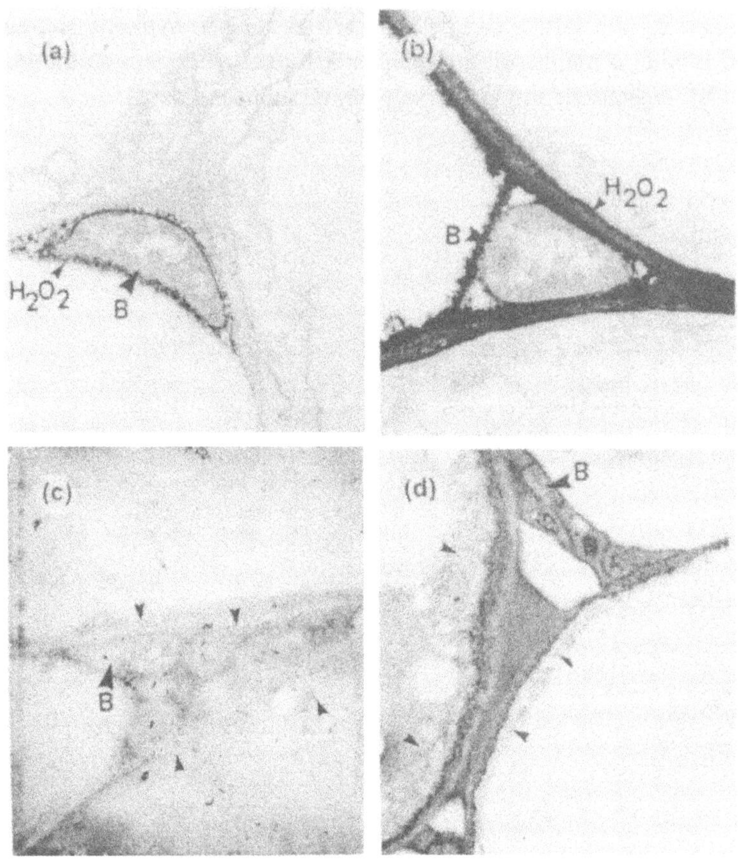

Figure 6. Co-localization of peroxidase and 42-kDa CBPRP in bacterial infection sites. (a) Staining for hydrogen peroxide using cerium chloride at an early stage in bacterial infection. (b) More extensive staining along the wall at a late stage in the defense response. (c) Immunogold localization of the 46-kDa peroxidase that generates the oxidative burst in French bean. The peroxidase has accumulated in the wall and extracellular matrix in the area of the bacteria, probably by directed secretion. (d) Immunogold localization of the 42-kDa CBPRP (arrows show examples of gold particles). The CBPRP has been secreted into developing papillae in the paramural deposits where it is cross-linked and associated with callose and phenolic deposition. For a more detailed description and analysis see Brown *et al* (1998). B, bacteria.

ACKNOWLEDGEMENTS

GPB thanks the BBSRC(UK) and the Leverhulme Trust for funding.

REFERENCES

Allen, A. K., Bolwell, G. P., Brown, D. S., Sidebottom, C., and Slabas, A. R., 1996, Potato lectin: a three-domain glycoprotein with novel hydroxyproline-containing sequences and sequence similarities to wheat-germ agglutinin, *Int. J. Biochem. Cell Biol.* **28**: 1285--1291.

Bolwell, G. P., 1986, Microsomal arabinosylation of polysaccharide and elicitor-induced carbohydrate-binding glycoprotein in French bean, *Phytochemistry* **25**: 1807–1813.

Bolwell, G. P., 1987, Elicitor induction of the synthesis of a novel lectin-like arabinosylated hydroxyproline-rich glycoprotein in suspension cultures of *Phaseolus vulgaris* L., *Planta* **172**: 184–191.

Bolwell, G. P., 1993, Dynamic aspects of the plant extracellular matrix, *Int. Rev. Cytol.* **146**: 261–324.

Bolwell, G. P., 1996, The origin of the oxidative burst in plants, *Biochem. Soc. Trans.* **24**: 438–441.

Brown, I., Trethowan, J., Kerry, M., Mansfield, J., and Bolwell G. P., 1998, Localisation of components of the oxidative cross-linking of glycoproteins and of callose synthesis in papillae formed during the interaction between non-pathogenic strains of *Xanthomonas campestris* and French bean mesophyll cells, *Plant J.* **15**: 333–345.

Du, H., Simpson, R. J., Clarke, A. E., and Bacic, A., 1996, Molecular characterisation of a stigma-specific gene encoding an arabinogalactan protein (AGP) from *Nicotiana alata*, *Plant J.* **9**: 313–323.

Millar, D. J., Slabas, A. R., Sidebottom, C., Smith, C. G., Allen, A. K., and Bolwell, G. P., 1992, A major stress-inducible M_r-42 000 wall glycoprotein of French bean (*Phaseolus vulgaris* L.), *Planta* **187**: 176–184.

Schultz, C. J., Hauser, K., Lind, J. L., Atkinson, A. H., Pu, Z., Anderson, M. A., and Clarke, A. E., 1997, Molecular characterisation of a cDNA sequence encoding the backbone of a style-specific 120 kDa glycoprotein which has features of both extensins and arabinogalactan proteins, *Plant Mol. Biol.* **35**: 833–845.

Sommer-Knudsen, J., Bacic, A., and Clarke, A. E., 1998, Hydroxyproline-rich glycoproteins, *Phytochemistry* **47**: 483–497.

Trethowan, J. B., 1997, Characterisation and expression of a novel chitin-binding protein involved in plant defence, Ph.D. Thesis, London University, UK.

Wojtaszek, P., Trethowan, J., and Bolwell, G. P., 1995, Specificity in the immobilisation of cell wall proteins in response to different elicitor molecules in suspension-cultured cells of French bean (*Phaseolus vulgaris* L.), *Plant Mol. Biol.* **28**: 1075–1087.

Wojtaszek, P., Trethowan, J., and Bolwell, G. P., 1997, Reconstitution *in vitro* of the components and conditions required for the oxidative cross-linking of extracellular proteins in French bean (*Phaseolus vulgaris* L.), *FEBS Lett.* **405**: 95–98.

Wojtaszek, P., Smith C. G., and Bolwell, G. P., 1999, Ultrastructural localisation and further biochemical characterisation of prolyl 4-hydroxylase from *Phaseolus vulgaris*: comparative analysis. *Int. J. Biochem. Cell Biol.* **31:** 463-477.

Zimmerlin, A., Wojtaszek, P., and Bolwell, G. P., 1994, Synthesis of dehydrogenation polymers of ferulic acid with high specificity by a purified cell wall peroxidase from French bean (*Phaseolus vulgaris* L.), *Biochem. J.* **299**: 747–753.

Chapter 9

Arabinogalactan-Proteins and Cell Development in Roots and Somatic Embryos

Clare G. Steele-King, William G.T. Willats and J. Paul Knox
Centre for Plant Sciences, University of Leeds, Leeds LS2 9JT, United Kingdom

1. ARABINOGALACTAN-PROTEINS AND PLANT CELL DEVELOPMENT

The cellular basis of plant development and the need for cell-to-cell interactions to direct the differentiation of cells is well established. However, our understanding of the molecular mechanisms of cell-to-cell interactions within a developing plant organ is limited. The occurrence of the structurally complex class of arabinogalactan-protein (AGP) proteoglycans at the surface of all plant cells has led to much speculation on their function(s) during development and their possible involvement in interactions between cells (Clarke *et al* 1979, Fincher *et al* 1983, Basile and Basile 1993, Knox 1995, Kreuger and van Holst 1996, Du *et al* 1996, Nothnagel 1997, Schultz *et al* 1998, Pennell 1998). Structural aspects of the complex carbohydrates of AGPs, the variability and diversity of the protein core and details of their location and attachment at plasma membranes and cell walls are discussed elsewhere (Clarke *et al* 1979, Fincher *et al* 1983, Du *et al* 1996, Nothnagel 1997, Schultz *et al* 1998). The biochemical complexity of AGPs, their heterogeneity and their range of physical properties have led to suggestions that they may function as lubricants, adhesion molecules or nutrients. It has also been proposed that they in some way encode information at cell surfaces that is utilized during development. Given such apparent complexity, several diverse experimental approaches to AGP function are applicable and useful. These varied approaches have begun to focus attention on AGPs in relation to particular cellular processes. The AGPs are now implicated in the three fundamental cellular processes that are coordinated to produce the plant body: cell proliferation, cell expansion and cell differentiation.

Cell and Developmental Biology of Arabinogalactan-Proteins
Edited by Nothnagel *et al.*, Kluwer Academic/Plenum Publishers, 2000

This chapter focuses on two approaches to understanding AGP function during cell development in roots and somatic embryos. The first of these approaches is the generation and use of monoclonal antibodies to the glycan components of AGPs. Such antibodies are powerful tools for investigating AGPs in a developmental context. Second, we consider aspects of the biological activities of the AGP-binding β-glycosyl Yariv reagents, which have emerged as a novel class of synthetic plant growth inhibitors. The term AGP is broad, as currently used, and is likely to include a range of structures and functions. Throughout, we indicate progress towards identifying AGPs associated with specific functions in root and somatic embryo development.

2. INSIGHTS FROM ANTI-AGP GLYCAN MONOCLONAL ANTIBODIES

The preparation of monoclonal antibodies to glycan components of AGPs has been discussed extensively elsewhere (Knox 1997, Nothnagel 1997). These antibodies have been widely used, and here we are predominantly concerned with recent observations. The important features of the anti-AGP monoclonal antibodies currently in use are that, first, they recognize the carbohydrate components of AGPs, and, second, they indicate that the carbohydrate components of AGPs are developmentally regulated within organs. In most cases this developmental regulation means that an antibody binds to the surface, most often the plasma membrane, of a subset of cells within a developing structure. The subset of cells relates to particular cell types or, in cases of very early cell development, to cell position. (See Knox [1997] for a detailed summary of AGP epitope occurrence in relation to cell development.)

The AGPs are widespread and are likely to occur in all plants. Most studies have focused on AGPs in higher plants, although considerable work has indicated a role for AGPs in controlling cell proliferation in liverworts (Basile and Basile 1993). The consideration of any differences between species in AGPs or the patterns of antibody binding is an important point, because they may indicate some taxonomic significance of the epitopes or antigens.

2.1 AGP Epitopes and Cell Development in Roots

In higher plants, primary roots have often been a focus for the study of the patterns of occurrence of AGP and other cell surface epitopes since the first report of a developmentally regulated plasma membrane AGP epitope (Knox *et al* 1989). Although such antibody studies have only been applied to a small number of species and families, some preliminary observations can now be drawn together to indicate trends. For example, the binding of the monoclonal antibody designated JIM13 was associated with xylem cells, but not phloem, in the roots of a range of dicotyledons, including carrot (Knox *et al* 1991), pea (Rae *et al* 1991, Casero *et al*

1998), *Arabidopsis* (Dolan *et al* 1995), and radish (Casero *et al* 1998), whereas the binding of the same antibody was associated with phloem cells, but not xylem, in the monocotyledons onion (Casero *et al* 1998) and maize (Šamaj *et al* 1998). Within the dicotyledons, however, there are subtle differences in the details of the occurrence of epitopes, indicating that the sequential appearance of an epitope does not always reflect developmental events in different species in the same way. For example, in radish and carrot the appearance of the JIM13 epitope correlates with xylem maturation, occurring first in the outer protoxylem and then appearing in association with inner metaxylem elements (Casero *et al* 1998). In contrast, in the *Arabidopsis* root, JIM13 binding occurs initially to a central xylem initial and then, as the root body develops, extends out to the protoxylem and parenchyma cells on either side of the xylem axis, and also ultimately to all the endodermal cells (Dolan *et al* 1995). In this latter case, the appearance of the JIM13 epitope reflects the emerging pattern of xylem cells rather than xylem cell maturation.

In roots, developmentally regulated AGP epitopes appear in files of cells during patterning events close to the meristem. The patterns mostly reflect differentiation outside the region of meristematic initials, and no AGP epitopes have yet been shown to be solely associated with cell proliferation at a root apical meristem. One possibility is that subsets of AGPs, as identified by epitopes, are involved in the control or suppression of cell proliferation (see Basile and Basile [1993], and work on liverworts), and the epitope patterns may reflect the transition from proliferation to differentiation. In several cases, patterns of AGP epitopes have been observed within the outer layer of the vascular tissue known as the pericycle. The pericycle is a meristematic tissue and is the site of origin of lateral roots. Root pericycle cells undergo defined patterns of cell division well before the emergence of lateral roots (Casero *et al* 1996). An epitope (recognized by a monoclonal antibody designated JIM4) appears sequentially in adjacent cells of the developing pericycle of carrot just before the cells undergo distinctive oblique divisions (Knox *et al* 1989, Casero *et al* 1998), suggesting a connection between an AGP and a tissue-specific proliferation. In this case, the JIM4 AGP epitope appears in two groups of cells opposite the xylem, and AGP epitope patterns within the pericycle generally reflect the radial symmetry of the vascular tissues. However, patterns of antibody binding to pericycle cells opposite the phloem or xylem do not appear to correlate with the sites of origin of lateral roots, which also occur in relation to the vascular pattern (Casero *et al* 1998).

In addition to extending the survey of epitope occurrence in a range of species, anti-AGP antibodies are also useful for the study of developmental mutants. For example, JIM13 has been used in a study of a mutation in *Arabidopsis* that causes an altered root anatomy. The mutation in the *SCARECROW* gene results in the loss of a cell layer between the epidermis and the pericycle of the *Arabidopsis* root (Laurenzio *et al* 1996). The one cell layer that replaces the cortex and the endodermis had cell surface characteristics of both cortical and endodermal cells, the binding of JIM13 being taken as an indicator of an endodermal character. These observations indicated that a heterogenous cell type was formed because of the absence of a crucial lineage-forming asymmetric division (Laurenzio *et al* 1996). Such observations also indicate the potential difficulties in ascribing causes and effects to AGP molecules. Is the JIM13 epitope in the *Arabidopsis* endodermis a marker of endodermis characteristics (i.e., associated with endodermis differentiation), or alternatively is it in some way a developmental marker of tissue or cell boundaries associated with the cell layer adjacent to the pericycle?

2.2 AGPs and Cell Dynamics During Carrot Somatic Embryogenesis

Primary roots have the advantage for experimental analysis of concurrent spatial and temporal development sequences within one organ. Carrot somatic embryogenesis is a powerful experimental system with the advantage that it is possible to dissect the development of tissue systems, several organs, and root and shoot meristems. Somatic embryo culture systems can be readily accessed and are open to manipulation. In this system, AGP epitopes are also developmentally regulated, reflecting the earliest cell distinctions within the embryo (Stacey *et al* 1990). Furthermore, AGPs have been shown to influence the progress of carrot somatic embryogenesis when added to the culture medium. Distinct populations of AGPs, sorted using monoclonal antibodies, have been shown to promote or inhibit somatic embryogenesis, suggesting that AGP epitopes are associated with distinct activities (Kreuger and van Holst 1995 1996). Several studies have focused on the occurrence of one epitope, recognized by a monoclonal antibody designated JIM8, in the cell cultures beginning at the early stage, when embryogenic potential is developing, and continuing through the formation of carrot somatic embryos. The JIM8 antibody binds to the outer surface of the cell wall of living cells in embryogenic cultures, and observations indicate that the occurrence of the JIM8 cell wall epitope is dynamic, occurring in only a proportion of the cell population (Toonen *et al* 1996, McCabe *et al* 1997). The epitope appears to be lost from the cell wall before embryo formation, but such non-JIM8-binding cells require the presence of JIM8-binding cells to develop (McCabe *et al* 1997). Furthermore, a soluble signal obtained from JIM8-binding cells can substitute for these cells, although whether this signal is one of the JIM8 antigens occurring in the active extracts is not yet determined. These observations indicate that differences in cell populations parallel differences in the occurrence of the JIM8 epitope and suggest the possible involvement of the JIM8 cell wall antigen in cell-to-cell interactions that underpin cell development (McCabe *et al* 1997).

2.3 AGP Glycan Epitope Structure

The monoclonal antibody JIM8 can bind to plasma membrane AGPs (Pennell *et al* 1991), but some JIM8-reactive material in the cell walls of carrot cells appears to be related to pectic polysaccharides (McCabe *et al* 1997). It is an often overlooked complication that the type of arabinogalactan occurring in AGPs, known as Type II arabinogalactan (Clarke *et al* 1979), may also occur as minor components and as side chains of pectic polysaccharides (Guillon and Thibault 1989, Renard *et al* 1991). The observations with JIM8 highlight a major obstacle to progress in studies with the anti-AGP glycan monoclonal antibodies, and this obstacle is the lack of understanding of AGP glycan epitope structure. To some extent the preparation of the monoclonal antibodies previously mentioned (JIM4, JIM13 and JIM8) was fortuitous in that they were selected subsequent to immunization with complex immunogens (see Knox [1997] for a full discussion). An attempt has been made to address the question of epitope structure for these antibodies, and the indications (from studies using hapten inhibition of antibody binding) are that some developmentally regulated epitopes contain arabinose or glucuronic acid and are likely to be side chains and terminal oligosaccharides on the (1→3)-β-(1→6)-β-galactan core structure of the AGP glycan (Yates *et al* 1996). However, we still have no knowledge as to whether the observed patterns of antibody binding reflect the highly regulated and directed insertion of AGPs at the cell surface or patterns of alteration and processing following incorporation at the cell surface. One possible mechanism for the latter could be the processing of membrane-bound forms into soluble forms by cleavage of glycosylphosphatidylinositol anchors (Schultz *et al* 1998). The unattached soluble forms of AGPs may not be as securely fixed in to the plant material by aldehyde fixatives as their membrane-bound counterparts prior to immunocytochemistry. However, this is perhaps unlikely, as most regulated epitopes appear during development, rather than disappear. We currently have few clues to synthetic or degradative enzymes involved in generating the oligosaccharides that form the epitopes and that could be targets for addressing epitope function. Furthermore, the biochemistry of the relationships and links between constitutive AGP epitopes (that occur at the surface of all cells in a developing organ) and developmentally regulated epitopes is still almost entirely unknown (Knox *et al* 1991, Smallwood *et al* 1996).

An alternative approach to understanding the developmental regulation of AGP glycan structure is the preparation of antibodies to defined oligosaccharide structures (which probably need to be at least three residues in size to ensure antibody specificity). A defined antisera has been made for a tetrasaccharide of the (1→6)-β-galactan of AGPs, and immunolocalization studies have indicated that the epitope occurs at all cell surfaces in radish and flax roots (Kikuchi *et al* 1993, Vicré *et al* 1998). However, the full range of oligosaccharide structures occurring in AGPs is not known, and many appropriate oligosaccharides would be difficult to obtain from plant material in amounts and purity sufficient for the covalent coupling to protein to produce immunogens. The preparation of defined anti-glycan probes for the pectic polysaccharides has progressed further than that for AGPs because of an increased availability of appropriate oligosaccharides for immunogen preparation, and monoclonal antibodies recognizing fully defined epitopes of pectic (1→4)-β-galactan and (1→5)-α-arabinan have been generated (Jones *et al* 1997, Willats *et al* 1998). The chemical synthesis of oligosaccharides occurring in AGPs is also time-consuming and difficult (Valdor and Mackie 1997).

3. INSIGHTS FROM STUDIES OF THE BIOLOGICAL ACTIVITIES OF β-GLUCOSYL YARIV REAGENT

The work with monoclonal antibodies is important in indicating the extensive developmental regulation of AGPs. It is also of considerable interest to test directly the function of particular AGP structures. The demonstration that β-glucosyl Yariv reagent (βGlcY), a synthetic phenylglycoside, could reversibly inhibit the proliferation of cultured plant cells (Serpe and Nothnagel 1994) led to a dramatic change in the way this class of compounds is viewed, adding biological action to an AGP-specific probe previously used for diagnostic, isolation and localization purposes.

The nature of the interaction of βGlcY with AGPs is not fully understood but probably requires a self-associated polymeric form of βGlcY, and a recent review is provided by Nothnagel (1997). A β-linkage of the glycoside to the phenyl group is known to be required for interaction with AGPs, and the non-AGP-binding α-galactosyl Yariv reagent (αGalY) is often used as a control Yariv reagent. Although a recent report indicates that βGlcY can also associate with non-AGP cell wall macromolecules, including cellulose, the interaction with cellulose requires high levels of βGlcY and has no stereoselectivity in that αGalY interacts in an equivalent manner to βGlcY (Triplett and Timpa 1997). It is therefore important for all studies involving the use of βGlcY to probe AGP function, that αGalY or another non-AGP-binding Yariv reagent is used as a control.

The initial observation of the biological activity of βGlcY on proliferation in cell cultures has now been extended to other systems, including tip-growing pollen tubes (Roy *et al* 1998). Here we focus on the use of βGlcY to probe AGP function during carrot somatic embryogenesis and during root growth in *Arabidopsis* seedlings. The action of βGlcY in these complex systems reveals more subtle and specific effects than those observed upon proliferating cell cultures, in that certain cells appear to be targeted by βGlcY with consequent disruptions of balanced growth patterns. These observations raise the possibility that in certain systems βGlcY interacts with specific populations of AGPs, probably restricted to the surface of a group of cells within an organ.

3.1 βGlcY and Carrot Somatic Embryogenesis

The application of βGlcY to proliferating rose cells in culture resulted in the cessation of proliferation, indicating a role for AGPs in cell proliferation (Serpe and Nothnagel 1994, Langan and Nothnagel 1997). How the binding of βGlcY to AGPs results in the reversible blockage of proliferation is not yet known. Such effects may be direct and act upon cell proliferation or indirect and disrupt other cell processes. In a cell culture system that can be switched from cell proliferation to cell elongation (by removal from the culture medium of the synthetic auxin 2,4-D), the application of βGlcY resulted in a blockage of cell elongation (Willats and Knox 1996). This observation could indicate that cell expansion is disrupted during the cell cycle or that βGlcY has a widespread, non-specific effect on cell functioning, albeit a reversible one. As indicated above, carrot somatic embryogenesis is a powerful experimental system, and studies indicate that βGlcY can perturb the developing embryos and does not have a global inhibitory effect.

The first of these observations is that the application of βGlcY during the early stages of somatic embryogenesis resulted in the disruption of root–shoot coordination. When βGlcY, at concentrations as low as 5 μM (~5 μg/ml), was supplied to the liquid medium at the time of transfer of embryogenic cells from maintenance medium (with 2,4-D) to induction medium (without 2,4-D), it resulted in the subsequent development of roots, rather than somatic embryos (Thompson and Knox 1998). A direct effect of βGlcY promoting root growth cannot be ruled out, but a more plausible interpretation is that βGlcY specifically blocks shoot development, and this allows or promotes extensive root growth. Shoot development was also blocked and root development continued when globular-stage embryos were transferred to medium containing βGlcY, although some radial expansion of the shoot apex occurred (Thompson and Knox 1998). It is known that roots of carrot somatic embryos elongate rapidly when the shoot apex end is surgically removed, an effect that can be inhibited by the application of auxin (Schiavone and Racusen 1990). If βGlcY binds to a subset of cells at the shoot end of developing globular embryos, preventing proliferation and development, then this may result in the disruption of auxin flow within the developing embryo. Possible candidates for the target cells and target AGPs in this system are the AGPs and cells identified by the monoclonal antibody JIM4. This antibody binds specifically to the plasma membrane of cells of the protoderm at the shoot apex end of a carrot globular embryo (Stacey *et al* 1990). It is possible that it is cell proliferation that is blocked by βGlcY at the shoot end of the embryos, as there is no extensive cell expansion occurring at this stage. When βGlcY was applied to later, torpedo-stage carrot somatic embryos, an inhibition of growth was observed, which appeared to be largely due to reduced cell expansion (Thompson and Knox 1998). It is also of interest that the effect of βGlcY on carrot somatic embryo development is similar to the effect of mutations in the *GURKE* gene on embryo development in *Arabidopsis*, in that they result in reduced or no shoot development but have no effect on root development (Torres-Ruiz *et al* 1996).

A further intriguing effect of βGlcY on embryogenic carrot cell cultures results in growth promotion. When clumps of proembryogenic masses, up to 1 mm in diameter, were isolated from an embryogenic culture and placed on solid media containing 5 μM βGlcY there was a three- to fourfold increase in the fresh weight of material when examined after 32 days and compared with cells placed upon media without βGlcY (Thompson and Knox 1998). This effect was not observed if 30 μM βGlcY was supplied in the media, nor if the media contained 2,4-D (i.e., if cells were grown on a maintenance rather than induction media [Thompson and Knox 1998]). So how does a low level of βGlcY have a growth promotion effect in this case? After 30 days, plant material grown on maintenance media was undifferentiated callus, whereas that grown on induction medium was largely plantlets and not callus. It therefore appears that the asymmetric application of 5 μM βGlcY to developing proembryogenic masses somehow resulted in a growth promotion of developing cells and the resulting plantlets, but had no effect on cells that were just proliferating. Plantlet number and the proportions of root, hypocotyl and shoot in the plantlets resulting from treatment and nontreatment have not yet been examined in detail, and this may provide some insight. It is likely that only a few cells of the proembryogenic masses were in direct contact with the surface of the media and accessible to βGlcY in this system. The growth promotion may indicate that the βGlcY disrupted some interactions, possibly inhibitory, between cells in the clusters, but that at the same time the flow of nutrients from the media to these cells was not impeded. As discussed above, subsets of AGPs (sorted by monoclonal antibodies) have been shown to have inhibitory or promotive effects on carrot somatic embryogenesis (Kreuger and van Holst 1996). The diverse responses to the application of different concentrations of βGlcY may reflect interaction with different populations of AGPs.

3.2 βGlcY and *Arabidopsis* Seedling Root Growth

The germination and growth of *Arabidopsis* seedlings in solid media containing 30 μM βGlcY results in a specific effect: germination is unaffected, proliferation within the root meristem appears to be unaffected, but cells proximal to the meristem do not elongate in a manner equivalent to wild type (Willats and Knox 1996). In addition, root epidermal cells bulge outwards in a manner equivalent to that observed in *reb* mutants (Baskin *et al* 1992). In βGlcY-treated roots all post-meristematic root body cells have longitudinal axes less than 50% of the length of those occurring in untreated seedlings. We have interpreted the bulging of epidermal cells as an indirect effect due to an inhibition of internal load-bearing cell layers, although clearly βGlcY can also access the outer cells (Willats and Knox 1996). For a discussion of the evidence that the load-bearing cell layer controlling root growth is not the epidermis but an internal layer, see Pritchard (1994). Although the βGlcY does not appear to occur beyond the root/hypocotyl boundary in treated seedlings, shoot growth is also reduced, most likely as an indirect effect of reduced root growth. These effects have been confirmed by Ding and Zhu (1997), who have interpreted the induced bulging of the epidermal cells as a direct effect. These authors have also shown that the *reb1-1* mutant, with bulging epidermal cells, has a reduced level and an altered profile of βGlcY-reactive AGPs occurring in its roots (Ding and Zhu 1997). Does βGlcY have this regulatory effect on root growth in other systems? In the authors' laboratory we have germinated carrot seedlings on artificial media in the presence of 30 μM βGlcY. In this case, highly βGlcY-reactive and abundant AGPs associated with the root cap bind to most of the βGlcY supplied, and there are no effects on root growth rate or epidermal cell morphology (S.E. Marcus and J.P. Knox, unpublished observations).

In our laboratory, we have used the defined effect of βGlcY on *Arabidopsis* root growth as the basis of a screen for mutants that have altered AGPs. Such mutants will be powerful tools to investigate AGP function. We have developed

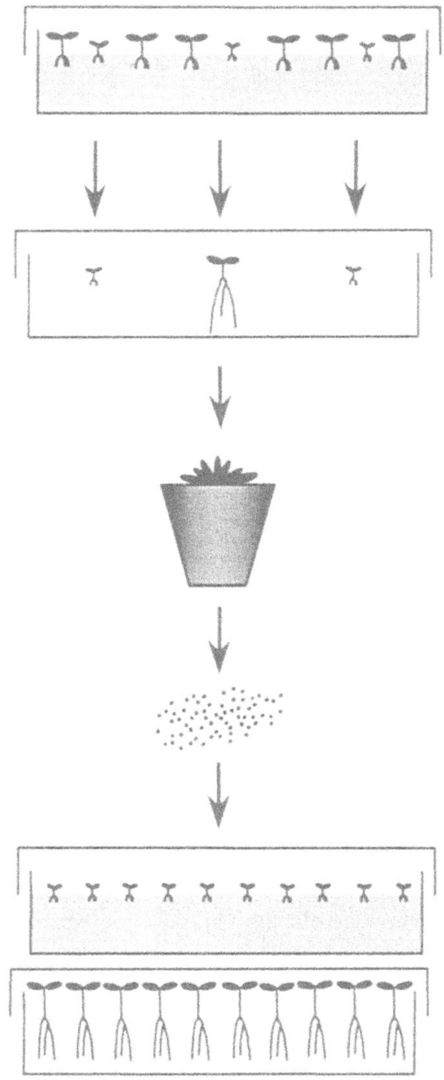

Figure 1. Schematic outline of a strategy used to screen for altered growth responses to βGlcY and to select mutant seedlings of *Arabidopsis thaliana* with altered AGPs.

procedures to screen populations of T-DNA-inserted mutants and EMS-treated seed for germination and growth on media containing βGlcY, and our strategy is shown schematically in Fig 1. The outlined procedures will be used to isolate mutant seedlings that appear to be insensitive or supersensitive to βGlcY. An important aspect of any supersensitive mutants will be the reversibility of any response to βGlcY, which will indicate that they are likely to be AGP mutants. Selected mutants may have directly altered AGPs that disrupt or enhance interaction with βGlcY. Alternatively, modified aspects of root structure or cell wall architecture may alter the ability of βGlcY to enter the root and interact with AGPs. In addition to any mutants selected from these screens, it is also of interest to examine AGPs in other established growth mutants of *Arabidopsis* using βGlcY and anti-AGP monoclonal antibodies. Although in most of these cases the genetic lesions are unlikely to affect AGP structure directly, they may provide insight into AGP function within pathways directing and controlling cell expansion and other aspects of growth and development. For example, an extreme dwarf phenotype results from mutations in the *DIMINUTIO* gene, which encodes an enzyme involved in brassinosteroid biosynthesis (Klahre *et al* 1998), and mutant plants have reduced levels of βGlcY-reactive AGPs in their hypocotyls (Takahashi *et al* 1995).

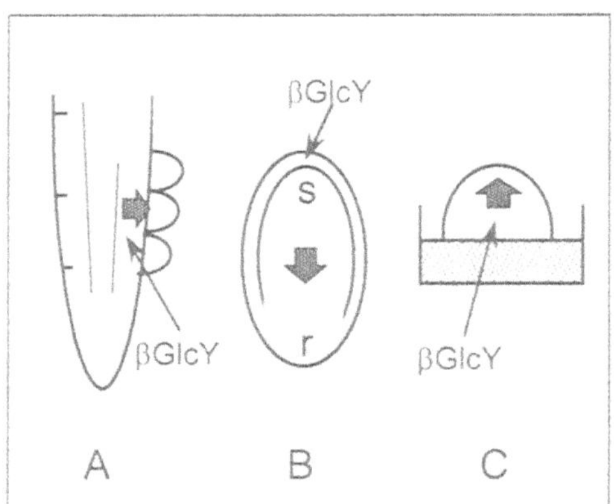

Figure 2. Diagrammatic representation of the proposed action of βGlcY resulting in indirect growth effects. A. Inhibition of cell elongation in the *Arabidopsis* seedling root resulting in bulging of epidermal cells. B. Inhibition of shoot (s) apex development in the carrot somatic embryo with consequent promotion of root (r) development. C. Growth promotion of proembryogenic masses placed on solid media containing βGlcY. The shaded arrow in each case indicates disruption/influence extending from the proposed site of action of βGlcY.

4. A WAY FORWARD

The AGPs are biochemically complex and appear to have patterns of expression at plant cell surfaces reflecting the cell patterning and cell differentiation events that occur in developing organs. An important next step is to characterize and dissect AGPs within a single developmental system in more detail. This will involve integration of the structural characterization of individual AGPs or populations of AGPs (protein core and carbohydrate analysis, immunoblotting and two dimensional gel analysis) with a precise knowledge of the developmental regulation of the structure. The disruption of the development of the system by βGlcY will also be a powerful tool with which to focus on a set or sets of AGPs and their involvement in a specific process (e.g., the βGlcY-binding AGPs and cell expansion of the *Arabidopsis* seedling root). In this chapter we have highlighted three experimental systems where βGlcY appears to disrupt specific cell processes and result in indirect effects, as shown schematically in Fig 2. In two cases, the response to βGlcY involves growth promotion, and this may reflect a disruption of cell-to-cell interactions. Systems such as these should provide the opportunity to focus on specific AGPs that are targets for βGlcY and also to determine the consequence of βGlcY binding on cell wall architecture and cell processes in general. Knowledge of individual AGPs, integrated with observations on AGPs in growth and developmental mutants of *Arabidopsis*, should lead to the placing of particular AGPs, and their associated functions, within pathways and networks that underlie plant cell development.

REFERENCES

Basile, D. V., and Basile, M. R., 1993, The role and control of the place-dependent suppression of cell division in plant morphogenesis and phylogeny, *Mem. Torrey Bot. Club* **25**: 63–84.

Baskin, T. I., Betzner, A. S., Hoggart, R., Cork, A., and Williamson, R. E., 1992, Root morphology mutants in *Arabidopsis thaliana*, *Aust. J. Plant Physiol.* **19**: 427–437.

Casero, P. J., Casimiro, I., and Lloret, P. G., 1996, Pericycle proliferation pattern during the lateral root initiation in adventitious roots of *Allium cepa*, *Protoplasma* **191**: 136–147.

Casero, P. J., Casimiro, I., and Knox, J. P., 1998, Occurrence of cell surface arabinogalactan-protein and extensin epitopes in relation to pericycle and vascular tissue development in the root apex of four species, *Planta* **204**: 252–259.

Clarke, A. E., Anderson, R. L., and Stone, B. A., 1979, Form and function of arabinogalactans and arabinogalactan-proteins, *Phytochemistry* **18**: 521–540.

Ding, L., and Zhu, J-K., 1997, A role for arabinogalactan-proteins in root epidermal cell expansion, *Planta* **203**: 289–294.

Dolan, L., Linstead, P., and Roberts, K., 1995, An AGP epitope distinguishes a central metaxylem initial from other vascular initials in the Arabidopsis root, *Protoplasma* **189**: 149–155.

Du, H., Clarke, A. E., and Bacic, A., 1996, Arabinogalactan-proteins: a class of extracellular proteoglycans involved in plant growth and development, *Trends Cell Biol.* **6**: 411–414.

Fincher, G. B., Stone, B. A., and Clarke, A. E., 1983, Arabinogalactan-proteins: structure, biosynthesis and function, *Annu. Rev. Plant Physiol.* **34**: 47–70.

Guillon, F., and Thibault, J-F., 1989, Methylation analysis and mild acid hydrolysis of the "hairy" fragments of sugar beet pectins, *Carbohyd. Res.* **190**: 85–96.

Jones, L., Seymour, G. B., and Knox, J. P., 1997, Localization of pectic galactan in tomato cell walls using a monoclonal antibody specific to (1→4)-β-D-galactan, *Plant Physiol.* **113**: 1405–1412.

Kikuchi, S., Ohinata, A., Tsumuraya, Y., Hashimoto, Y., Kaneko, Y., and Matsushima, H., 1993, Production and characterization of antibodies to the β(1→6)-galactotetrasyl group and their interactions with arabinogalactan-proteins, *Planta* **190**: 525–535.

Klahre, U., Noguchi, T., Fujioka, S., Takatsuto, S., Yokota, T., Nomura, T., Yoshida, S., and Chua, N. H., 1998, The Arabidopsis *DIMINUTO/DWARF1* gene encodes a protein involved in steroid synthesis, *Plant Cell* **10**: 1677–1690.

Knox, J. P., 1995, Developmentally-regulated proteoglycans and glycoproteins of the plant cell surface, *FASEB J.* **9**: 1004–1012.

Knox, J. P., 1997, The use of antibodies to study the architecture and developmental regulation of plant cell walls, *Int. Rev. Cytol.* **171**: 79–120.

Knox, J. P., Day, S., and Roberts, K., 1989, A set of cell surface glycoproteins forms a marker of cell position, but not cell type, in the root apical meristem of *Daucus carota.* L., *Development* **106**: 47–56.

Knox, J. P., Linstead, P. J., Peart, J., Cooper, C., and Roberts, K., 1991, Developmentally-regulated epitopes of cell surface arabinogalactan-proteins and their relation to root tissue pattern formation, *Plant J.* **1**: 317–326.

Kreuger, M., and van Holst, G.-J., 1995, Arabinogalactan-protein epitopes in somatic embryogenesis of *Daucus carota* L., *Planta* **197**: 135–141.

Kreuger, M., and van Holst, G.-J., 1996, Arabinogalactan-proteins and plant differentiation, *Plant Mol. Biol.* **30**: 1077–1086.

Langan, K. J., and Nothnagel, E. A., 1997, Cell surface arabinogalactan-proteins and their relation to cell proliferation and viability, *Protoplasma* **196**: 87–98.

Laurenzio, L. D., Wysocka-Diller, J., Malamy, J. E., Pysh, L., Helariutta, Y., Freshour, G., Hahn, M. G., Feldmann, and Benfy, P. N., 1996, The *SCARECROW* gene regulates an asymmetric cell division that is essential for generating the radial organization of the Arabidopsis root, *Cell* **86**: 423–433.

McCabe, P. F., Valentine, T. A., Forsberg, L. S., and Pennell, R. I., 1997, Soluble signals from cells identified at the cell wall establish a developmental pathway in carrot, *Plant Cell* **9**: 2225–2241.

Nothnagel, E. A., 1997, Proteoglycans and related components in plant cells, *Int. Rev. Cytol.* **174**: 195–291.

Pennell, R. I., 1998, Cell walls: structures and signals, *Curr. Opin. Plant Biol.* **1**: 504–510.

Pennell, R. I., Janniche, L., Kjellbom, P., Scofield, G. N., Peart, J. M., and Roberts, K., 1991, Developmental regulation of a plasma membrane arabinogalactan-protein epitope is oilseed rape flowers, *Plant Cell* **3**: 1317–1326.

Pritchard, J., 1994, The control of cell expansion in roots, *New Phytol.* 127: 3–26.

Rae, A. L., Perotto, S., Knox, J. P., Kannenberg, E. L., and Brewin, N. J., 1991, Expression of extracellular glycoproteins in the uninfected cells of developing pea nodule tissue, *Mol. Plant Micr. Interact.* **4**: 563–570.

Renard, C. M. G. C., Voragen, A. G. J., Tibault, J.-F., and Pilnik, W., 1991, Studies on apple protopectin V: structural studies on enzymatically extracted pectins, *Carbohyd. Polymers* **16**: 137–154.

Roy, S., Jauh, G. Y., Hepler, P. K., and Lord, E. M., 1998, Effects of Yariv phenylglycoside on cell wall assembly in the lily pollen tube, *Planta* **204**: 450–458.

Šamaj, J., Baluska, F., and Volkmann, D., 1998, Cell-specific expression of two arabinogalactan protein epitopes recognized by monoclonal antibodies JIM8 and JIM13 in maize roots, *Protoplasma* **204**: 1–12.

Schiavone, F. M., and Racusen, R. H., 1990, Microsurgery reveals regional capabilities for pattern re-establishment in somatic carrot embryos, *Dev. Biol.* **141**: 211–219.

Schultz, C., Gilson, P., Oxley, D., Youl, J., and Bacic, A., 1998, GPI-anchors on arabinogalactan-proteins: implications for signalling in plants, *Trends Plant Sci.* **3**: 426–431.

Serpe, M. D., and Nothnagel, E. A., 1994, Effects of Yariv phenylglycosides on Rosa cell suspensions: evidence for the involvement of arabinogalactan-proteins in cell proliferation, *Planta* **193**: 542–550.

Smallwood, M., Yates, E. A., Willats, W. G. T., Martin, H., and Knox, J. P., 1996, Immunochemical comparison of membrane-associated and secreted arabinogalactan-proteins in rice and carrot, *Planta* **198**: 452–459.

Stacey, N. J., Roberts, K., and Knox, J. P., 1990, Patterns of expression of the JIM4 arabinogalactan protein epitope in cell cultures and during somatic embryogenesis in *Daucus carota* L., *Planta* **180**: 285–292.

Takahashi, T., Gasch, A., Nishizawa, N., and Chua, N.-H., 1995, The *DIMINUTO* gene of Arabidopsis is involved in regulating cell elongation, *Genes Dev.* **9**: 97–107.

Thompson, H. J. M., and Knox, J. P., 1998, Stage-specific responses of embryogenic carrot cell suspension cultures to arabinogalactan protein-binding β-glucosyl Yariv reagent, *Planta* **205**: 32–38.

Toonen, M. A. J., Schmidt, E. D. L., Hendriks, T., Verhoeven, H. A., Van Kammen, A., and de Vries, S. C., 1996, Expression of the JIM8 cell wall epitope in carrot somatic embryogenesis, *Planta* **200**: 167–173.

Torres-Ruiz, R. A., Lohner, A., and Jürgens, G., 1996, The *GURKE* gene is required for normal organization of the apical region in the Arabidopsis embryo, *Plant J.* **10**: 1005–1016.

Triplett, B. A., and Timpa, J. D., 1997, β-Glucosyl and α-galactosyl Yariv reagents bind to cellulose and other glucans, *J. Agric. Food. Chem.* **45**: 4650–4654.

Valdor, J. F., and Mackie, W., 1997, Synthesis of a trisaccharide repeating unit related to arabinogalactan-protein (AGP) polysaccharides, *J. Carbohyd. Chem.* **16**: 429– 440.

Vicré, M., Jauneau, A., Knox, J. P., and Driouich A., 1998, Immunolocalization of β(1→4)- and β(1→6)-D-galactan epitopes in the cell wall and Golgi stacks of developing flax root tissues, *Protoplasma* **203**: 26–34.

Willats, W. G. T., and Knox, J. P., 1996, A role for arabinogalactan-proteins in plant cell expansion: evidence from studies on the interaction of β-glucosyl Yariv reagent with seedlings of *Arabidopsis thaliana*, *Plant J.* **9**: 919–925.

Willats, W. G. T., Marcus, S. E., and Knox J. P., 1998, Generation of a monoclonal antibody specific to (1→5)-α-L-arabinan, *Carbohyd. Res.* **308**: 149–152.

Yates, E. A., Valdor, J-F., Haslam, S. M., Morris, H. R., Dell, A., Mackie, W., and Knox, J. P., 1996, Characterization of carbohydrate structural features recognized by anti-arabinogalactan-protein monoclonal antibodies, *Glycobiology* **6**: 131–139.

Chapter 10

Effect of Arabinogalactan-Proteins and Chitinases on Somatic Embryogenesis

Marc Kreuger,[1] Arjon van Hengel[2,3] and Sacco de Vries[2]
[1]Novartis Seeds BV., Plant Propagation Technology, P.O. Box 2, 1600 AA Enkhuizen, The Netherlands; [2]Laboratory of Molecular Biology, Wageningen University, Dreijenlaan 3, 6703 HA Wageningen, The Netherlands; [3]Present address: Department of Cell Biology, John Innes Institute, Colney, Norwich NR4 7UH, United Kingdom

1. INTRODUCTION

The process of somatic embryogenesis is interesting in several ways. Because it takes place in liquid cultures, it provides an opportunity to study the development of embryos completely free from surrounding structures such as the seed. Although differences in morphology are quite frequently observed, it has been established that somatic embryos are quite similar, in molecular and physiological aspects, to zygotic embryos in developing seeds. For means of plant propagation, somatic embryos offer a way of cloning that can be applied on a large scale and at relatively low cost. Both aspects have attracted scientists over the years, and a wealth of literature is available on the subject.

Making plant cells from every crop embryogenic in tissue culture is a wish of many businessmen in the propagation industry. As many scientists in the same industry know, this is not always easy to achieve. Classical growth regulators, especially auxin, stress and/or other factors are well known to induce embryogenic capacity. What is also known is that these factors alone do not guarantee successful induction of embryogenic potential. Other factors, such as the state of the explant or the medium composition, can also be decisive and are much harder to grasp. This difficulty may have contributed to the fact that today still only a few crops are considered as model systems for somatic embryogenesis. The search for "regeneration powder", the "Holy Grail" of somatic embryologists, is still underway but never accomplished. Of the many non-classical growth regulators known to influence somatic embryogenesis, arabinogalactan-proteins (AGPs) are an attractive candidate to perhaps at least fill up the long-sought Holy Grail.

In this chapter, several of the experiments that suggest an important role of AGPs in somatic embryogenesis will be discussed. The system that has been studied most frequently thus far is carrot suspension cultures. Important effects of AGPs on this system have been observed by at least three different laboratories, all using different sources, preparations, cultures and assay procedures. Thus, it appears now well established that AGPs are important molecules in somatic embryogenesis. Unfortunately, their mode of action, as well as whether the reported effects are to be attributed to single AGPs, mixtures of AGPs, the entire intact molecule or AGP fragments only, remains unclear.

2. EMBRYOGENESIS IN PLANTS

Before embarking on a discussion of the role of AGPs, it appears appropriate to sketch a few aspects of embryogenesis to provide a backdrop for the particular aspects of embryogenesis in which AGPs may be involved. In seed plants, pollen transfer from anther to stigma initiates sexual reproduction. One of the two sperm cells carried by the pollen grain fertilizes the egg cell in the flower's carpel, giving rise to a fertilized egg cell or zygote. Zygotic embryos develop through a series of characteristic morphological stages, which, in dicots, are the globular, heart, torpedo, and bent-cotyledon stages. During this development, all distinct organs and tissues present in the seedling are arranged in their proper positions, a process called pattern formation. Along the apical-basal or longitudinal axis, the pattern consists of the shoot apical meristem, cotyledons (embryonic leaves), hypocotyl (embryonic stem) and radicle (embryonic root), including the root cap and root meristem. Along the radial axis, another pattern is apparent as a concentric arrangement of tissue types from outside to inside: the epidermis, ground tissue, and central vascular system. In the model plant *Arabidopsis* (wall cress), the sequence of cell divisions during zygotic embryogenesis is highly invariant, so the origin of seedling organs and tissues appears traceable to specific cells in the early embryo. However, except for the early epidermal cell fate, no clonally transmitted lineages appear to be instrumental in pattern formation. Currently, numerous studies focus on the molecular events underlying early plant embryo development, with the final aim of understanding the molecular mechanisms that plants employ to generate axes of polarity. Several studies involving mutants have emphasized that signaling between the apical and basal compartments of the early zygotic embryo may be an important mechanism employed by plants to set up and maintain an axis of polarity. This hypothesis is of special relevance, because certain studies hint at a role of AGPs in such apical-basal signaling in somatic embryos, resulting in change and/or maintenance of different cell fates.

Like many, but not all, plant species, carrot (*Daucus carota* L.) and *Arabidopsis* always form the apical-basal axis of the embryo in the same orientation relative to the surrounding embryo sac. This orientation can be dependent on the morphological polarity of the unfertilized egg cell, or imposed by the polar organization of the surrounding maternal tissues. Maize zygotes formed by *in vitro* fusion of isolated egg and sperm cells acquire normal polarity in the absence of maternal tissue (Breton *et al* 1995). Egg cells of certain plant species appear apolar, or their polarity is reversed upon fertilization (Johri 1984), suggesting that axes of polarity in plant cells do not generally require strict maternal control.

Examples of the ability in plants to reset axis formation are the formation of secondary embryos from suspensor cells if the primary embryo is aborted or arrested in development (Schwartz *et al* 1994, Vernon and Meinke 1994, Yadegari *et al* 1994, Yadegari and Goldberg 1997). This observation suggests that the apical part of the embryo normally inhibits an "embryo" fate in suspensor cells. In *twin2* (*twn2*) mutants, the development of the apical daughter cells of the zygote arrests after one or two zygotic divisions, and subsequently multiple embryos are formed from suspensor cells (Zhang and Somerville 1997). In *twin1* (*twn1*) mutants, aberrant divisions occur in the embryo, and subsequently a secondary embryo is formed from the suspensor (Vernon and Meinke 1994). Strikingly, the axis of polarity of the secondary embryo is either in the same or in the reverse orientation as that of the primary embryo. This raises the possibility that during normal embryo development, the position of the suspensor relative to the early embryo proper is instrumental in establishing the basal embryo pole. Alternatively, the basal pole of suspensor-derived embryos may be established at random, because in contrast to normal embryos, these embryos do not originate from a polarized zygote.

Thus, the available evidence suggests that the formation of the apical-basal axis of the embryo is not fixed before fertilization, requires segregation of cell fates after the first division, and may be completely stabilized only later in embryo development. It has not been unequivocally demonstrated whether the maternal tissues surrounding the early embryo influence the orientation of the apical-basal axis. It might be that maternal tissues produce a class of hypothetical polarity-inducing signals prior to or after fertilization. Alternatively, it may be the physical attachment of the embryo to the suspensor that is instrumental in establishing the basal embryo pole.

3. AGPS AND PLANT DEVELOPMENT

In recent years a number of reviews have been published concerning the molecular and physiological data on AGPs (Du *et al* 1996, Kreuger and van Holst 1996, Nothnagel 1997). A view that comes forward is that AGPs appear to change prior to differentiation of cells. Monoclonal antibodies have been particular useful in this respect. The AGPs contain epitopes that are spatially highly regulated (sometimes even cell specific), and epitopes may appear or disappear just before or during differentiation (Knox *et al* 1989 1991, Pennell and Roberts 1990, Stacey *et al* 1990, Pennell *et al* 1991 1992, Schindler *et al* 1995). Although AGP-encoding DNA sequences do not appear to be very conserved (Du *et al* 1996), similar AGP epitopes have been shown to be present in different species.

One way to visualize the variability of AGPs is crossed electrophoresis (van Holst and Clarke 1986). This technique shows the distribution of size and charge in an AGP population. Each tissue has its own AGP profile, which changes with the development of the tissues (van Holst and Clarke 1986). Peaks arise or disappear, indicating a change in size or charge. The population of AGPs present in a tissue changes in composition during differentiation. On the basis of the strikingly specific patterns of AGP epitopes, many authors have suggested that AGPs have a function in cell signaling and cell identity. Although in intact plants this may only be shown conclusively after identification of the appropriate mutants, *in vitro* cultures are more accessible and therefore easier to manipulate. This accessibility is the main reason that the use of cell cultures is a popular tool for the study of the potential role of AGPs in plant cell differentiation and somatic embryogenesis.

Analysis of AGPs secreted in the medium of embryogenic cell lines of carrot shows a changing AGP profile when the cell line ages (Kreuger and van Holst 1993). This change coincides with the increase and subsequent decrease of embryogenic potential of a carrot cell line during a period of about one year (Pennell *et al* 1992). Cell lines of other crops, like cyclamen, which tend to loose their embryogenic potential much slower, show a more continuous AGP pattern (i.e., both embryogenic potential and AGP pattern remain unaltered for a long period [Kreuger *et al* 1995]). The AGP pattern may be a reflection of the cell types present in the cell line, suggesting that a pattern that does not change in time may be linked to the cell-type composition, including embryogenic cells, of a given line. An unanswered question of course is whether the observed changes in AGP composition are cause or effect. To help answer this point, AGPs have been added to cell cultures of various sources, composition and differentiation state.

4. THE ADDITION OF AGPS TO EXPLANT CULTURES

The AGPs can be purified by precipitation with the β-glucosyl Yariv reagent (Yariv *et al* 1962 1967). Subsequently the Yariv reagent can be broken down, releasing the AGPs from the precipitate (Kreuger and van Holst 1995). A purity check shows that no additional protein is present.

The addition of AGPs isolated from the conditioned medium of old, non-embryogenic carrot cell lines to explant cultures showed some interesting effects. In these experiments the explant, a sliced etiolated carrot hypocotyl, was cultured at a very low density of 1 hypocotyl/10 ml of liquid medium containing the auxin 2,4-D (normally 0.4 g sliced hypocotyl/10 ml medium is used). The use of liquid medium is important since a constant exponential growth is possible, and it avoids limitation of growth due to diffusion. In the control cultures (no AGPs added) loosely clustered, small cells were present after 19 days, the time the explant normally was removed. These cells were round, often plasmolyzed and probably dead. In contrast, when AGPs from the non-embryogenic cell lines were added, all cells were large and vacuolated, and not many dead cells were observed (Kreuger and van Holst 1993). This culture very much resembled a non-embryogenic cell line and did not contain pro-embryogenic masses.

One might conclude that to get an embryogenic cell line from an explant it is inevitable that dead cells are present. Much of the explant is released in the medium and falls apart, but most of these cells seem to be inactive. The part of the explant that generates the competent cells, the central cylinder (Guzzo *et al* 1994 1995), also produces the cells that appear to be dead. A small portion of the cells in the central cylinder acquire the capability to form competent cells, and this event is accompanied by the expression of a somatic embryogenesis receptor kinase, SERK (Schmidt *et al* 1997). The use of the monoclonal antibody JIM8 showed that there might be a link between dead cells and competent cells (Pennell *et al* 1992, McCabe *et al* 1997). The epitope recognized by JIM8 may very well be an AGP (McCabe *et al* 1997). Cells on the path of embryo formation are daughter cells of JIM8 positive cells, which themselves are on the path of apoptosis. In between the first SERK expression and the formation of an embryo, dead cells, which contain certain AGP epitopes, may be formed. The dead cells and/or the AGP epitope may be essential to obtain competent cells. These AGP epitopes are excreted, as was shown with nurse cultures, and without these excreted products embryo formation is impossible (McCabe *et al* 1997). Inhibition of cell death might therefore inhibit the release of the AGP epitopes and subsequently the formation of competent cells. This sequence is exactly what happens when AGPs isolated from non-embryogenic cell lines are added to explant cultures. This change in AGP complement would result in a totally different environment for the cells able to become competent. Because of the imposed imbalance in AGPs, the culture sets off in a different direction. No dying cells are observed, and subsequently, no pro-embryogenic masses are formed (Kreuger and van Holst 1993). However, this change in fate may also be caused by the improved growth of the cells that would normally not form competent cells. This enhanced growth would change the ratio between different cell types, resulting in a totally different culture (i.e., the competent cells would be outgrown by the non-embryogenic cells). A similar effect can be observed in very diluted cell lines. This low density induces cell death and stops growth; the addition of the non-embryogenic AGPs prevents cell death and restores growth.

It is clear that during the initiation phase of a cell line, timing of the different events is crucial. The amount and duration of auxin supply has to be within certain limits (Guzzo *et al* 1994, Kreuger *et al* 1995). Any imbalance (e.g., too brief auxin supply or too low concentration) will cause a different morphogenic path; in most cases only root formation is observed. To obtain an embryogenic cell line from an explant, many prerequisites must be fulfilled — obviously a correct auxin supply is one of them. The AGPs and the SERK product may very well be others. Even more may be unknown, but it is clear that these complex events can easily be disturbed if one of the components is altered.

5. AGPS IN EMBRYOGENIC CELL SUSPENSIONS

The localization of AGP epitopes in suspension cultures showed some interesting results. The use of the monoclonal antibody JIM4 showed that the epitope is restricted to a few cells at the surface of pro-embryogenic masses. During embryo development the epitope is also restricted to the outer surface (Stacey *et al* 1990). In contrast, in roots the epitope was found only in two segments of the developing vascular tissue (Knox *et al* 1989). This localization pattern correlates with epidermal differentiation and vascular development.

The JIM8 epitope was localized to single cells in suspension cultures, and its presence seemed to correlate with the ability of the cell line to produce embryos (Pennell *et al* 1992). Later it was shown by cell tracking that the JIM8-positive single cells did not produce embryos themselves (Toonen *et al* 1996). However, a preparation of compounds secreted into the medium by JIM8-positive cells was found to rescue embryo formation in a JIM8 negative cell population that was previously unable to produce embryos (McCabe *et al* 1997). This preparation included compounds that bound the JIM8 antibody, although it was not ascertained that these were the active components. A "nursing" function of the cells producing the JIM8 compound would explain the apparent conflicting results, and a role for the JIM8-containing AGPs was suggested in cell–cell communication.

Removal of AGPs by the addition of the AGP binding compound, the Yariv reagent (Yariv *et al* 1962 1967), caused cell enlargement and cessation of growth of *Rosa* cell suspensions (Serpe and Nothnagel 1994). We observed the same effects after addition of the monoclonal antibody ZUM18 to carrot cell suspensions. The AGPs, in general, therefore seem indispensable for growth of plant cells in liquid medium. Once secreted they are able to direct the development of other cells present in the medium.

The addition of AGPs isolated from dry carrot seeds to a carrot cell population smaller than 100 μm resulted in an increased formation of small cytoplasmic cells, designated embryogenic cells (Kreuger and van Holst 1993). Additional purification of carrot seed AGPs with the use of the monoclonal antibody ZUM18 further increased the relative amount of these cells. The active concentration was 0.1 mg/l, which is in the nanomolar range. When another antibody (ZUM15) was used, the relative amount of embryogenic cells decreased. The dose response curve of AGP fractions could be changed by further removal of AGPs containing different epitopes. This effect was possible because of cross-reactivity of the two antibodies. The AGPs containing only the ZUM18 epitope and not the ZUM15 epitope showed a different dose response curve than AGPs containing both epitopes (Kreuger and van Holst 1995). This observation indicates that balance between epitopes directly influences the activity of an AGP population.

The changes in the relative amount of small cytoplasmic cells after AGP treatment were also reflected in the number of embryos formed after transfer of the cells to hormone-free medium (Kreuger and van Holst 1995). In contrast, the addition of these AGPs to pro-embryogenic masses, consisting of the same cell type, in hormone-free medium had much less effect (Kreuger and van Holst 1993) or no effect (Toonen *et al* 1997). Auxin and AGPs may therefore have a synergistic effect on cells. The latter authors also added the ZUM18 AGPs to cell fractions smaller than 22 μm, in the presence of auxin. Again no increase in the number of embryos was observed (Toonen *et al* 1997). This observation reflects both the importance and the difference between the two systems used. To detect activity of AGPs, rapidly dividing cells give the best result. In the single-cell population smaller than 22 μm only a small percentage of the cells actually divide. Besides this, the treatment of the cell lines and cells used in the experiments was also different, thereby possibly inducing changes in the susceptibility of the cells. Precisely defined test systems are clearly essential to compare and validate these results.

6. CHITINASES AND PLANT EMBRYOGENESIS

Apart from AGPs, many different proteins secreted by cell cultures accumulate in the medium, a process that contributes to the conditioning of the medium. The initiation of somatic embryogenesis results in major changes in the pattern of the extracellular proteins (De Vries *et al* 1988). Some of the secreted proteins are thought to be related to the formation of embryogenic cells and somatic embryos. A causal relationship between secreted proteins and embryogenic potential has been determined for the extracellular protein 3 (EP3), identified as a chitinase. The EP3 was originally purified as a protein capable of rescuing somatic embryos in the mutant carrot cell line ts11 at the non-permissive temperature (De Jong *et al* 1992). The acidic endochitinase EP3 was found to be a member of a small family of class IV chitinase genes (Kragh *et al* 1996). Two of these proteins, EP3-1 and EP3-3, were purified and shown to have subtly different effects on the formation of somatic embryos in newly initiated ts11 embryo cultures (Kragh *et al* 1996).

Since the effect of the chitinases was mimicked by *Rhizobium*-produced Nod factors, it was proposed that the chitinases are involved in the generation of signal molecules essential for embryogenesis in ts11 (De Jong *et al* 1993). The EP3 proteins produced by ts11 at the permissive as well as at the non-permissive temperature did not show any difference in biochemical characteristics compared with the ones produced in wild-type cultures, and they were also capable of rescuing ts11 somatic embryos. It was also shown that the sensitivity of ts11 to chitinases coincided with a transient decrease in the abundance of this otherwise functional set of proteins (De Jong *et al* 1995).

The roots of leguminous plants are known to produce chitinases, and during the interaction with *Rhizobium* these plant-produced chitinases have been suggested to control the biological activity of Nod factors by cleaving and inactivating them. In this way, chitinases are proposed to have the potential to control plant morphogenesis and cell division (Staehelin *et al* 1994).

The use of whole-mount *in situ* hybridization revealed that a subset of the morphologically distinguishable cell types in embryogenic and non-embryogenic suspension cultures, including ts11, express *EP3* genes. No expression was found in somatic embryos. In carrot plants *EP3* genes are expressed in the inner integument cells of young fruits and in a specific subset of cells located in the middle of the endosperm of mature seeds. No expression was found in zygotic embryos. These results support the hypothesis that the EP3 endochitinase has a "nursing" function during zygotic embryogenesis and that this function can be mimicked by suspension cells during somatic embryogenesis.

One of the major unknowns remaining of course is the identity of the natural plant substrates for chitinases. In the absence of high M_r chitin in plant cell walls, a search was initiated to identify plant compounds with at least three adjacent N-acetylglucosamine (GlcNAc) residues capable of being cleaved by endochitinases of the EP3 class IV type. Labeling suspension-culture cells with D-[1-^{14}C] glucosamine and N-acetyl-D-[1-^{14}C] glucosamine, followed by Yariv precipitation of conditioned culture medium, showed that a subpopulation of AGPs secreted by suspension cultured carrot cells contained GlcNAc and glucosamine (GlcN). The [1-^{14}C] GlcN and [1-^{14}C] GlcNAc were found only in AGPs from embryogenic but not in non-embryogenic cultures, perhaps one of the reasons why the presence of these GlcNAc-containing AGPs was not previously detected. The AGPs isolated from developing seeds were shown to contain oligosaccharides susceptible to EP3 class IV endochitinase activity. During seed development the native electrophoretic mobility of total seed AGPs increased, whereas AGP epitopes disappeared, suggesting increased processing. Seed AGPs occur predominantly in the endosperm. Together with the previously reported presence of the EP3 endochitinase enzyme in the endosperm, these results suggest that GlcNAc-containing AGPs are a natural substrate for plant chitinases.

Finally, a novel assay system was established to evaluate the possible oligosaccharides released from the interaction between endochitinases and AGPs. Both endochitinases and AGPs extracted from immature carrot seeds can increase the number of somatic embryos formed from protoplasts obtained from wild-type embryogenic suspension cultures. The immature seed AGPs were much more active than the chitinases. Pre-treatment of AGPs with EP3 endochitinases or N-acetylhexosaminidase resulted in optimal somatic embryo-forming activity. The carbohydrate part of the AGPs was responsible for the embryo-promoting effect, demonstrating that activation of immature seed AGPs by hydrolytic enzymes is an important component of embryogenesis. Apart from the increase in embryogenesis, AGPs appear to activate a subpopulation of otherwise non-dividing protoplasts but are only capable of promoting embryogenesis during a short period preceding cell wall formation (van Hengel 1998).

7. REGENERATION POWDER, OR DO AGPS CONTROL AND MAINTAIN PLANT CELL IDENTITY?

Carrot cell lines tend to lose their pro-embryogenic masses gradually and simultaneously their embryogenic capacity (Halperin 1966, Bayliss 1975). During this phase the growth rate of the cell line increases, the pro-embryogenic masses become more translucent, and a higher proportion of the cell volume seems to be taken by the vacuole. The resulting culture consists mainly of large, single, vacuolated cells and aggregates thereof. The addition of carrot seed AGPs to such cultures resulted in the formation of cell clumps consisting of smaller cells. Further subculturing of these clumps, thereby removing a lot of vacuolated cells, eventually resulted in restoration of embryo formation (Kreuger and van Holst 1993). A similar effect was observed after the addition of the ZUM18 AGPs to small cell clusters. Large vacuolated cells seemed to form small cells by division, resulting in cytoplasm-rich cells (Kreuger and van Holst 1995). To reinduce embryogenic potential, removal of excessive vacuolated cells seemed to be necessary (Toonen *et al* 1997). Only then could the added AGPs do their job. Apparently these vacuolated cells have a negative influence on the desired cell type. A similar conclusion was drawn from the experiments with the explant cultures. A balance between different cell types may therefore greatly influence the cell line. Nursing effects between cells and cell types will therefore always influence experiments with cell populations. It appears that AGPs, endochitinases and the resulting cleavage products are a component of such a nursing system.

The AGP-mediated promotion of somatic embryogenesis might also provide more insight into the processes that occur during zygotic embryogenesis, especially since both the AGPs that were shown to be very effective in the protoplast bioassay and the EP3 endochitinases that can activate the immature seed AGPs were present in the endosperm. This coincidence in localization suggests that EP3 endochitinases may provide a way of signaling between maternal tissues, endosperm and embryo. The result of this signaling most likely resides in the characteristics and the function of AGPs. Therefore, changing AGPs by means of hydrolytic enzymes like EP3 might affect plant development by changing the identity of plant cells or by generating signal molecules.

We suggest that the EP3 endochitinase-mediated release of oligosaccharides from endosperm AGPs or, alternatively, the dissociation of AGP complexes in the endosperm might have an effect on the cells of the developing embryo that are in the immediate vicinity of the endosperm. An intriguing puzzle that remains to be solved is the difference in timing, since "active" AGPs and EP3 endochitinases are present around globular stage zygotic embryos, whereas in the protoplast assay they promote somatic embryogenesis mainly by affecting single protoplasts. Regarding the effect of AGPs on the formation of somatic embryos, we propose that the biological effect of (activated) GlcNAc-containing AGPs is to maintain the "embryo identity" of both the somatic and zygotic embryo.

REFERENCES

Bayliss, M. W., 1975, The effects of growth *in vitro* on the chromosome complement of *Daucus carota* (L.) suspension cultures, *Chromosoma (Berl.)* **51**: 401–411.

Breton, C., Faure, J. E., and Dumas, C., 1995, From *in vitro* fertilization to early embryogenesis in maize, *Protoplasma* **187**: 3–12.

De Jong, A. J., Cordewener, J., Lo Schiavo, F., Terzi, M., Vandekerckhove, J., van Kammen, A., and De Vries, S.C., 1992, A carrot somatic embryo mutant is rescued by chitinase, *Plant Cell* **4**: 425–433.

De Jong, A. J., Heidstra, R., Spaink, H. P., Hartog, M. V., Meijer, E. A., Hendriks, T., Lo Schiavo, F., Terzi, M., Bisseling, T., van Kammen, A., and De Vries, S. C., 1993, *Rhizobium* lipo-oligosaccharides rescue a carrot somatic embryo mutant, *Plant Cell* **5**: 615–620.

De Jong, A. J., Hendriks, T., Meijer, E. A., Penning, M., Lo Schiavo, F., Terzi, M., van Kammen, A., and De Vries, S. C., 1995, Transient reduction in secreted 32 kD chitinase prevents somatic embryogenesis in the carrot (*Daucus carota* L.) variant *ts11, Dev. Genet.* **16**: 332–343.

De Vries, S. C., Booij, H., Janssens, R., Vogels, R., Saris, L., Lo Schiavo, F., Terzi, M., and van Kammen, A., 1988, Carrot somatic embryogenesis depends on the phytohormone-controlled presence of correctly glycosylated extracellular proteins, *Genes Dev.* **2**: 462–476.

Du, H., Clarke, A. E., and Bacic, A., 1996, Arabinogalactan-proteins: a class of extracellular matrix proteoglycans involved in plant growth and development, *Trends Cell Biol.* **6**: 411–414.

Guzzo, F., Baldan, B., Mariani, P., Lo Schiavo, F., and Terzi, M., 1994, Studies on the origin of totipotent cells in explants of *Daucus carota* L., *J. Exp. Bot.* **45**: 1427–1432.

Guzzo, F., Baldan, B., Levi, M., Sparvoli, E., Lo Schiavo, F., Terzi, M., and Mariani, P., 1995, Early cellular events during induction of carrot explants with 2,4-D, *Protoplasma* **185**: 28–36.

Halperin, W., 1966, Alternative morphogenetic events in cell suspensions, *Am. J. Bot.* **53**: 443–453.

Johri, B. M., 1984, *Embryology of Angiosperms*, Springer-Verlag, Berlin.

Knox, J. P., Day, S., and Roberts, K., 1989, A set of surface glycoproteins forms an early marker of cell position, but not cell type, in the root apical meristem of *Daucus carota* L., *Development* **106**: 47–56.

Knox, J. P., Linstead, P. J., Peart, J., Cooper, C., and Roberts, K., 1991, Developmentally regulated epitopes of cell surface arabinogalactan proteins and their relation to root tissue pattern formation, *Plant J.* **1**: 317–326.

Kragh, K. M., Hendriks, T., De Jong, A. J., Lo Schiavo, F., Bucherna, N., Hojrup, P., Mikkelsen, J. D., and De Vries, S. C., 1996, Characterization of chitinases able to rescue somatic embryos of the temperature-sensitive carrot variant *ts11*, *Plant Mol. Biol.* **31**: 631–645.

Kreuger, M., and van Holst, G-J., 1993, Arabinogalactan-proteins are essential in somatic embryogenesis of *Daucus carota* L., *Planta* **189**: 243–248.

Kreuger, M., and van Holst, G-J., 1995, Arabinogalactan-protein epitopes in somatic embryogenesis of *Daucus carota* L., *Planta* **197**: 135–141.

Kreuger, M., and van Holst, G-J., 1996, Arabinogalactan-proteins and plant differentiation, *Plant Mol. Biol.* **30**: 1077–1086.

Kreuger, M., Postma, E., Brouwer, Y., and van Holst, G-J., 1995, Somatic embryogenesis of *Cyclamen persicum* in liquid medium, *Physiol. Plant.* **94**: 605–612.

McCabe, P. F., Valentine, T. A., Forsberg, L. S., and Pennell, R. I., 1997, Soluble signals from cells identified at the cell wall establish a developmental pathway in carrot, *Plant Cell* **9**: 2225–2241.

Nothnagel, E. A., 1997, Proteoglycans and related components in plant cells, *Int. Rev. Cytol.* **174**: 195–291.

Pennell, R. I., and Roberts, K., 1990, Sexual development in the pea is presaged by altered expression of arabinogalactan protein, *Nature* **344**: 547–549.

Pennell, R. I., Janniche, L., Kjellbom, P., Scofield, G. N., Peart, J. M., and Roberts, K., 1991, Developmental regulation of a plasma membrane arabinogalactan protein epitope in oilseed rape flowers, *Plant Cell* **3**: 1317–1326.

Pennell, R. I., Janniche, L., Scofield, G. N., Booij, H., De Vries, S. C., and Roberts, K., 1992, Identification of a transitional cell state in the developmental pathway to carrot somatic embryogenesis, *J. Cell Biol.* **119**: 1371–1380.

Schindler, T., Bergfeld, R., and Schopfer, P., 1995, Arabinogalactan proteins in maize coleoptiles: developmental relationship to cell death during xylem differentiation but not to extension growth, *Plant J.* **7**: 25–36.

Schmidt, E. D. L., Guzzo, F., Toonen, M., and De Vries, S. C., 1997, A leucine-rich repeat containing receptor-like kinase marks somatic plant cells competent to form embryos, *Development* **124**: 2049–2062.

Schwartz, B. W., Yeung, E. C., and Meinke, D. W., 1994, Disruption of morphogenesis and transformation of the suspensor in abnormal suspensor mutants of *Arabidopsis*, *Development* **120**: 3235–3245.

Serpe, M. D., and Nothnagel, E. A., 1994, Effects of Yariv phenylglycosides on *Rosa* cell suspensions: Evidence for the involvement of arabinogalactan-proteins in cell proliferation, *Planta* **193**: 542–550.

Stacey, N. J., Roberts, K., and Knox, J. P., 1990, Patterns of expression of the JIM4 arabinogalactan-protein epitope in cell cultures and during somatic embryogenesis in *Daucus carota* L., *Planta* **180**: 285–292.

Staehalin, C., Schultze, M., Kondorosi, E., Mellor, R. B., Boller, T., and Kondorosi, A., 1994, Structural modifications in *Rhizobium meliloti* Nod factors influence their stability against hydrolysis by root chitinases, *Plant J.* **5**: 319–330.

Toonen, M., Schmidt, E. D. L., Hendriks, T., Verhoeven, H. A., van Kammen, A., and De Vries, S. C., 1996, Expression of the JIM8 cell wall epitope in carrot somatic embryogenesis, *Planta* **200**: 167–173.

Toonen, M., Schmidt, E. D. L., van Kammen, A., and De Vries, S. C., 1997, Promotive and inhibitory effects of arabinogalactan-proteins on *Daucus carota* somatic embryogenesis, *Planta* **203**: 188–195.

van Hengel, A. J., 1998. Chitinases and arabinogalactan proteins in somatic embryogenesis, Ph.D. Thesis, Wageningen Agricultural University, Wageningen, The Netherlands.

van Holst, G. J., and Clarke, A. E., 1986, Organ specific arabinogalactan-proteins of *Lycopersicon peruvianum* (Mill) demonstrated by crossed electrophoresis, *Plant Physiol.* **8**: 786–789.

Vernon, D. M., and Meinke, D. W., 1994, Embryogenic transformation of the suspensor in *twin*, a polyembryonic mutant of *Arabidopsis*, *Dev. Biol.* **165**: 566–573.

Yadegari, R., and Goldberg, R.B., 1997, Embryogenesis in dicotyledonous plants, In *Cellular and Molecular Biology of Plant Seed Development* (B.A. Larkins and I.K. Vasil, eds.), Kluwer Academic Publishers, Dordrecht, The Netherlands, pp. 3-52.

Yariv, J., Rapport, M. M., and Graf, L., 1962, The interaction of glycosides and saccharides with antibody to the corresponding phenylazo glycoside, *Biochem. J.* **85**: 383–388.

Yariv, J., Lis, H., and Katchalski, E., 1967, Precipitation of arabic acid and some seed polysaccharides by glycosylphenylazo dyes, *Biochem. J.* **105**: 1c–2c.

Zhang, J. Z., and Somerville, C. R., 1997, Suspensor-derived polyembryony caused by altered expression of valyl-tRNA synthetase in the *twn2* mutant of Arabidopsis, *Proc. Natl. Acad. Sci. USA* **94**: 7349–7355.

Chapter 11

Arabinogalactan-Proteins in Reproductive Tissues of Flowering Plants

A Historical Perspective of Work from the Plant Cell Biology Research Centre, University of Melbourne, Victoria, Australia

Adrienne E. Clarke,[1] G. Currie,[1] P. Gilson,[1] S-L. Mau,[2] D. Oxley,[1] C.J. Schultz,[2] J. Sommer-Knudsen,[2] and A. Bacic[1]
[1]Plant Cell Biology Research Centre and [2]Cooperative Research Centre for Industrial Plant Biopolymers, School of Botany, University of Melbourne, Parkville, Victoria 3052, Australia

1. INTRODUCTION

This paper is intended to give a historical account of the work on arabinogalactan-proteins (AGPs) that came from our research group between 1975 and 1998. Although some work of others is mentioned, the paper is not a comprehensive review. For a more extensive discussion of the subject, the reader is referred to Clarke *et al* (1979), Fincher *et al* (1983), Knox (1995), Du *et al* (1996a), Nothnagel (1997), Sommer-Knudsen *et al* (1998) and Schultz *et al* (1998).

The history of our work on AGPs from 1975 onwards has been intertwined with our research on fertilization in higher plants. In the early period, from 1975–1984, we used *Gladiolus gandavensis* to study both fertilization and AGPs. We chose this model because flowers were readily available from the local florist at a time when glasshouse facilities were not available to us. We needed a ready supply of plant material to establish methods for handling the very small amounts of proteins and glycoconjugates present on the stigma surface and within the stylar canals.

In the second period, from 1984 to the present, our work on fertilization focused on a search for the product of the *S*- (self-incompatibility) gene. We had chosen an ornamental tobacco, *Nicotiana alata*, as our experimental model (for reviews see Clarke and Newbigin 1993, Matton *et al* 1994, Dodds *et al* 1996). This

Cell and Developmental Biology of Arabinogalactan-Proteins
Edited by Nothnagel *et al.*, Kluwer Academic/Plenum Publishers, 2000

plant species was particularly suitable, as we had plants bearing combinations of five different *S*-alleles available for study. The plant was readily propagated vegetatively, and we anticipated that it would be readily transformable (based on experience with *Nicotiana tabacum*). Furthermore, the pistil was of a size comparable to that of *Gladiolus* so the methods we had developed for isolating particular components from *Gladiolus* could be used with *N. alata.*

2. THE EARLY PERIOD, 1975–1984

In 1975, in a collaboration with Dr. Michael Jermyn and Professors Bruce Stone and Bruce Knox, we decided to use the β-glucosyl Yariv reagent (Yariv *et al* 1962) as a histochemical reagent. The reasoning was that if the Yariv reagent (a red dye) precipitated AGPs from tissue extracts, we would be able to use it as a histochemical reagent to localize AGPs within tissues. Using this histochemical test, we found AGPs in a range of plants and tissues. In some cases, the staining was restricted to the intercellular spaces; in other cases staining was associated with secretory ducts, plasma membranes, cell walls and vesicles (Clarke *et al* 1975 1978). As we were working on *Gladiolus* for our research program on fertilization, we examined flowers for staining with the Yariv reagent. We found intense staining on the surface and in the cells lining the secretory canals, as well as the mucilage secreted into these canals. Later, we found that pistil extracts from all major families of flowering plants contained AGPs, based on their interaction with the Yariv reagent (Hoggart and Clarke 1984). Thus AGPs seemed to be a class of molecules associated with the female sexual tissues that had been conserved during the evolution of flowering plants.

We decided to focus on analyzing the composition of the *Gladiolus* stylar mucilage AGP in detail. We were aware of the possibility that the Yariv reagent might precipitate a number of closely related molecules, either different glycoforms of a particular AGP or separate but related molecules, from tissue extracts. We devised a method for purification based on the interaction with a galactose-binding lectin (Tridacnin) and compared preparations isolated by this method and by Yariv precipitation. The results of analyses of the two samples were essentially identical. The material was then passed through a Sepharose 4B column, and we analyzed samples across the symmetrical elution profile. Again, the results for all samples were the same. Results were typical of AGPs: content of approximately 90% carbohydrate and 3% protein, with galactose and arabinose in the ratio 6:1 as the only monosaccharides. Methylation analysis showed that the major linkages were 3-linked Gal, 3, 6-linked Gal, and t-Ara. The component was polydisperse in the M_r range of 150,000–400,000. This characteristic was typical of AGPs as we knew them at the time, and, with the methods available at the time, it seemed to be a single molecule. We tried to dissociate any co-purifying material by carrying out the final chromatographic step in 6 M urea. This step did not change the results, indicating the unlikelihood of a significant amount of other material being bound to the AGP (Gleeson *et al* 1979, Gleeson and Clarke 1979).

We speculated on the possible structure/function relationships of these stylar AGPs. We had found differences in the degree of branching of the carbohydrates of stigma and style AGP preparations, which led us to consider whether fine differences in the AGPs might reflect their role as markers of organ identity (Gleeson and Clarke 1979). The second function we considered was a nutritional role. We had observed fewer AGPs in our freeze-fracture studies of pollinated styles than in unpollinated styles (Clarke *et al* 1977). Labarca and Loewus (1972 1973), in pioneering experiments, had observed that radioactively labelled style exudate, when injected into the hollow stylar canals of *Lilium longiflorum,* was incorporated into pollen tube walls growing through the canal and that the label appeared in low-molecular-weight material after pollination. Together, these observations suggested a nutritive function. The third suggested function was that AGPs might act as informational molecules by binding to other molecules such as plant lectins or flavanol glycosides. Jermyn noted that flavanol glycosides were the only molecules of a number tested that inhibited the Yariv reagent–AGP interaction (Jermyn 1978). Thus, the potential for binding of other molecules to AGPs existed, but with the available information, we were unable to design experiments that would enable us to test the various hypotheses on function.

3. THE SECOND PERIOD: 1984–PRESENT

In 1986, we developed a crossed electrophoresis technique using the Yariv reagent in the second gel to give AGP profiles of tissue extracts (van Holst and Clarke 1986). We found different AGPs in the stigmas and styles of *N. alata* (Gell *et al* 1986) and another solanaceous plant, *Lycopersicon peruvianum* (van Holst and Clarke 1986). It was also clear that the nature of the AGPs changed during development. These observations provided more evidence for a role for AGPs as markers of identity and a role in development. It also suggested that the heterogeneity, which could be due to differences in either the protein or glycosyl components or both, probably involved the protein backbones, because the glycosyl residues detected were all neutral (Bacic *et al* 1988).

At this time, in our research program on self-incompatibility, we were considering how *S*-genotype-specific components might be detected in *N. alata* styles. We embarked on a program to make monoclonal antibodies from style extracts of different *S*-genotypes (Anderson *et al* 1984). We found, to our surprise and initial disappointment, that most of the hybridomas secreted antibodies directed to AGPs. Of these AGP-specific antibodies, some were directed primarily to terminal arabinosyl residues; others primarily to terminal galactosyl residues.

We could not define the epitopes precisely because of the lack of commercially available, chemically defined oligosaccharides. We did, however, define one antibody that detected a style-specific epitope, which further emphasized the potential of AGPs as markers of identity. All the AGP-specific antibodies have proved to be extremely useful in our studies of particular AGPs, although they had no value for us in our quest for the product of the *S*-gene in styles.

At this time we made two decisions. Using our model plant *N. alata*, we decided to identify and analyze all the major components of style secretions, then clone the genes encoding the protein backbone of AGPs. We were acutely aware of the difficulties. In particular, isolating a single AGP in pure form was a challenge because of the different glycoforms and the experience that closely related but distinct AGPs co-purified even after several different steps designed to separate individual AGPs.

Recovery of the protein backbone after removal of the carbohydrate was another major technical challenge. In many cases, the protein accounted for only 1% (w/w) of the total molecule, and only microgram amounts of AGP could be isolated from a single style. We then had to deal with problems of preparing and sequencing peptides from the recovered protein backbones. Finally, if we did get information, we recognized that most of the protein backbone was composed of amino acids with multiple codons, Ser, Thr, Ala and Pro. This composition foreshadowed the extreme difficulty in finding a suitable region for making DNA primers for PCR-based cloning.

We adopted the strategy of taking a model system for which there was plenty of material available to develop the techniques. The model we chose was the extracellular culture filtrate from suspension-cultured cells of *Pyrus communis* (Chen *et al* 1994). This choice was based on the availability of large batches of these cells from another research program and on the knowledge that these cells secreted AGPs into the culture medium. The availability of large amounts of AGPs from this source made it practical for us to establish methods for the deglycosylation and recovery of the protein backbone. We could also practice and refine methods for fragmenting the backbone and sequencing the recovered peptides. In this we were helped immeasurably by Richard Simpson, an experienced protein chemist at the Ludwig Institute for Cancer Research located near our laboratory. We found that material, which on several chromatographic systems appeared to be a single AGP, when deglycosylated resulted in protein backbones that on one chromatographic system appeared as a single peak but could be resolved into two or more peaks with use of multiple chromatographic systems. Eventually, the technical difficulties were overcome, and a full-length cDNA was obtained (Chen *et al* 1994). The sequence and its derived amino acid sequence showed a profile with a classic signal sequence, a central domain rich in Pro/Hyp, Ala, Ser/Thr and a hydrophobic C-terminal domain that corresponded to a transmembrane helix. This suggested that the molecule could be anchored within the plasma membrane.

We then embarked on the task of obtaining a cDNA encoding the backbone of the *N. alata* style AGP. The hydropathy plot of the deduced amino acid sequence from this cDNA clone also showed three domains: a hydrophobic domain at the N-terminus corresponding to a signal sequence; a central domain rich in Hyp/Pro, Ser and Ala; and a hydrophobic C-terminal domain corresponding to a transmembrane helix, very similar to that obtained for the pear AGP (Du *et al* 1994). Similar techniques were used to identify a stigma-specific gene encoding a different AGP (Du *et al* 1996b). This cDNA had a different profile: it encoded a protein with three domains: an N-terminal secretion signal, a Pro-rich domain and a C-terminal domain containing all eight Cys residues. No C-terminal domain corresponding to a transmembrane helix was present. This cDNA encodes a "non-classical" AGP. The amino acid composition of the deduced mature protein is rich in Pro, Asx, Glx, and Ala, in contrast to the "classical" AGPs, which consist mainly of Hyp/Pro, Ala and Ser.

The questions of position, linkage and structure of the oligosaccharide chains attached to these backbones are not resolved. Because the carbohydrate is such an overwhelming proportion of the total molecule, we would expect that the protein would effectively be covered by carbohydrate. However, it is theoretically possible that the carbohydrate could be present in one or a few large chains, leaving regions of the protein backbone "bare". Understanding the arrangement of carbohydrate chains on these protein backbones is one of the next important challenges.

The decision to investigate all the major components of the style mucilage led us to isolate some extremely interesting molecules. In the high-molecular-weight range of style mucilage components, we found three groups of components that contain significant amounts of arabinose and galactose. In addition to the AGPs, a glycoprotein with apparent molecular weight of 120 kDa was present, as was a molecule known as the galactose-rich style glycoprotein (GaRSGP), which appeared disperse on the SDS gel in the range of 45–120 kDa. We isolated, purified and analyzed both the 120-kDa glycoprotein (Lind *et al* 1994, Schultz *et al* 1997) and GaRSGP (Sommer-Knudsen *et al* 1996). These two molecules are classified as Pro/Hyp-rich glycoproteins (P/HRGPs), rather than AGPs. Neither of these glycoproteins react significantly with the Yariv reagent, although GaRSGP does bind slightly. Although both proteins contain significant amounts of Hyp in the protein moiety and galactose and arabinose in the carbohydrate moiety, they differ with respect to the carbohydrate content: the 120-kDa glycoprotein has 35% carbohydrate, and the GaRSGP has 65–75% carbohydrate (w/w). Monosaccharide ratios of Ara to Gal also differ: 1:0.64 for the 120-kDa glycoprotein and 1:12 for the GaRSGP. The comparative analyses are shown in Table 1.

Table 1. Hyp-rich glycoproteins (HRGPs) isolated from sexual tissues of *Nicotiana alata*.

Name	GaRSGP[a]		120 kDa glycoprotein[b]		AGP RT25[c]		AGP RT35[c]	
Plant tissue	Transmitting tract of style		Transmitting tract of style		Style		Stigma	
Protein (% w/w)	25–35		~65		< 10		< 10	
Carbohydrate (%w/w)	65–75		~35		> 90		> 90	
Most abundant amino acids (mol %)	Lys	14	Hyp	21	Ala	20	Asx	15
	Pro	12	Pro	12	Hyp	18	Glx	12
	Hyp	9	Ser	8	Ser	15	Ala	11
	Val	8	Lys	7	Gly	9	Thr	7
	Ser	7	Ala	7	Thr	7	Hyp	5
	Thr	7	Leu	6	Val	6	Pro	4
Carbohydrate (mol %)	Gal	83	Gal	37	Gal	~60	Gal	52
	Ara	7	Ara	63	Ara	~30	Ara	45
	GlcNAc	4					Rha	2
	Man	4					Glc	1
	Xyl	2					Xyl	<1
	Glc	1						
Apparent M_r (kDa)								
Native	45–120		120		> 90		> 90	
Deglycosylated	28–30		78		–		–	
pI	≥ 10		10		–		–	
Corresponding clone	*NaPRP4*		*NaPRP5*		*NaAGP1*		*NaAGP3*	
Class	P/HRGP		P/HRGP		AGP		AGP	

[a]Sommer-Knudsen *et al* 1996
[b]Lind *et al* 1994, Schultz *et al* 1997
[c]Du *et al* 1994, Du *et al* 1996a

The protein backbones of the 120-kDa glycoprotein and GaRSGP were isolated, and N-terminal and internal peptide sequences obtained. The cDNAs for both molecules show domain structures. The predicted mature backbone of the 120-kDa glycoprotein can be divided into three domains (Fig 1). The N-terminal domain contains 45% Pro and many Ser-Pro_{2-7} motifs. The central domain is also Pro-rich (44% Pro). The dominant motifs in this domain are Pro_4-Ala and Pro_{2-3}. The C-terminal is relatively Pro-poor and contains all six of the Cys residues.

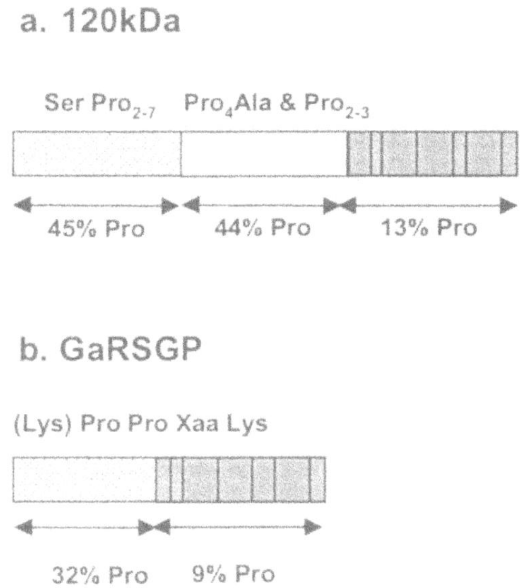

Figure 1. The domain structure of the encoded backbones of two style-specific glycoproteins: the 120-kDa glycoprotein and the galactose-rich style glycoprotein (GaRSGP). a. The predicted mature backbone of the 120-kDa glycoprotein can be divided into three domains: the N-terminal domain (residues 50–150) contains 45% Pro and many Ser-Pro_{2-7} motifs; the central domain (residues 151-320) is also Pro-rich with 44% Pro and predominant motifs of Pro_4-Ala and Pro_{2-3}; the C-terminal domain (residues 321–461) is relatively Pro-poor (13% Pro) and contains all six of the Cys residues (vertical lines) in the 120-kDa backbone. The dominant motifs are written above the appropriate domains and the % Pro of each domain is below. b. The predicted backbone of the GaRSGP contains a C-terminal domain with 57% amino acid identity to the C-terminal domain of the 120-kDa backbone. The GaRSGP also has six Cys residues (vertical lines) in the C-terminal domain (residues 117–252). The N-terminal domain of GaRSGP (residues 1–95) is Hyp/Pro-rich (32% Pro) but does not contain Ser-Pro_n motifs. It includes the motif Lys-Pro-Pro-Xaa-Lys (Sommer-Knudsen *et al* 1996). Xaa represents any amino acid.

We were able to make some predictions about the likely distribution of extensin- and AGP-like carbohydrate substituents on the 120-kDa glycoprotein. The residues are t-Ara, 2-Ara, t-Gal, 6-Gal, 3-Gal and 3,6-linked Gal in the ratio 9:12:6:3:1:3. Therefore, if AGP-like 3,6 branched galactosyl substituents that terminate in Ara were found, then t-Gal and 2-linked Ara would remain to be accounted for in other structures. These residues could be arranged in configurations characteristic of extensins. The N-terminal domain contains several Ser-Pro_{2-7} motifs (typical of extensins) that could bear extensin-like carbohydrate substituents. The positions of the presumptive AGP-like substituents are not known; the substituents could be scattered along the entire protein backbone or confined to a single domain.

The backbone of GaRSGP has two domains. The N-terminal region (residues 1-95) is Pro/Hyp-rich (32%), does not contain Ser-Pro_{2-7} motifs, but does contain a Lys-Pro-Pro-Xaa-Lys motif. The C-terminal (residues 96-232) is Pro/Hyp-poor (9%). The alignment between the C-terminal domains of the protein backbones of the 120-kDa glycoprotein and GaRSGP shows that the number and position of the Cys residues is conserved, and the overall identity is 57% at the amino acid level.

This similarity between the two backbones is consistent with the finding that they share a common antigenic epitope. Indeed, a stretch of five identical amino acids (TDKNG) in the C-terminal region is predicted to be an antigenic peptide, although whether this is the common antigen is not proven. Another stretch, PSAKPPVKPP, is also common (with the exception of the single alanine residue) between the two polypeptides.

The GaRSGP is heavily glycosylated (65–75% carbohydrate), has two N-glycan glycosylation sites present in the C-terminal region and must have a number of branched chains and t-Gal residues to account for the methylation data.

The common domain structure, the common peptide sequence and the common monosaccharides of GaRSGP and the 120-kDa molecule explain the complex immunological specificities found during our attempts to prepare specific antibodies both to the glycosylated and deglycosylated molecules (Sommer-Knudsen *et al* 1996). An antibody to the deglycosylated 120-kDa molecule binds both intact and deglycosylated 120-kDa protein but interestingly does not bind the intact GaRSGP, although it does bind to the deglycosylated GaRSGP. When used on tissue sections containing the intact molecules, the antibody would be predicted to bind the 120-kDa glycoprotein but not GaRSGP.

This antibody was used in immunogold electron microscopy to show that in unpollinated styles the 120-kDa molecule is evenly distributed in the extracellular matrix of the style transmitting tissue. It is not associated with the walls of the transmitting tract cells. In pollinated pistils, labelling of the pollen tube tip occurred, particularly in the cytoplasm, very close to the apical zone (Lind *et al* 1996). Further back from the tip, large vesicle-like structures appeared in the pollen cytoplasm, which were heavily labelled. The structures were membrane bound and contained a mixture of components with different electron opacities. This region of the cytoplasm stained darkly and contained large lipid droplets. In the zone of P-particles, antibody binding occurred in regions of the cytoplasm that were rich in secretory vesicles. Antibody did not bind to *in vitro*-grown pollen tubes. However, when the 120-kDa glycoprotein was incubated with *in vitro*-grown tubes and then examined, gold labelling was observed in both the cytoplasm and cell walls of the tubes (Lind *et al* 1994). These experiments show that in some unknown way the 120-kDa molecule is taken into the pollen tubes. Whether the molecule is processed in the tubes is not known, but at least the antigenic epitope(s) is intact within the tube. The uptake and incorporation into the wall does not necessarily indicate a primary role in pollen tube nutrition. Indeed the role of any of these molecules, the AGPs, the 120-kDa glycoprotein or GaRSGP, is still not defined.

The suggestions we made in 1979 for nutritional or informational roles are still considered, but 20 years later, no definitive molecular information to ascribe a particular role to any AGP is available. Many key questions remain. Why does the plant synthesize so many closely related molecules? What is the purpose of generating such diversity of structures? How are the different domains of these molecules related biosynthetically? Unravelling the complexity will certainly require a focus on the detailed structures. The capacity to design experiments in which individual molecules of this class can be followed is essential to moving forward. Looking back over 25 years of AGP research, we see that the technologies of the time: electrophoresis, histochemistry, immunology, monoclonal antibody technology, immunogold electron microscopy, carbohydrate chemistry, protein chemistry and cDNA cloning—have all been applied to further our understanding of this group of molecules. The key questions of diversity and structure/function relationships still, however, remain to be unraveled in the next 25 years.

ACKNOWLEDGEMENTS

Our work is funded by a Special Research Centre grant from the Australian Research Council and a Cooperative Research Centre grant to the Cooperative Research Centre for Industrial Plant Biopolymers from the Australian Government.

REFERENCES

Anderson, M. A., Sandrin, M. S., and Clarke, A. E., 1984, A high proportion of hybridomas raised to a plant extract secrete antibody to arabinose or galactose, *Plant Physiol.* **75**: 1013–1016.

Bacic, A., Gell, A. C., and Clarke, A. E., 1988, Arabinogalactan-proteins from stigmas of *Nicotiana alata, Phytochemistry* **27**: 679–684.

Chen, C-G., Pu, Z-Y., Moritz, R. L., Simpson, R. J., Bacic, A., Clarke, A. E., and Mau, S-L., 1994, Molecular cloning of a gene encoding an arabinogalactan-protein from pear (*Pyrus communis*) cell suspension culture, *Proc. Natl. Acad. Sci. USA* **91**: 10305–10309.

Clarke, A. E., and Newbigin, E. N., 1993, Molecular aspects of self-incompatibility in flowering plants, *Annu. Rev. Genet.* **27**: 257–279.

Clarke, A. E., Knox, R. B., and Jermyn, M. A., 1975, Localization of lectins in legume cotyledons, *J. Cell Sci.* **19**: 157–167.

Clarke, A. E., Considine, J. A., Ward, R., and Knox, R. B., 1977, Mechanism of pollination in *Gladiolus*: roles of the stigmatic papillae and pollen-tube guide, *Ann. Bot.* **41**: 15–20.

Clarke, A. E., Gleeson, P. A., Jermyn, M. A., and Knox, R. B., 1978, Characterization and localization of ß-lectins in lower and higher plants, *Aust. J. Plant Physiol.* **5**: 707–722.

Clarke, A. E., Anderson, R. L., and Stone, B. A., 1979, Form and function of arabinogalactans and arabinogalactan proteins, *Phytochemistry* **18**: 521–540.

Dodds, P. N., Clarke, A. E., and Newbigin, E. J., 1996, A molecular perspective on pollination in flowering plants, *Cell* **85**: 141–144.

Du, H., Simpson, R. J., Moritz, R. L., Bacic, A., and Clarke, A. E., 1994, Isolation of the protein backbone of an arabinogalactan-protein from styles of *Nicotiana alata* and characterization of a corresponding cDNA, *Plant Cell* **6**: 1643–1653.

Du, H., Clarke, A. E., and Bacic, A., 1996a, Arabinogalactan-proteins: a class of extracellular matrix proteoglycans involved in plant growth and development, *Trends Cell Biol.* **6**: 411–414.

Du, H., Simpson, R. J., Clarke, A. E., and Bacic, A., 1996b, Molecular characterization of a stigma-specific gene encoding an arabinogalactan-protein (AGP) from *Nicotiana alata, Plant J.* **9**: 313–323.

Fincher, G. B., Stone, B. A., and Clarke, A. E., 1983, Arabinogalactan-proteins: structure, biosynthesis and function, *Annu. Rev. Plant Physiol.* **34**: 47–70.

Gell, A. C., Bacic, A., and Clarke, A. E., 1986, Arabinogalactan proteins of the female sexual tissue of *Nicotiana alata.* I. Changes during flower development and pollination, *Plant Physiol.* **82**: 885–887.

Gleeson, P. A., and Clarke, A. E., 1979, Structural studies on the major component of *Gladiolus* style mucilage, an arabinogalactan-protein, *Biochem. J.* **181**: 607–621.

Gleeson, P. A., Jermyn, M. A., and Clarke, A. E., 1979, Isolation of an arabinogalactan protein by lectin affinity chromatography on Tridacnin-Sepharose 4B, *Anal. Biochem.* **92**: 41–45.

Hoggart, R. M., and Clarke, A. E., 1984, Arabinogalactans are common components of angiosperm styles, *Phytochemistry* **23**: 1571–1573.

Jermyn, M. A., 1978, Isolation from the flowers of *Dryandra praemorsa* of a flavanol glycoside that reacts with beta lectins, *Aust. J. Plant Physiol.* **5**: 697–705.

Knox, J. P., 1995, Developmentally regulated proteoglycans and glycoproteins of the plant cell surface, *FASEB. J.* **9**: 1004–1012.

Labarca, C., and Loewus, F., 1972, The nutritional role of pistil exudate in pollen tube wall formation in *Lilium longiflorum*. 1. Utilisation of injected stigmatic exudate, *Plant Physiol.* **50**: 7–14.

Labarca, C., and Loewus, F., 1973, The nutritional role of pistil exudate in pollen tube wall formation in *Lilium longiflorum*. 2. Production and utilisation of exudate from stigma and stylar canal, *Plant Physiol.* **52**: 87–92.

Lind, J. L., Bacic, A., Clarke, A. E., and Anderson, M. A., 1994, A style-specific hydroxyproline-rich glycoprotein with properties of both extensins and arabinogalactan proteins, *Plant J.* **6**: 491–502.

Lind, J. L., Bonig, I., Clarke, A. E., and Anderson, M. A., 1996, A style-specific 120-kDa glycoprotein enters pollen tubes of *Nicotiana alata* in vivo, *Sex. Plant Reprod.* **9**: 75–86.

Matton, D., Nass, N., Clarke, A. E., and Newbigin, E., 1994, Self-incompatibility: How plants avoid illegitimate offspring, *Proc. Natl. Acad. Sci. USA* **91**: 1992–1997.

Nothnagel, E. A., 1997, Proteoglycans and related components in plant cells, *Int. Rev. Cytol.* **174**: 195–291.

Schultz, C. J., Hauser, K., Lind, J. L., Atkinson, A. H., Pu, Z., Anderson, M. A., and Clarke, A. E., 1997, Molecular characterisation of a cDNA sequence encoding the backbone of a style-specific 120 kDa glycoprotein which has features of both extensins and arabinogalactan-proteins, *Plant Mol. Biol.* **35**: 833–846.

Schultz, C., Gilson, P., Oxley, D., Youl, J., and Bacic, A., 1998, GPI-anchors on arabinogalactan-proteins: implications for signalling in plants, *Trends Plant Sci.* **3**: 426–431.

Sommer-Knudsen, J., Clarke, A. E., and Bacic, A., 1996, A galactose-rich, cell-wall glycoprotein from styles of *Nicotiana alata, Plant J.* **9**: 101–113.

Sommer-Knudsen, J., Bacic, A., and Clarke, A. E., 1998, Hydroxyproline-rich glycoproteins, *Phytochemistry* **47:** 483-497.

van Holst, G-J., and Clarke, A. E., 1986, Organ-specific arabinogalactan-proteins of *Lycopersicon peruvianum* (Mill) demonstrated by crossed-electrophoresis, *Plant Physiol.* **80**: 786–789.

Yariv, J., Rapport, M. M., and Graf, L., 1962, The interaction of glycosides and saccharides with antibody to the corresponding phenylazo glycosides, *Biochem J.* **85**: 383–388.

Chapter 12

Transcriptional, Post-Transcriptional and Post-Translational Regulation of a *Nicotiana* Stylar Transmitting Tissue-Specific Arabinogalactan-Protein

Alice Y. Cheung,[1,2,3] Xiao-yan Zhan,[4] Eric Wong,[1] Hong Wang,[5] and Hen-ming Wu[1,2]
[1]*Department of Biochemistry and Molecular Biology,* [2]*Molecular and Cell Biology Graduate Program, and* [3]*Plant Biology Graduate Program, University of Massachusetts, Amherst, MA 01003*
[4]*Current address: School of Public Health, University of California, Berkeley, CA 94720*
[5]*Department of Molecular Biophysics and Biochemistry, Yale University, New Haven, CT 06511*

1. INTRODUCTION

Nicotiana tabacum, one of its progenitors *N. sylvestris* and its self-incompatible relative *N. alata* have structurally very similar pistils. They have an elongated style connecting the stigma and the ovary in which the ovules develop and fertilization takes place (Fig 1). Pollen grains hydrate and germinate on the stigmatic papilla, each grain extruding a pollen tube. Pollen tubes elongate directionally in the extracellular matrix (ECM) of the stigmatic and the stylar transmitting tissues until they reach the ovary. In the ovary, pollen tubes turn from their basal growth orientation to enter the ovules and penetrate the embryo sacs to deliver the male gametes for fertilization. Pistillate tissues along the pollen tube growth pathway are packed with highly secretory cells, and they produce an ECM enriched with sugar- and lipid-containing compounds through which pollen tubes elongate. Pollen tube growth in the pistil is extremely efficient (e.g., reaching a rate of 1 cm/hour in maize). In *N. tabacum*, pollen tubes elongate at an approximate rate of 0.15 cm/hour under normal greenhouse conditions (Wang *et al* 1996). However, *N. tabacum* pollen tubes elongated at a rate of 235 μm/hour under what was considered to be the most optimal *in vitro* conditions reported (Read *et al* 1993). The pistil-secreted ECM is thus believed to aid in both the efficiency and the directionality of the pollen tube growth process. [For reviews on pollen tube growth, see Lord and Sanders (1992), Mascarenhas (1993), Cheung (1996a,b), Taylor and Hepler (1997)].

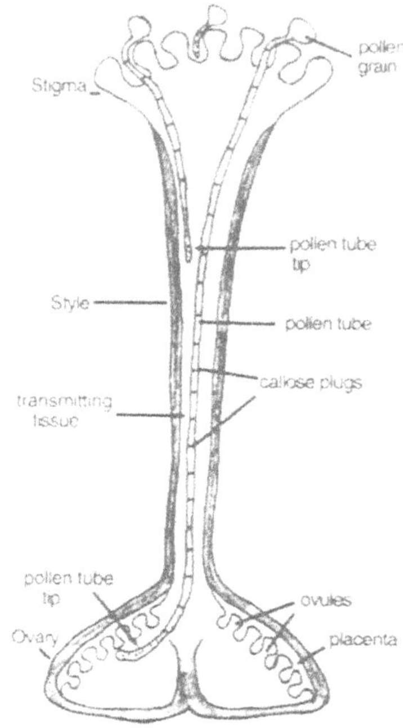

Figure 1. Pollen tube elongation in the style. Reprinted with permission from Cheung 1996a, Pollen–pistil interactions during pollen tube growth, *Trends Plant Sci.* **1**: 45–51, Fig 1A, Elsevier Science.

The proteins TTS and NaTTS are virtually identical arabinogalactan-proteins (AGPs) from *N. tabacum* and *N. alata*, respectively (Cheung *et al* 1995, Cheung and Wu 1999, Wu *et al* 2000). The amino acid sequences of their backbone polypeptides are approximately 97% homologous (Cheung *et al* 1993, Chen *et al* 1993). Similar proteins are also present in *N. sylvestris* and a number of other Solanaceous plants (Cheung and Wu 1999). Cyto-immunodetection using antibodies against a partial TTS polypeptide backbone showed that both TTS and NaTTS proteins are located exclusively in the stylar transmitting tissue ECM (Wang *et al* 1993, Wu *et al* 2000). Both of these AGPs promote pollen tube growth *in vitro* and attract pollen tubes in a semi-*in vivo* pollen tube growth system (Cheung *et al* 1995, Wu *et al* 2000). The TTS protein is also important for the *in vivo* pollen tube growth process, since transgenic *N. tabacum* plants with reduced levels of this AGP showed reduced pollen tube growth rates, sometimes resulting in a detectable decrease in reproductive efficiency (Cheung *et al* 1995). The TTS protein interacts intimately with pollen tubes, binding to the pollen tube tip and surface and incorporating into the pollen tube wall. In turn, the sugar moieties on TTS protein are hydrolyzed by pollen tube-bound enzymes (Wu *et al* 1995). *Nicotiana tabacum* and *N. alata* pollen tubes hydrolyze TTS and NaTTS proteins interchangeably (H.-M. Wu and A.Y. Cheung, unpublished data), suggesting little species specificity on this level of pollen–TTS protein interaction. The sugar moieties on TTS protein appear to be important for its pollen tube growth-stimulating and pollen tube-attracting activities (Cheung *et al* 1995). Analysis of *TTS* mRNA, TTS and NaTTS proteins led to the revelation of multiple regulatory mechanisms that operate on the transcriptional, post-transcriptional and post-translational levels. Together, these regulatory mechanisms lead to the stylar transmitting tissue-specific production of mature, fully glycosylated forms of these AGPs.

2. TRANSCRIPTIONAL REGULATION OF *TTS* GENES

2.1 The Bimodular Structure of TTS Protein is Encoded by Two Exons in the *TTS-1* Gene

Genomic blot analysis (Cheung *et al* 1993) and a 5' mRNA fragment mapping method (Wang *et al* 1996) showed that TTS protein in *N. tabacum* is encoded by two genes, *TTS-1* and *TTS-2*. *Nicotania tabacum* has apparently inherited these genes, one from each of its two progenitors *N. sylvestris* and *N. tomentosiformis* (Fig 2). The *TTS-1* and *TTS-2* genes both appear to be expressed to a similar extent during pistil development and in response to pollination (Cheung *et al* 1993, Wang *et al* 1996). One of these genes, *TTS-1*, has been isolated and characterized (Fig 3) (Zhan 1995).

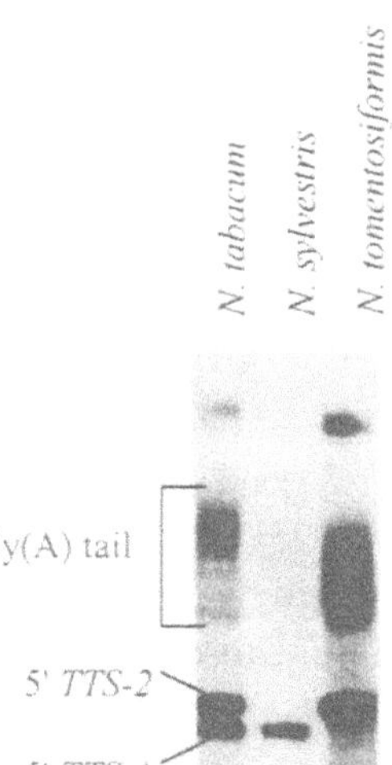

Figure 2. Nicotiana tabacum derives its two TTS genes, *TTS-1* and *TTS-2*, each from one of its progenitors, *N. sylvestris* and *N. tomentosiformis*, respectively. An mRNA 5'-end mapping technique (Wang *et al* 1996) was used. Total stylar RNA from *N. tabacum*, *N. sylvestris* and *N. tomentosiformis* was hybridized with an oligonucleotide complementary to an identical region of *TTS-1* and *TTS-2* mRNA. The RNA/DNA hybrids were digested with RNase H. The resulting RNA samples were electrophoresed on a denaturing urea-acrylamide gel, electro-transferred to Zetaprobe membrane, and hybridized with ^{32}P-labeled *TTS* DNA. The RNase H cleaved the *TTS-1* and *TTS-2* mRNAs at the RNA-oligonucleotide hybrid region into discrete-sized 5' end fragments (indicated as 5' *TTS-1* and 5' *TTS-2*) and heterogeneous 3'-end fragments that appeared as the smears that trailed the discrete 5'-end fragments. *Nicotiana sylvestris* expresses only *TTS-1* mRNA, *N. tomentosiformis* expresses only *TTS-2* mRNA, and *N. tabacum* expresses both.

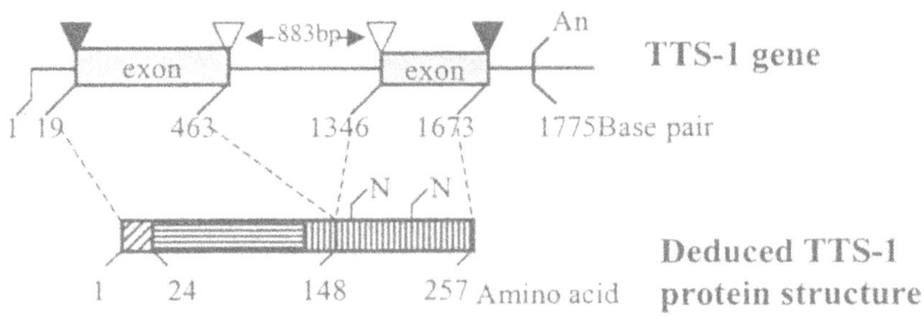

Figure 3. The *TTS-1* gene structure and its deduced bimodular TTS-1 protein. The *TTS-1* gene encodes a protein with a signal peptide (▨), a Pro-rich domain (▤) and a Cys-rich domain (▥). Numbers underneath the *TTS-1* gene indicate nucleotide positions. 1, Transcription initiation site; open arrowheads, exon/intron junctions; [An], polyadenylation signal. Numbers underneath the deduced TTS-*1* protein structure indicate amino acid positions. 24, Putative N-terminal amino acid of mature TTS-1 protein.

The deduced primary sequences of TTS and NaTTS proteins have a bimodular structure (Cheung *et al* 1993, Chen *et al* 1993). The N-terminal half of these proteins is Hyp-rich and extensively O-glycosylated. The C-terminal half has six cysteine residues and is N-glycosylated (Wang *et al* 1993, Cheung *et al* 1995). Nucleotide sequence analysis of the *TTS-1* gene revealed that the coding region of these two structural domains is interrupted by an 833-bp intron. Presence of an intron separating the Pro-rich domain from another recognizable structural domain is a relatively constant feature among Pro-rich proteins (e.g., Wu *et al* 1993, Li and Showalter 1996). Both transcriptional and post-transcriptional regulatory mechanisms are employed in the expression and accumulation of the *TTS* mRNAs.

2.2 5' Regulatory Region of *TTS-1* Directs Transmitting Tissue-Specific Expression

An approximately 3-kbp region upstream of the *TTS-1* gene has been used in the analysis of its transcriptional regulation (Fig 4A) (Zhan 1995). A series of 5'-deleted and internally deleted fragments of this region were fused with the reporter gene encoding β-glucuronidase (β-GUS) (Fig 4B). The β-GUS expression directed from these chimeric genes was analyzed in transgenic *N. tabacum* plants. An upstream region of 1242 bp was adequate to direct a high level of transmitting tissue-specific expression of β-GUS, whereas sequences between –569 and –1242 bp upstream of the transcription initiation site had an enhancing effect on the *TTS-1* promoter (Fig 4C, Figs 5B,C,D). Sequences between –1242 bp to about –3 kbp did not affect the qualitative (data not shown) or quantitative properties of the *TTS-1* promoter. An internal deletion between -192 and –569 (F2) resulted in the abolition of *TTS-1* promoter activity (Fig 4C). This observation suggests the presence of important regulatory elements in this 377-bp region, in addition to the CAAT and TATA boxes found between –192 bp and the transcription initiation site (Fig 4A).

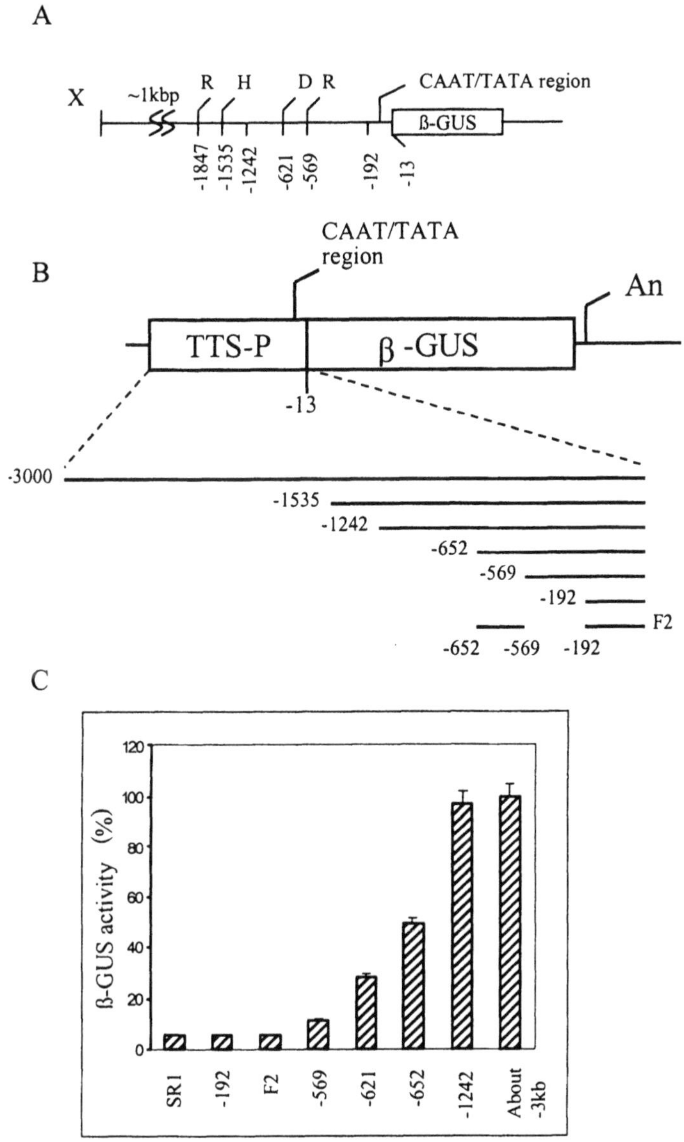

Figure 4. A. The 5'-upstream regulatory region of the *TTS-1* gene fused with the β-GUS reporter gene. R, EcoRI; H, HindIII; D, DdeI; X, XbaI. B. Chimeric constructs between the longest (~3000), various 5'-deleted, and an internal-deleted (F2) *TTS-1* 5' upstream region and the β-GUS reporter gene. These constructs were introduced into *N. tabacum* plants via Agrobacterium Ti plasmid transformation. C. Comparison of β-GUS activity from each of the chimeric genes shown in Fig 4B. The β-GUS activity from various constructs is expressed as percentage (%) of that expressed by the TTS-p(-1242)-β-GUS construct (set at 100%). Data from each construct were obtained from 10 separately analyzed independent transgenic plants. SR1, wild type as nontransgenic control.

D

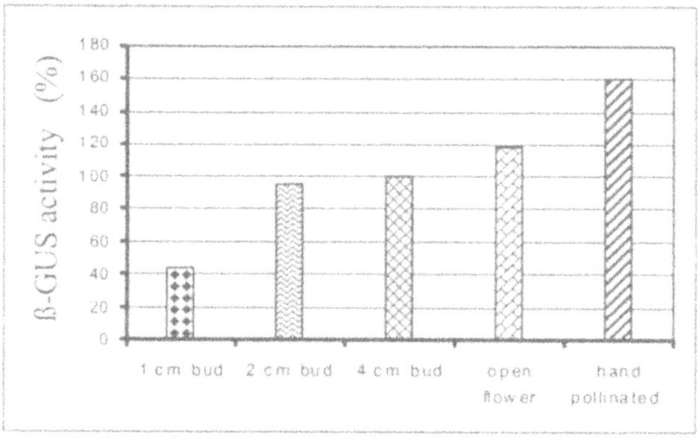

E

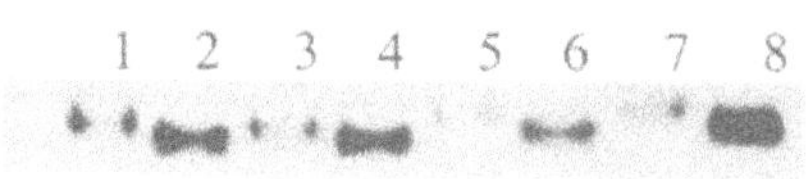

Figure 4. (continued) D. Pistil development- and pollination-regulated *TTS-1* promoter activity. For each construct, the styles from 10–20 flowers at the indicated developmental stages were used to prepare protein extracts for β-GUS activity assays. The hand-pollinated sample was obtained by hand-pollinating styles of emasculated 2-cm flowers. The response was similar when emasculated 4-cm flowers were hand-pollinated (data not shown). E. Pollination stimulates *TTS* mRNA levels and induces *TTS* mRNA shortening. Odd-numbered lanes are for unpollinated control styles; even-numbered lanes are for their corresponding pollinated samples. Lanes 1–4 were RNA from the upper half of the styles; lanes 5–8 were from the lower half of the styles. Pollinated samples were either 1 day (lanes 2,6) or 2 days (lanes 4,8) after pollination. Unpollinated samples were either 1 day (lanes 1,5) or 2 days (lanes 3,7) after emasculation of styles developmentally similar to those used for pollination. The zig-zagging of RNA bands represented an average shortening of about 180 bases at the 3' end of the *TTS* mRNA in the pollinated styles (see Wang *et al* 1996). Figure 4E adapted with permission from Wang *et al* 1993, Development and pollination-regulated accumulation and glycosylation of a stylar transmitting tissue-specific proline-rich protein, *Plant Cell* **5**: 1639–1650, Fig 8A, American Society of Plant Physiologists.

The *TTS-1* promoter was activated in relatively young pistils in these transgenic plants. Its activity increased with pistil development, and pollination provided a slight stimulation of the *TTS-1* promoter activity (Fig 4D). The observed transcriptional activity of the *TTS-1* promoter paralleled closely the mRNA accumulation pattern reported previously for the *TTS* mRNAs (Fig 4E) (Cheung *et al* 1993, Wang *et al* 1996).

2.3 The *NAG1* (*N. tabacum Agamous* gene) -Activated Pistil Developmental Pathway Induces the *TTS-1* Promoter

Transgenic plants ectopically expressing the *Arabidopsis Agamous* gene or its orthologs from other plant species, including *NAG1* from *N. tabacum*, by the CaMV35S promoter showed a morphological transformation of sepals into carpelloid tissues (Mandel *et al* 1992, Kempin *et al* 1993, Cheung *et al* 1996). These carpelloid sepal tissues in transgenic *N. tabacum* plants expressing the *CaMV35S-NAG1* chimeric gene accumulated at least three normally stylar-transmitting tissue-specific mRNAs: TTS, β-(1,3)-glucanase and MG15 [another stylar transmitting tissue-specific AGP (De Graaf *et al* 1998)] (Cheung *et al* 1996). This observation suggests that the *CaMV35S-NAG1* gene activated a pistil-specific gene expression program in the sepals of these transgenic plants and that this program controls even genes that are believed to operate late in pistil development during the pollen tube growth process.

The *CaMV35S-NAG1* gene was crossed into transgenic *N. tabacum* plants harboring the chimeric *TTS-p(-1535)-β-GUS* and *TTS-p(-621)-β-GUS* genes. The doubly transformed plants showed β-GUS activities in their stigmatoid sepals (Fig 5E). Transgenic *N. tabacum* plants expressing the *CaMV35S-AGL5* (a downstream target of the *Arabidopsis Agamous* gene) gene also showed similar morphological and molecular changes in their sepal tissues (data not shown). These results indicate transcriptional activation of the *TTS-1* promoter by the *NAG1*- and *AGL5*-controlled pistil developmental pathway and suggest that *TTS-1* was either directly or indirectly controlled by these two MADS box genes.

3. POST-TRANSCRIPTIONAL REGULATION OF *TTS* mRNA IN RESPONSE TO POLLINATION

Pollination slightly stimulates the activity of the *TTS-1* promoter (Fig 4D), whereas RNA blot analysis showed that pollination results in a three- to fivefold increase in the level of *TTS* mRNA; concomitantly, the length of *TTS* mRNA became shorter (Fig 4E) (Wang *et al* 1993 1996). Using a 3' mRNA mapping technique, Wang *et al* (1996) showed that the poly(A) tail of *TTS* mRNA was reduced upon pollination by an average of approximately 180 bases. The shortening of *TTS* mRNA was detectable in both the upper and lower halves of the styles, between 3 to 6 hours after pollination, when pollen tubes were just elongating past the stigmatic-transmitting tissue junction. This suggests a relatively fast-moving signal that traveled ahead of the pollen tubes to induce the *TTS* mRNA shortening process. Ethylene and okadaic acid were shown to be capable of inducing similar *TTS* mRNA shortening in unpollinated styles, suggesting a signal transduction pathway involving this gaseous hormone and a phosphorylation cascade.

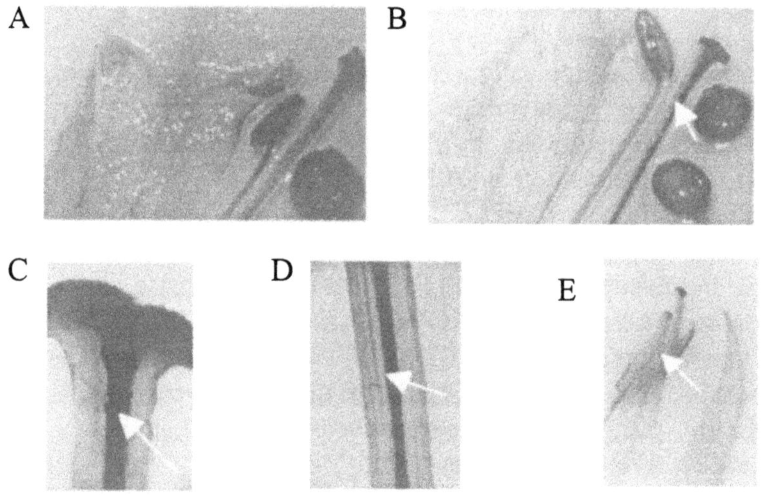

Figure 5. A,B. Histochemical staining for β-GUS activity (arrow) in wild-type (A) and *TTS-p(-1242)-β-GUS* transformed (B) floral parts. From left to right are sepal, petal, stamen, stigma/style and ovary tissues. C,D. Higher magnification of stigma/style (C) and style (D) regions of a *TTS-p-1242)-β-GUS*-transformed style, showing transmitting tissue-specific β-GUS activity. Qualitatively, the stylar transmitting tissue expression specificity was the same for all the deletions that retained adequate promoter activity for β-GUS detection (data not shown). E. The β-GUS expression in a carpelloid sepal of a *CaMV35S-NAG1*, *TTS-p(-1535)-GUS* doubly transformed plant (left side) compared to a control wild-type sepal (right side). Qualitatively similar results were obtained from *CaMV35S-NAG1*, *TTS-p(-621)-β-GUS* plants (data not shown).

Pollination also induces cell deterioration and death within the transmitting tissue (Wang *et al* 1996). Shortening of *TTS* mRNA preceded observable histological changes in the pollinated tissues. Pollination-induced transmitting tissue cell death suggests significant overall cellular deterioration and degradation among cellular biochemical components. Shortening of mRNA appears to be a common degradative event among transmitting tissue mRNAs, since at least two additional mRNAs, MG15 and β-(1,3)-glucanase mRNAs, were also detectably shortened after pollination (Wang *et al* 1996). However, in contrast to the pollination-induced enhancement of *TTS* mRNA levels (Fig 4E), the levels of MG15 and β-(1,3)-glucanase mRNAs were greatly reduced by 48 hours after pollination, consistent with the general observation that poly(A)-tail trimming often precedes mRNA degradation (Sachs 1993). A threshold level of TTS protein is apparently necessary for optimal reproductive success, since pollen tube growth rate was reduced in transgenic *N. tabacum* plants in which the level of TTS protein was significantly reduced by antisense suppression or co-suppression, sometimes resulting in reduced female fertility (Cheung *et al* 1995). The pollination-induced shortening of the *TTS* mRNA poly(A) tail suggests that *TTS* mRNA, like other cellular components, is sensitive to the deteriorating biochemical conditions within the pollinated transmitting tissue cells. The pollination-enhanced *TTS* mRNA levels, despite an overall RNA degradative environment, suggests a need to ensure the presence of adequate amounts of TTS protein in the pollinated styles. Mechanisms apparently evolved to accomplish this enhancement of TTS expression (Wang *et al* 1993).

4. REGULATION OF TTS PROTEINS AT THE POST-TRANSLATIONAL LEVEL

4.1 A *NAG1*-controlled Pistil Developmental Program Regulates Post-Translational Modification of TTS Protein

On immunoblots and by histo-immunostaining, TTS protein was detectable only in the stylar transmitting tissue (Fig 6A,B) (Wang *et al* 1993, Cheung *et al* 1996). Transgenic *N. tabacum* plants harboring a chimeric *CaMV35S-TTS* gene produced TTS polypeptides constitutively (Fig 6A). In the wild-type styles, TTS protein was found most predominantly in the molecular weight range between 50 and 100 kDa. However, all of the ectopically produced TTS protein in the *CaMV35S-TTS* transgenic plants was underglycosylated and showed molecular weights below 45 kDa (Fig 6A) (Cheung *et al* 1996). These protein molecules represented TTS polypeptides that had been N- and O-glycosylated to extents significantly lower than that of the predominant TTS protein species found in the stylar transmitting tissue (see Fig 6D). These results suggest that activities to fully glycosylate TTS protein were either limiting or absent in all non-stylar transmitting tissue, including the adjacent stylar cortex and the ovary. However, the extent of glycosylation of the ectopically expressed TTS polypeptides did not change among transgenic plants expressing either trace amounts or high levels of these molecules in non-stylar tissues (H.-M. Wu and A.Y. Cheung, unpublished data). Thus, the presence of a glyco-linkage(s) in TTS protein requiring an enzyme(s) uniquely active in the stylar transmitting tissue would be a preferred hypothesis to explain the observed transmitting tissue-specific glycosylation of TTS polypeptides.

Interestingly, in the *CaMV35S-NAG1* (Cheung *et al* 1996) and *CaMV35S-AGL5* (D. Vu and A.Y. Cheung, unpublished data) transgenic plants, *TTS* mRNA was induced and was translated in the sepals of the transformed plants. The carpelloid sepal-produced TTS polypeptides became fully glycosylated as in the wild-type stylar tissue (Fig 6D). These observations indicate that the activities that fully glycosylate TTS backbone polypeptides were also activated by the *NAG1*- and *AGL5*-controlled pistil developmental program.

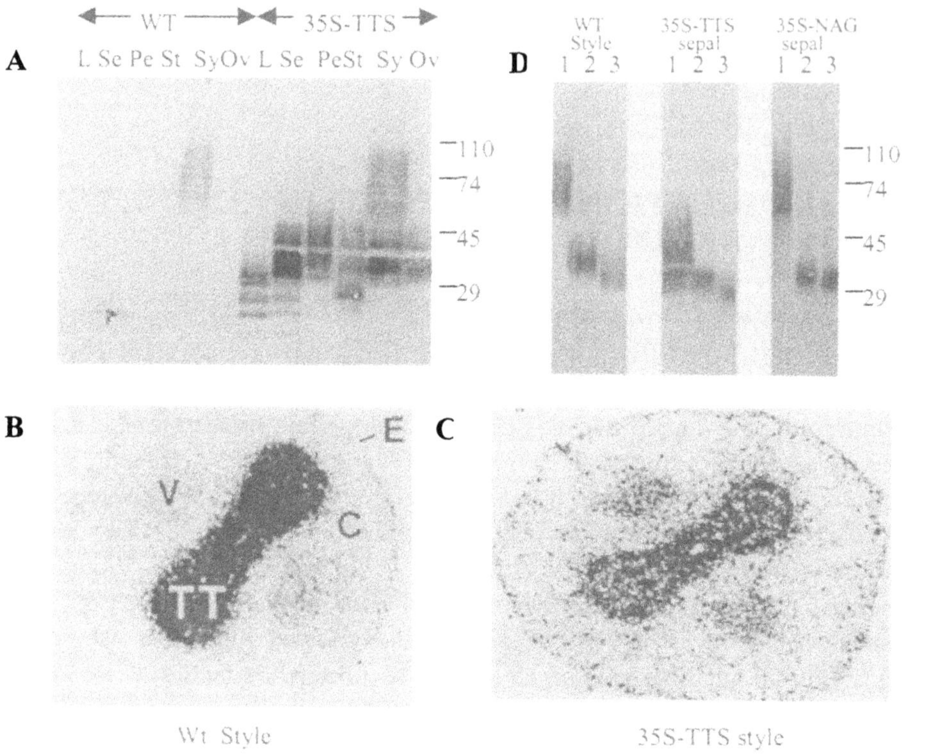

Figure 6. Immunoblot (A) and histo-immunostaining of wild-type (B) and *CaMV35S-TTS*-transformed (C) stylar tissues by TTS antibodies. In (A), proteins from wild-type (WT) and *CaMV35S-TTS* (35S-TTS)-transformed leaf (L) and floral tissues (Se, sepal; Pe, petal; St, stamen; Sy, stigma and styles; Ov, ovary) were compared. B,C. Cross-sections of wild-type and *CaMV35S-TTS*-transformed styles stained with TTS antibodies. (E, epidermis; V, vascular tissue; TT, transmitting tissue). D. Immunoblotting and TTS antibody staining of proteins isolated from WT styles, *CaMV35S-TTS* (35S-TTS)-transformed sepals and *CaMV35S-NAG1* (35S-NAG)-transformed sepals. Lane 1 in the three panels shows proteins directly after extraction from the designated tissue. The same proteins are shown after chemical deglycosylation (lane 2) or after chemical and enzymatic deglycosylation (lane 3). Adopted with permission from Cheung *et al* 1996, Organ-specific and *Agamous*-regulated expression and glycosylation of a pollen tube growth-promoting protein, *Proc. Natl. Acad. Sci. USA* **93**: 3853–3858, Fig 2, National Academy of Sciences, U.S.A.

4.2 Glycosylation of TTS and NaTTS Polypeptide Backbones Is Highly Modulated, Resulting in a Broad Spectrum of Sugar Modifications That Affect How These Proteins Interact with the Cell Wall Matrix

The predominant amounts of TTS and NaTTS proteins have sizes between 50 and 105 kDa. They are loosely associated with the transmitting tissue cell wall matrix and are released from it under a broad range of buffer conditions (Fig 7) (Wang *et al* 1993, Cheung *et al* 1995, Wu *et al* 2000). A careful analysis of the extractability of TTS and NaTTS proteins from the transmitting tissue ECM revealed that the predominant amounts of these proteins were solubilized from the ECM by low salt buffers. High salt conditions were needed to extract the remaining, smaller amounts of TTS and NaTTS proteins (Fig 7B; Wu *et al* 2000). Immunoblot analysis revealed that the low salt-extractable, predominant population of TTS and NaTTS proteins spanned a molecular weight spectrum of 50–110 kDa. However, most of the high salt-extractable TTS and NaTTS proteins spanned a somewhat lower molecular weight spectrum of 45–100 kDa. Furthermore, the high salt-extractable protein sample from *N. alata* contained significant amounts of NaTTS protein molecules in the 30–45 kDa range. Although proteins in this range were also present in the high salt-extractable TTS protein sample from *N. tabacum*, the amount of these proteins was significantly lower than that found in the *N. alata* sample. This observation suggests that the *N. tabacum* style has a capacity to fully glycosylate more of the TTS polypeptide backbones than the *N. alata* style. A no-salt, mildly acidic buffer solubilized the entire spectrum of TTS and NaTTS proteins. The differential solubilization of more- and less-glycosylated TTS and NaTTS proteins by low and high salt buffers, respectively, suggests that the level of glycosylation affects how these AGPs interact with the cell wall matrix. Glycosylation acidifies the TTS and NaTTS polypeptide backbones, modifying the highly basic polypeptides (with pIs of approximately 10) to proteins of pIs ranging between 7.5 and 8.5 (Fig 8C) (Wu *et al* 1995 2000). These observations together indicate that the less glycosylated, more basic TTS and NaTTS protein molecules associate more tightly with the ECM than the more highly glycosylated, relatively more acidic molecules.

Purified NaTTS proteins isolated by low salt extraction of transmitting tissue or by high salt-extraction of the low salt-washed transmitting tissue have been analyzed on protein blots by antibodies against the TTS polypeptide backbone, by β–glucosyl Yariv reagent and by a monoclonal antibody (JIM13) specific for arabinogalactan moieties. These revealed dramatic variability in the quantity and quality of glyco-modifications on the NaTTS polypeptide backbone (Wu *et al* 2000). Antibodies against the C-terminal domain of the TTS polypeptide backbone showed that the low salt-extractable NaTTS protein ranges from 45–110 kDa, whereas the high salt-extractable NaTTS protein ranges from 30-100 kDa, similar to what is shown in Fig 7B. The AGP-diagnostic probes β-glucosyl Yariv reagent and JIM13 (Knox *et al* 1991) reacted efficiently with a NaTTS protein that was larger than 55 kDa and 40 kDa, respectively. These observations imply that either the sugar moieties recognized by these two reagents were absent or were not present at detectable levels in the less glycosylated forms of NaTTS protein. In addition, the β-glucosyl Yariv reagent reacted more weakly with the high salt-extractable NaTTS protein than with the low salt-extractable counterpart. This observation suggests either a qualitative or a quantitative difference in the sugar moieties between these two populations of NaTTS proteins.

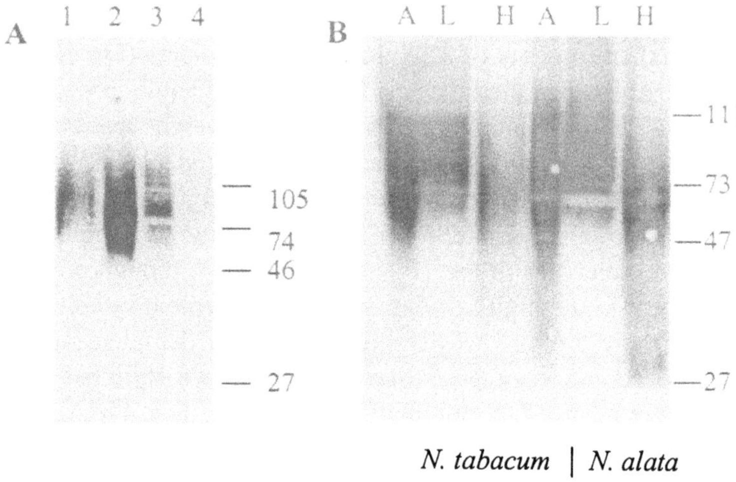

Figure 7. Immunoblots of TTS proteins isolated by various buffer conditions. A. Protein samples are: lane 1, total proteins extracted from pulverized stylar tissues by boiling in an SDS-containing protein extraction buffer; lane 2, proteins eluted from isolated transmitting tissue by an isotonic buffer conditioned not to rupture transmitting tissue cells; lane 3, proteins extracted from pulverized transmitting tissue after it had been washed by the isotonic buffer to obtain the sample shown in lane 2; lane 4, total protein extracted from pulverized pistil tissues from which the transmitting tissues had been removed (thus containing almost exclusively epidermal, cortical and vascular tissues). Adopted with permission from Wang *et al* 1993, Development and pollination-regulated accumulation and glycosylation of a stylar transmitting tissue-specific proline-rich protein, *Plant Cell* **5**: 1639–1650, Fig 4B, American Society of Plant Physiologists. B. Lanes A, stylar transmitting tissue ECM proteins isolated by a mildly acidic, no-salt wash (84 mM citric acid, pH 3); lanes L, proteins isolated by a low salt wash (100 mM NaCl, 20 mM Tris-HCl, pH 7.5); lanes H, proteins isolated by a high salt wash (400 mM NaCl, 20 mM Tris-HCl, pH 7.5) of previously low salt-washed tissue. All buffers contained 2 mM $Na_2S_2O_4$. Buffers that differed slightly from those described here yielded similar results (Wu *et al* 2000). Because these different extraction conditions recovered vastly different amounts of proteins, samples were loaded on a constant volume per fresh tissue weight basis. The TTS antibodies, raised against a partial backbone polypeptide, reacted more efficiently with less glycosylated TTS proteins than with the most highly glycosylated forms of these AGPs.

Carbohydrate modifications are known to play key roles in modulating recognition and interactions between biological molecules. The broad range of glyco-modification observed for TTS and NaTTS proteins underscores the complexity of the biochemical and biological properties of these AGPs, allowing tremendous versatility for them to interact with other molecules in the transmitting tissue ECM environment.

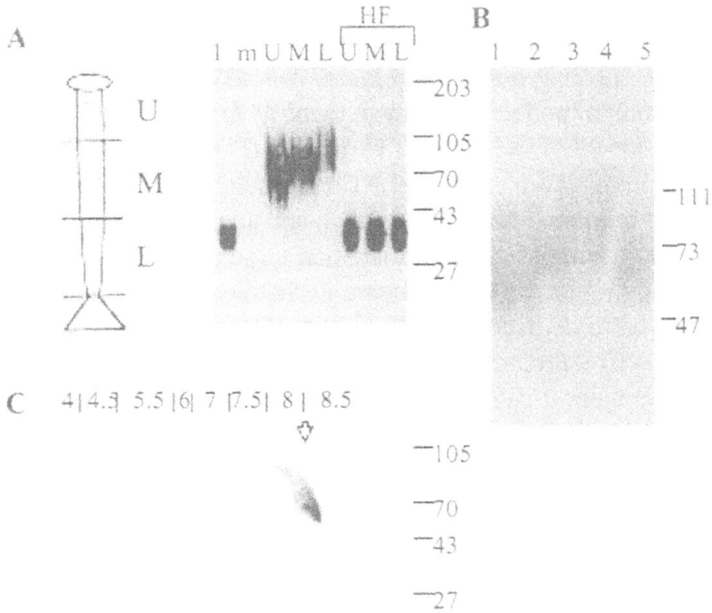

Figure 8. A. Immunoblot showing TTS protein isolated from the upper (U), middle (M) and lower (L) segments of the *N. tabacum* styles. HF, chemically deglycosylated samples. Lane 1, chemically deglycosylated total stylar proteins; lane m, molecular weight markers. B. Immunoblot for the corresponding TTS protein from the styles of *N. sylvestris*. Because of the long styles, proteins were isolated from four stylar segments between the stigma and the ovary. Lanes 1–4 are for proteins isolated from the top, second, third and the lowest segments, respectively, of the styles. Lane 5 shows the same material as in lane 1, but at one-fifth the load, to obtain comparable TTS antibody-stainable material as that loaded in lane 4. This comparison accentuates the differences in molecular weight range of the *N. sylvestris* TTS protein from the top and the bottom stylar segments and shows that the amount of proteins loaded did not affect the mobility of these proteins. The NaTTS protein also shows a similar glycosylation level gradient (Wu *et al* 2000). C. Immunoblot of a two-dimensional-SDS-PAGE showing that higher-molecular-weight TTS protein molecules have lower pIs. The NaTTS protein shows a similar phenomenon (Wu *et al* 2000). Panels A and C adapted with permission from Wu *et al* 1995, A pollen tube growth stimulatory glycoprotein is deglycosylated by pollen tubes and displays a glycosylation gradient in the flower, *Cell* **82**: 395-403, Fig. 6, Cell Press. Panel B adapted with permission from Cheung and Wu, 1999, Arabinogalactan proteins in plant sexual reproduction, *Protoplasma* **208**: 87-98, Fig. 3, Springer.

4.3 Differential Glycosylation of TTS Protein along the Length of the Style Results in a Gradient of Increasing TTS Protein Glycosylation Levels in the Direction of Pollen Tube Growth

The stylar transmitting tissue of at least three different *Nicotiana* plants—*N. tabacum*, *N. alata* and N. *sylvestris*—show differential glycosylation of their corresponding TTS proteins from the stigmatic to the ovarian ends (Fig 8) (Wu *et al*

1995, Cheung and Wu 1999, Wu *et al* 2000). The TTS protein from the ovarian end of the styles showed a molecular weight spectrum significantly higher than that of TTS protein isolated from the stigmatic end of the styles (Fig 8A). This differential glycosylation of the TTS protein results in an increasing TTS protein-bound sugar gradient in the direction of pollen tube growth in the *N. tabacum* style. Similar gradients were also observed for the *N. sylvestris* TTS protein (Fig 8B) (Cheung and Wu 1999) and for the NaTTS protein (Wu *et al* 2000). These results and the observation that TTS and NaTTS proteins promote pollen tube growth and attract pollen tubes (Cheung *et al* 1995, Wu *et al* 2000) prompt us to speculate that these protein-bound sugar gradients play a role in keeping pollen tubes advancing in the direction of the ovary. When the ability of pollen tubes to deglycosylate TTS protein (Wu *et al* 1995), NaTTS protein and the *N. sylvestris* TTS protein (H.-M. Wu and A.Y. Cheung, unpublished data) is brought into consideration, it can be further envisaged that the advancing pollen tubes would produce a local, self-sharpening effect on the TTS protein-bound sugar gradient, as each pollen tube would leave behind TTS protein molecules that were less glycosylated and would encounter its next substrates in the form of more glycosylated TTS protein molecules (Wu *et al* 1995).

5. DISCUSSION

The TTS protein in *N. tabacum* and its counterparts in *N. alata* and *N. sylvestris* are major glycoprotein constituents in the stylar transmitting tissue of these plants.

The multiple mechanisms that confer a transmitting tissue-specific accumulation of highly glycosylated forms of these proteins can be envisaged to ensure the presence of these pollen tube growth-promoting and -attracting AGPs in the pollinated styles, presumably to facilitate pollen tube growth. Transcriptional control confers transmitting tissue-specific accumulation of *TTS* mRNAs. The post-transcriptional regulation of maintaining the stability of 3'-end shortened *TTS* mRNA after pollination assures the presence of high levels of *TTS* gene products. Should the control of transmitting tissue expression of *TTS* genes lose its stringency and low levels of *TTS* mRNA accumulate in non-stylar transmitting tissue, the post-translational control of transmitting tissue-specific glycosylation of TTS protein backbone precludes the accumulation of fully glycosylated TTS protein in non-transmitting tract tissues. This level of regulation may be important, especially at the cortical/transmitting tissue junction, so that pollen tubes do not elongate non-productively into the cortical tissue. The exclusion of *TTS* mRNA and fully glycosylated TTS protein in the ovary is more intriguing since pollen tubes continue to elongate in the ovary. However, a need to turn from a basally oriented elongation direction along the placental surface to enter the ovules may be better served in the absence of TTS protein so that pollen tubes may be more receptive to guidance signals emanating from the ovules.

The range of glycosylation observed for TTS protein and other *Nicotiana* counterparts provides extraordinary opportunity to modulate the biochemical properties and thus influence the biological activities of these AGPs. The differentially glycosylated TTS proteins interact with the transmitting tissue ECM with different affinities, possibly affecting their availability to interact with the traversing pollen tubes. This observation is especially intriguing if considered in the light of the observed gradient of increasing TTS protein glycosylation levels in the styles of these *Nicotiana* plants. The more glycosylated TTS protein molecules closer to the ovary should be more readily recruited by pollen tubes since they are associated with the transmitting tissue ECM with lower affinity. If true, this further strengthens the speculation that the gradient of increasing TTS protein-bound sugars in the direction of pollen tube growth plays a crucial role in the pollen tube growth process. Further work towards perturbing this gradient will be critical toward defining its functional significance. Moreover, the *Nicotiana* pistil is enriched with multiple glycoproteins in addition to TTS protein (Cheung and Wu 1999). How they individually and/or together contribute to female organ development and the reproductive processes of pollination and fertilization need to be explored more globally.

ACKNOWLEDGEMENTS

Research in our laboratory was supported by grants from the US Department of Agriculture, the US Department of Energy and the National Institutes of Health.

REFERENCES

Chen, C.-G., Mau, S.-L., and Clarke, A. E., 1993, Nucleotide sequence and style-specific expression of a novel proline-rich protein gene from *Nicotiana alata*, *Plant Mol. Biol.* **21**: 391–395.

Cheung, A. Y., 1996a, Pollen–pistil interactions during pollen tube growth, *Trends Plant Sci.* **1**: 45–51.

Cheung, A. Y., 1996b, The pollen tube growth pathway: its molecular and biochemical contributions and responses to pollination, *Sex. Plant Reprod.* **9**: 330–336.

Cheung, A. Y., and Wu, H.-M., 1999, Arabinogalactan proteins in plant sexual reproduction, *Protoplasma* **208**: 87-98..

Cheung, A. Y., May, B., Kawata, E. E., Gu, Q., and Wu, H.-M., 1993, Characterization of cDNAs for stylar transmitting tissue-specific proline-rich proteins in tobacco, *Plant J.* **3**: 151–160.

Cheung, A. Y., Wang, H., and Wu, H.-M., 1995, A floral transmitting tissue-specific glycoprotein attracts pollen tubes and stimulates their growth, *Cell* **82**: 383–393.

Cheung, A. Y., Zhan, X.-Y., Wang, H., and Wu, H.-M., 1996, Organ-specific and *Agamous*-regulated expression and glycosylation of a pollen tube growth-promoting protein, *Proc. Natl. Acad. Sci. USA* **93**: 3853–3858.

De Graaf, B. H. J., Knuiman, B., and Mariani, C., 1998, The PELPs in the transmitting tissue of *Nicotiana tabacum* are translocated through the pollen walls in vivo, XVth International Congress on Sexual Plant Reproduction, Wageningen, The Netherlands, August 1998 (abstr.), p.29.

Kempin, S. A., Mandel, M. A., and Yanofsky, M. F., 1993, Conversion of the tobacco perianth into reproductive organs by ectopic expression of the tobacco floral homeotic gene *NAG1*, *Plant Physiol.* **103**: 1041–1046.

Knox, J. P., Linstead, P. J., Peart, J., Cooper, C., and Roberts, K., 1991, Developmentally regulated epitopes of cell surface arabinogalactan proteins and their relation to root tissue pattern formation, *Plant J.* **1**: 317–326.

Li, S.-X., and Showalter, A. M., 1996, Cloning and developmental/stress regulated expression of a gene encoding a tomato arabinogalactan protein, *Plant Mol. Biol.* **32**: 641–652.

Lord, E. M., and Sanders, L. C., 1992, Roles for the extracellular matrix in plant development and pollination: a special case of cell movement in plants, *Dev. Biol.* **153:** 16–28.

Mandel, M. A., Bowman, J. L., Kempsin, S. A., Ma, H., Meyerowitz, E. M., and Yanofsky, M. F., 1992, Manipulation of flower structure in transgenic tobacco, *Cell* **71**: 133–143.

Mascarenhas, J. P., 1993, Molecular mechanism of pollen tube growth and differentiation. *Plant Cell* **5**: 1303–1314.

Read, S. M., Clarke, A. E., and Bacic, A., 1993, Requirement for division of the generative nucleus in cultured pollen tubes of *Nicotiana*, *Protoplasma* **175**: 126–130.

Sachs, A. B., 1993, Messenger RNA degradation in eucaryotes, *Cell* **74**: 413–421.

Taylor, L. P., and Hepler, P. K., 1997, Pollen germination and tube growth, *Annu. Rev. Plant Physiol. Plant Mol. Biol.* **48**: 461–491.

Wang, H., Wu, H.-M., and Cheung, A. Y., 1993, Development and pollination-regulated accumulation and glycosylation of a stylar transmitting tissue-specific proline-rich protein, *Plant Cell* **5**: 1639–1650.

Wang, H., Wu, H.-M., and Cheung, A. Y., 1996, Pollination induces mRNA poly(A) tail-shortening and cell deterioration in flower transmitting tissue, *Plant J.* **7**: 715–727.

Wu, H.-M., Zou, J.-T., Mat, B., Gu, Q., and Cheung, A. Y., 1993, A tobacco gene family for flower cell wall proteins with a proline-rich and a cysteine-rich domain, *Proc. Natl. Acad. Sci. USA* **90**: 6829–6833.

Wu, H.-M., Wang, H., and Cheung, A. Y., 1995, A pollen tube growth stimulatory glycoprotein is deglycosylated by pollen tubes and displays a glycosylation gradient in the flower, *Cell* **82**: 393–403.

Wu, H.-M., Wang, E., Ogdahl, J., and Cheung, A. Y., 2000, A pollen tube growth-promoting arabinogalactan protein from *Nicotiana alata* is similar to the tobacco TTS protein, *Plant J.* (in press).

Zhan, X.-Y., 1995, Molecular studies on plant male and female reproductive tissue-specific genes, Ph.D. Thesis, Zhejiang Agricultural University, Peoples Republic of China.

Chapter 13

Characterization of Arabinogalactan-Proteins and a Related Oligosaccharide in Developing Rice Anthers

Kentaro Kawaguchi[1] and Naoto Shibuya
[1]*Hokkaido National Agricultural Experiment Station, Memuro, Hokkaido 082-0071; and National Institute of Agrobiological Resources, Tsukuba, Ibaraki 305-8602, Japan*

1. INTRODUCTION

Male sterility induced by chilling temperature (12–18°C) at the booting stage causes serious loss of grain yield in rice, one of the most serious problems for Asian agriculture. To resolve this problem, it is important to understand the physiological mechanisms of chilling injury during pollen development. Male gametogenesis is a very important step in plant reproduction and is known to be very sensitive to various forms of environmental stress, including chilling. Meiosis and mitosis are important events in gametogenesis, as are rapid turnovers of pollen walls that occur through interactions between microspores and tapetal cells. Although many anther-specific genes have been identified, the physiological or biochemical basis of male gametogenesis is not well understood. Thus, the physiological and biochemical characterization of these processes is essential for understanding agronomically important phenomena such as chilling injury in rice.

2. POLLEN FORMATION IN EARLY STEPS OF ANTHER DIFFERENTIATION

Since chilling injury in rice is known to be an organ- and stage-specific phenomenon (Satake and Hayase 1970), it is important to precisely describe the development of rice anthers. In this study, anther development was divided into six stages: (1) pollen mother cell, (2) tetrad, (3) early microspore, (4) middle microspore, (5) late microspore

Cell and Developmental Biology of Arabinogalactan-Proteins
Edited by Nothnagel *et al.*, Kluwer Academic/Plenum Publishers, 2000

and (6) uninucleate microspore stages (Fig 1). These stages can be distinguished through microscopic observation. It has been shown that the chilling-sensitive stages are the tetrad and early microspore stages, which together are also called the young microspore stage.

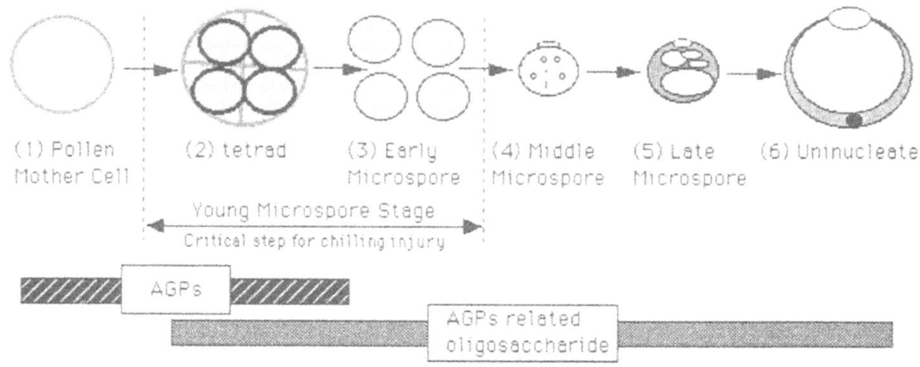

Figure 1. Schematic diagram of pollen morphogenesis in immature rice anthers. Pollen mother cells (1) develop to tetrads (2) through two successive meiotic divisions. Four early microspores (3) are released from a tetrad by dissolution of the callosic wall. Formation of a germ pore and fine vacuoles can be observed in middle microspores (4). Late microspores (5) have multiple large vacuoles and a thick wall. Uninucleate microspores (6) have a single large vacuole and a noticeable nucleus located diametrically opposite the germ pore. Bars indicate the accumulation pattern of arabinogalactan-proteins (AGPs) and the oligosaccharide in developing rice anthers. The AGPs accumulate only during two stages: pollen mother cell and tetrad. The tetrasaccharides accumulate from tetrad to uninucleate.

3. ANALYSIS OF FREE SUGARS IN DEVELOPING RICE ANTHERS

Free sugars in developing rice anthers were analyzed by high-performance anion-exchange chromatography using a PA-1 column attached to a Dionex HPLC system. These results showed that glucose, fructose, sucrose and an unknown oligosaccharide were the main low-molecular-weight carbohydrates that accumulated in developing anthers. The unknown oligosaccharide was abundant and accumulated transiently during microsporogenesis (stages 2–6). Furthermore, the concentration of the unknown oligosaccharide was specifically decreased by a chilling treatment (12°C, 4 days at booting stage), which can induce male sterility in the rice anther. This effect was especially more evident in the chilling-sensitive cultivars than in the tolerant cultivars. Other sugars did not show such stage-specific accumulation and chilling response. These results indicated that the oligosaccharide might have some important roles in pollen development and chilling sensitivity in rice anther.

4. STRUCTURAL ANALYSIS OF THE UNKNOWN OLIGOSACCHARIDE IN RICE ANTHERS

The structure of the purified oligosaccharide was analyzed by methylation, mass spectrometry/mass spectrometry, and nuclear magnetic resonance spectrometry (Kawaguchi *et al* 1996). Its structure was β-L-Ara*f*-(1→3)-α-L-Ara*f*-(1→3)-β-D-Gal*p* -(1→6)-D-Gal, which is very similar to the terminal sequence of the glycan chain of the AGPs from *Acacia senegal* (Defaye and Wong 1986). The β-configuration of the non-reducing terminal of the tetrasaccharide was unique, compared with the α-configuration of the acacia gum.

5. SPECIFIC ACCUMULATION OF AGPS IN DEVELOPING RICE ANTHER

The single radial gel diffusion method (van Holst and Clarke 1985) with β-glucosyl Yariv reagent showed that AGPs appeared transiently during only two stages: pollen mother cell and tetrad. Both AGPs and the tetrasaccharide accumulate during microsporogenesis, and their concentrations changed drastically at the critical step for chilling injury in rice (Fig 1). The sequential appearances of the AGPs and then the tetrasaccharide, when considered together with their structural similarity, suggested that they have some important roles for anther development.

6. LOCALIZATION AND POSSIBLE FUNCTION OF AGPS AND THE TETRASACCHARIDE IN RICE ANTHER

Antibody was produced against the tetrasaccharide. If the rice AGPs have a glycan chain that contains structural units similar to the tetrasaccharide, then the anti-tetrasaccharide antibody was expected to react not only with the tetrasaccharide but also with the AGPs. In fact, the accumulation pattern of AGPs obtained by ELISA using this antibody showed a pattern similar to that obtained with Yariv reagent. Although this was only circumstantial evidence, it revealed that the antibody also reacted with the rice AGPs. Immunocytochemical analysis using this antibody localized the epitope at the cell surfaces of the tapetal layer and the microspores. The results suggested that AGPs or the tetrasaccharide have important roles in both the development of the rice anther and its response to chilling. The AGPs have been indicated to be important components of the extracellular matrix of plant cells (Du *et al* 1996). They might be a kind of signal or marker for the communication between tapetal cells and microspores in the developing anther. Elucidation of the physiological functions of AGPs and the tetrasaccharide in developing rice anther will be the subject of our future study.

REFERENCES

Defaye, J., and Wong, E., 1986, Structural studies of gum arabic, the exudate polysaccharide from *Acacia senegal*, *Carbohyd. Res.* **150**: 221–231.

Du, H., Clarke, A. E., and Bacic, A., 1996, Arabinogalactan-proteins: a class of extracellular matrix proteoglycans involved in plant growth and development, *Trends Cell Biol.* **6**: 411–414.

Kawaguchi, K., Shibuya, N., and Ishii, T., 1996, A novel tetrasaccharide, with a structure similar to the terminal sequence of an arabinogalactan-protein, accumulates in rice anthers in a stage-specific manner, *Plant J.* **9**: 777–785.

van Holst, G.-J., and Clarke, A. E., 1985, Quantification of AGP in plant extracts by single radial gel diffusion, *Anal. Biochem.* **148**: 446–450.

Satake, T., and Hayase, H., 1970, Male sterility caused by cooling treatment at the young microspore stage in rice plants. V. Estimations of pollen developmental stage and the most sensitive stage to coolness, *Proc. Crop Sci. Soc. Japan* **39**: 468–473.

Chapter 14

Arabinogalactan-Proteins in Pollen Tube Growth

E. M. Lord,[1] T. Holdaway-Clarke,[2] S. J. Roy,[1,2] G. Y. Jauh,[3] and P. K. Hepler[2]
[1]*Department of Botany and Plant Sciences, University of California, Riverside, CA 92521-0124;* [2]*Biology Department, University of Massachusetts, Amherst, MA; and* [3]*Institute of Biological Chemistry, Washington State University, Pullman, WA*

Arabinogalactan-proteins (AGPs) were associated with reproductive tissues in the early days of their discovery (Nothnagel 1997). In particular, the stigmas and styles of flowering plants were abundant sources of AGPs, as revealed by Yariv phenylglycoside staining [(β-D-glucosyl) Yariv phenylglycoside [(β-D-Glc)$_3$]], a red-colored reagent thought to bind AGPs (Yariv *et al* 1962). Several species were examined in detail using the (β-D-Glc)$_3$ to precipitate the stigma and stylar AGPs (Clarke *et al* 1979, Gleeson and Clarke 1979 1980, Fincher *et al* 1983, Miki-Hirosige *et al* 1987). Only recently has DNA sequencing provided us with knowledge of the protein component of some of these large proteoglycans, and the first AGPs sequenced were from stylar transmitting tract tissues (Du *et al* 1994, Mau *et al* 1995). The carbohydrate branches of AGPs consist mainly of arabinose and galactose, and the protein core is typically rich in Hyp/Pro, Ala and Ser, but the entire molecular structure of any AGP is unknown (Nothnagel 1997). The exciting recent discovery that some plasma membrane AGPs have glycosylphosphatidylinositol anchors has suggested the possibility of their involvement in signaling cascades at the cell surface (Youl *et al* 1998, Schultz *et al* 1998, Svetek *et al* 1999). Polyclonal and monoclonal antibodies to AGPs have been produced with polysaccharide epitopes of mostly unknown composition (Pennell *et al* 1991, Knox 1992). There is good circumstantial evidence that they bind AGPs. These studies have confirmed that AGPs are abundant in stigmas, styles, and pollen tubes of many species. The antibodies have revealed the diversity of expression patterns for AGPs in development and have ignited interest in their roles in plant biology.

The AGPs are ubiquitous in plants and are no doubt filling important roles in cell–cell interactions during development. Here we focus on their localization patterns and possible roles in pollination, with an emphasis on pollen tube growth.

1. AGPS IN POLLINATION

In 1992, Li and colleagues used monoclonal antibodies (MAbs) to AGPs (JIM 8 and MAC 207) on tobacco pollen tubes and discovered a striking banding pattern of AGP deposition in the wall. This observation confirmed some earlier work suggesting that AGPs occurred in pollen tube walls (Harris *et al* 1987, Anderson *et al* 1987). In Li and colleagues' 1992 and 1995 studies on tobacco, there was no evidence of AGPs in the pollen tube tip where Golgi vesicles fuse with the plasma membrane and deposit wall materials for growth of the pollen tube. The authors showed that AGPs and callose co-localized in the pollen tube wall with AGPs at the interface between the outer pectin wall and the inner callosic wall. The AGPs were also seen on the plasma membrane and in secretory vesicles back from the tip. When similar MAbs (JIM 13-16, MAC 207) were used on lily pollen tubes, AGPs were also localized to the wall of the tube, the plasma membrane and vesicles but now including the tube tip (Fig 1). Several species show this pollen tube tip localization of AGPs using MAb JIM 13, and others, like tobacco, do not (Mollet *et al* [page 284, this volume]) (Table 1).

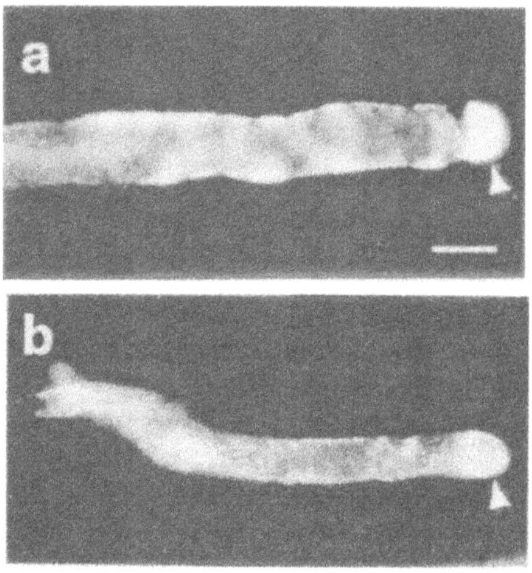

Figure 1. Immunolocalization of AGPs (MAb JIM 13) on lily pollen tubes grown *in vitro* 28 hours. (a) and *in vivo* 24 hours (b). Pollen tubes pretreated with pectinase to expose AGP epitopes back from the tips (arrowheads). Reproduced with permission from Jauh and Lord 1996, Localization of pectins and arabinogalactan-proteins in lily (*Lilium longiflorum* L.) pollen tube and style, and their possible roles in pollination, *Planta* **199**: 251–261, Fig 2E,F, Springer-Verlag. Bar = 10 μm.

Table 1. Pollen tubes that secrete AGPs at their tip are arrested by $(\beta\text{-D-Glc})_3$

Species	JIM 13 staining at tip	Growth arrested by 30 μM $(\beta\text{-D-Glc})_3$
Lilium longiflorum (Liliaceae)	+	+
Zea maize (Graminae)	+	+
Annona cherimoya (Annonaceae)	+	+
Nicotiana tabacum (Solanaceae)	–	–
Lycopersicon esculentum (Solanaceae)	–	–
Aquilegia eximia (Ranunculaceae)	–	–

In lily, as in many other species, the transmitting tract of the style is rich in AGPs, as revealed by several means including immunogold labeling with MAb JIM 13 (Fig 2). The transmitting tract epidermis in lily has a convoluted transfer cell-like wall at the outer face, topped by a continuous wall layer. With chemical fixation, much of the secreted matrix rich in AGPs is lost, but the matrix secretions are evident en route to the surface in these Spurrs-resin-embedded sections when we use MAb JIM 13 (Fig 2a). The copious secretions that fill the hollow style and surround the pollen tubes traveling on the transmitting tract epidermis are rich in AGPs (Fig 2b). These secretions can be retained and detected by immunocytochemistry by use of high pressure freezing/freeze substitution techniques for sample preparation (Roy *et al* 1997, Lennon *et al* 1998). Binding of (β-D-Glc)$_3$ to the pollen and stylar components recognized by MAb JIM 13 provides evidence that they are AGPs (Fig 3), but apparently the two probes bind different epitopes (Roy *et al* 1998).

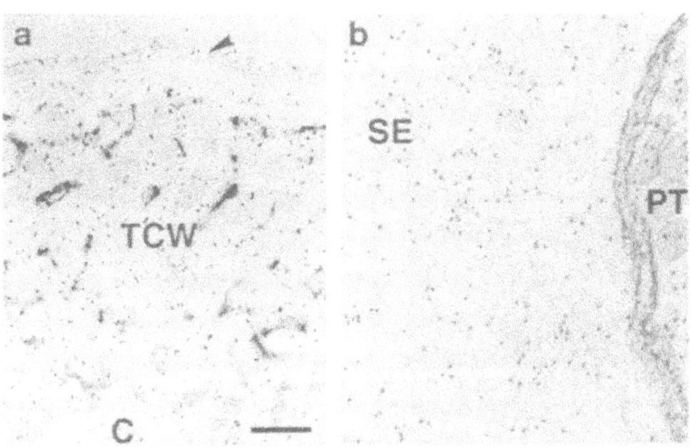

Figure 2. (a) Ultrastructure and immunolocalization of AGPs on lily stylar transmitting tract outer epidermal cell wall. Cells are labelled with MAb JIM 13 conjugated with 18 nm gold. Arrowhead indicates surface of cell where pollen tubes adhere. C, cytoplasm, TCW, transfer cell wall. Reproduced with permission from Jauh and Lord 1996, *Planta* **199**: 251–261, Fig 4C, Springer-Verlag. (b) Ultrastructure and immunolocalization of lily stylar secretions (SE) that surround *in vivo* pollen tubes (PT). Labeling is with MAb JIM 13 conjugated with 18 nm gold. Tissue was high pressure frozen and freeze-substituted, then embedded in Epon Araldite. Bar = 0.5 μm.

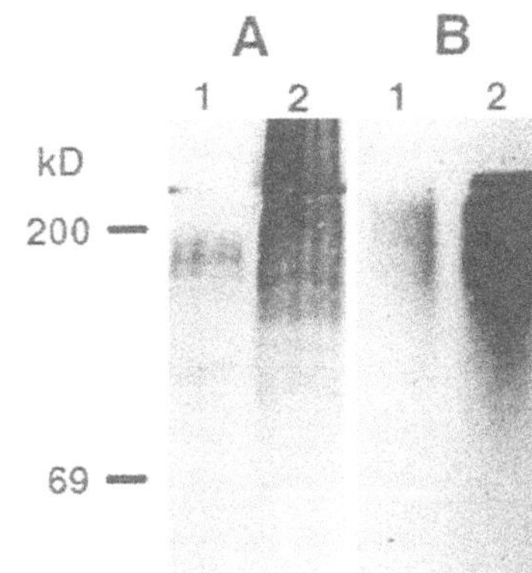

Figure 3. Immunoblot of total proteins from lily styles and *in vitro*-grown pollen tubes. Ten micrograms of total proteins from pollen tubes (*lane 1*) and 20 μg from style (*lane 2*) were fractionated by 10% SDS-PAGE and either immunoblotted and probed with MAb JIM 13 (A) or stained with 50 μM (β-D-Glc)$_3$ (B). Reproduced with permission from Jauh and Lord 1996, *Planta* **199**: 251–261, Fig 5, Springer-Verlag.

Examination of pollen tubes of lily by TEM show further that AGPs and esterified pectins are among the components of the Golgi vesicles secreted at the pollen tube tip (Fig 4) (Jauh and Lord 1996, Roy *et al* 1998). In lily, AGPs localize to the pollen tube cell plasma membrane, to the Golgi vesicles that fuse with the tip and, to a lesser degree, to the cell wall. Early biochemical studies on the vesicle contents of lily show evidence of AGPs as well (Van der Woude *et al* 1971). In tobacco, AGPs localize to the cell wall but are absent at the tip, both from the vesicles in the cytoplasm and from the plasma membrane or wall (Li *et al* 1995). In our model of tip growth in lily we envision the Golgi vesicles, lined with membrane-bound AGPs and containing esterified pectins, fusing at the tip. There the AGPs are exposed on the surface but are bound to the plasma membrane, perhaps by the GPI anchor recently discovered in several membrane-associated AGPs (Schultz *et al* 1998, Svetek *et al* 1999). The esterified pectins are also confined mainly to the tip, de-esterification occurring here and evident in the pectins along the wall back from the tip (Li *et al* 1995, Jauh and Lord 1996, Roy *et al* 1997 1998). The transmitting tract epidermis of the lily style is rich in AGPs and pectins, both of which are candidate adhesion molecules, binding the pollen tubes to the stylar extracellular matrix (ECM) (Park *et al* 2000).

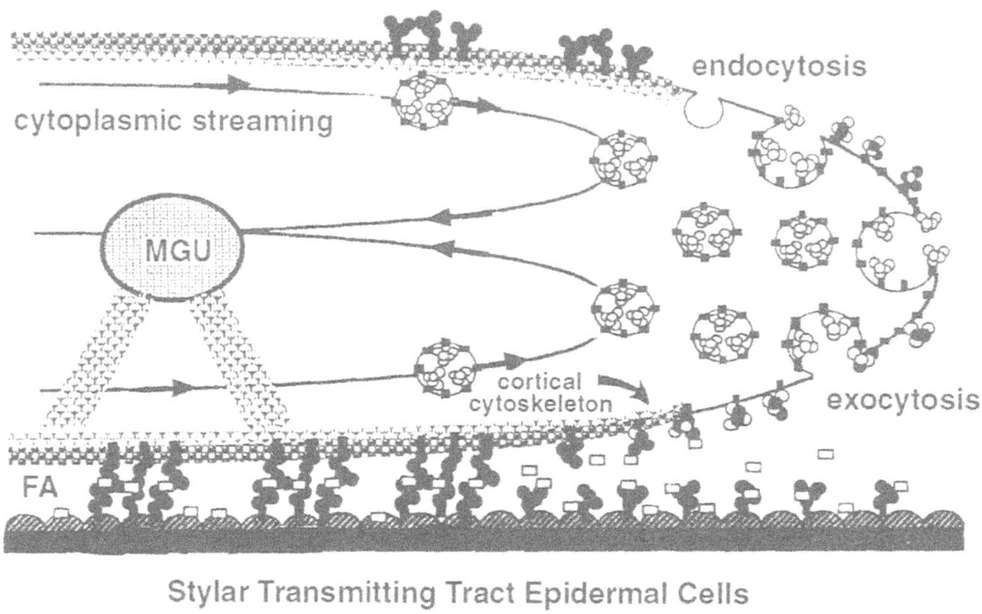

Figure 4. A model for adhesion and cell movement during pollination in lily. The pollen tube tip is illustrated as it progresses along the stylar transmitting tract ECM. The cell wall of the pollen tube is thin at the tip and thicker and more complex back from the tip. Vesicles containing AGPs and esterified pectins fuse at the tip. The mail germ unit (MGU) is composed of the tube cell nucleus and the generative cell (or two sperm). Components are not drawn to scale. FA, focal adhesion; ■ membrane-bound AGP; □ stylar secreted adhesion molecules; ●, unesterified pectins; ○, esterified pectins. Adapted with permission from Lord *et al* 1996, Cell adhesion in plants and its role in pollination, In *Membranes: Specialized Functions in Plants* (Smallwood *et al*, eds), Bios Scientific Publishers, Oxford.

2. THE EFFECT OF YARIV REAGENT [(β-D-GLC)$_3$] ON POLLEN TUBE GROWTH

Serpe and Nothnagel (1994) observed an arrest in growth of rose cells in culture after treatment with (β-D-Glc)$_3$. Adding (β-D-Glc)$_3$ to *in vitro* grown lily pollen tubes causes an arrest of pollen tube growth (Fig 5). Serpe and Nothnagel (1994) observed resumption of growth when (β-D-Glc)$_3$ was removed from the rose cell culture medium, and the same result occurred in lily pollen tubes in culture (Fig 5). Sixty percent of the arrested pollen tubes regenerated new tube tips from the flanks of the arrested tip when (β-D-Glc)$_3$ was removed. These new sites of pollen tube tip growth could be observed in treated pollen after many hours of culture (Fig 6a,b). They are evident due to the red color of the (β-D-Glc)$_3$ reagent which binds both the tube tip and the site of a flanking renewal tip that is also blocked in its emergence. An early effect of (β-D-Glc)$_3$ binding (within minutes) is the production of callose at the extreme tube tip (Fig 6c,d). Normally callose is produced only in the secondary wall back from the tip (Meikle *et al* 1991). The continued secretion of wall materials into the arrested tube tip is evidenced by MAb JIM 13 staining for AGPs (Fig 6e,f). After (β-D-Glc)$_3$ is removed from the medium, an early indicator of renewed tip growth on the flanks of a (β-D-Glc)$_3$-treated pollen tube is secretion of AGPs at the renewal site (Fig 6g,h), as visualized by use of MAb JIM 13 (Mollet *et al* [page 284, this volume]).

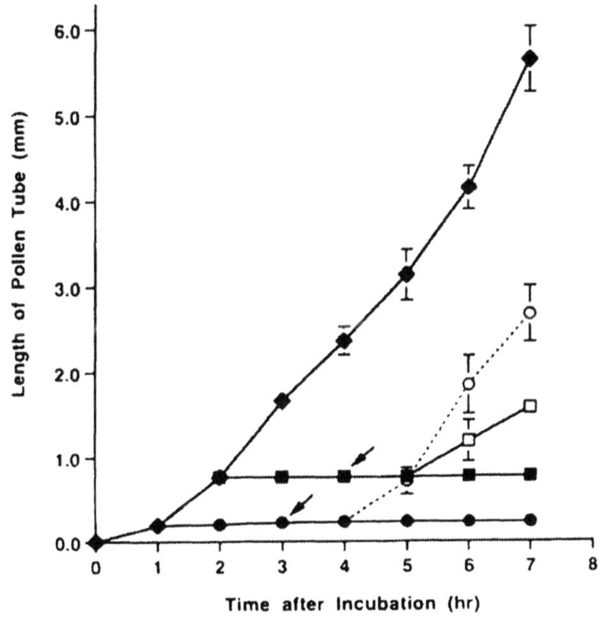

Figure 5. The effects of (β-D-Glc)$_3$ on the growth of lily pollen tubes. *In vitro*-grown pollen tubes were treated with either growth medium only (◆) or 30 μM (β-D-Glc)$_3$ after 1 hour (●) and 2 hours (■) of germination. Growth was arrested in 10 min. Arrows indicate when the (β-D-Glc)$_3$-treated pollen tubes were transferred to fresh growth medium. A total of 62% of those pollen tubes treated with (β-D-Glc)$_3$ after 1 hour in growth medium regenerated (○); for those treated with (β-D-Glc)$_3$ after 2 hours in growth medium, regeneration was 30% (□), and growth rates were diminished. The (β-D-Man)$_3$ Yariv phenylglycoside, which does not bind to AGPs, does not inhibit pollen tube growth (data not shown).

When the (β-D-Glc)$_3$-treated lily pollen tube tip is prepared for TEM observation using plunge freezing methods, a thickened wall or periplasm is evident at the tip (Fig 7). The original wall remains intact, but its expansion is arrested and a new extra-cellular matrix forms beneath it because of continued secretion by the tube cell.

Vesicles merge with the plasma membrane at the tip, even though pollen tube elongation is arrested; the result is wall ingrowths at the sites of secretion. In addition to vesicle fusion, callose depositions occur and a complex periplasm is created that resides outside the plasma membrane at the tip and inside the original pectin wall (Roy *et al* 1998). Use of MAbs to AGPs (Fig 8) and pectins, and a callose probe (Fig 9) reveal the organizational structure of this unusual ECM. In addition, what appears to be an aggregate of the (β-D-Glc)$_3$ (osmiophilic inclusions) can also be seen in these preparations in electron-translucent islands in the expanded periplasm (Figs 8a, b). The AGPs are localized in these islands as well. Labeling of AGPs in the cytoplasm of the tube cell is evident in the Golgi and its secretory vesicles (Fig 8c).

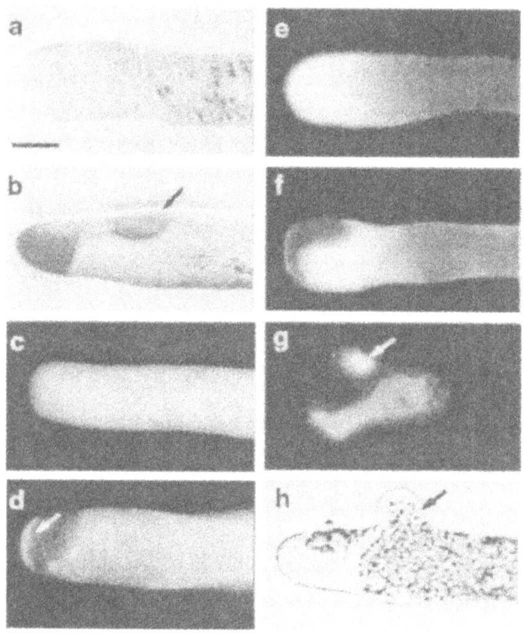

Figure 6. The effect of (β-D-Glc)$_3$ on lily pollen tube morphology (a, b), callose deposition (c, d, stained with aniline blue) and immunolocalization of AGPs by MAb JIM 13 (e, f). Pollen tubes were germinated for 2 hours, then transferred to fresh germination medium for 3 hours (a, c, e) or to medium containing 30 μM (β-D-Glc)$_3$ for 1 h (b, d, f). The (β-D-Glc)$_3$ caused a red wall to accumulate at the tip in 1 hour (b) and callose (arrow) (d) in 5 min after treatment. New centers of secretion on the flanks of the arrested tip are also stained by (β-D-Glc)$_3$ (b) (arrow). The AGP (localization by JIM 13) presages a new center for tip growth on the flanks of the (β-D-Glc)$_3$ inhibited tube tip (arrow) (g). The effects of (β-D-Glc)$_3$ on pollen tube tip growth are reversible. When (β-D-Glc)$_3$ was removed from the germination medium, a new tip (arrow) regenerates from the flanks of the old pollen tube tip (h). Bar = 11 μm.

Callose labeling occurs in these same electron-translucent islands where AGPs reside and at the plasma membrane in the region where the translucent islands appear to initiate (Fig 9). The bulk of the periplasm stains darkly (Figs 7, 8, 9) and contains pectins both esterified and unesterified, as shown by MAb localizations using JIM 5 and JIM 7 (Roy *et al* 1998).

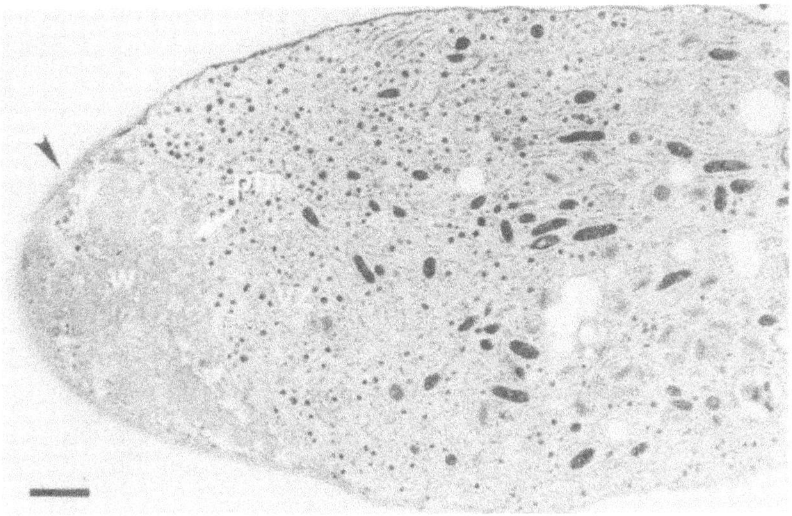

Figure 7. A transmission electron micrograph (TEM) of plunge frozen/freeze-substituted (β-D-Glc)$_3$-treated lily pollen tube. Beneath the original pectin wall (arrow), new wall material accumulates at the tip outside the plasma membrane of the tube cell. pm, plasma membrane; vz, vesicle zone; w, wall or expanded periplasm. Bar = 0.5 µm. Reproduced with permission from Roy *et al* 1998, Effects of Yariv phenylglycoside on cell wall assembly in the lily pollen tube, *Planta* **204**: 450-458, Fig 3, Springer Verlag.

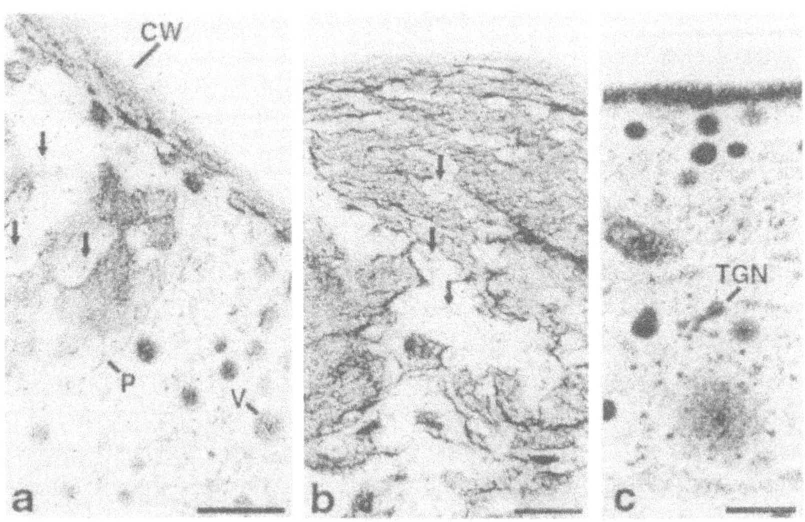

Figure 8. Localization of AGPs in (β-D-Glc)$_3$-treated lily pollen tubes by MAb JIM 13 with 18 nm gold. (a) The AGPs are detected in the secretory vesicles, on the plasmalemma and in the electron-translucent areas within the expanded periplasm. Note that the labeling is associated with the osmiophilic inclusions (arrows). (b) Expanded periplasm at the tip. Note that the labeling is restricted to the electron-translucent areas and associated with the osmiophilic inclusions (arrows). (c) Labeling within the cytoplasm is seen in the *trans*-Golgi network and the secretory vesicles. CW, cell wall; P, plasmalemma; TGN, *trans*-Golgi network; V, secretory vesicle. Bars = 0.4 µm. Reproduced with permission from Roy *et al* 1998, *Planta* **204:** 450-458, Figs 15–17, Springer-Verlag.

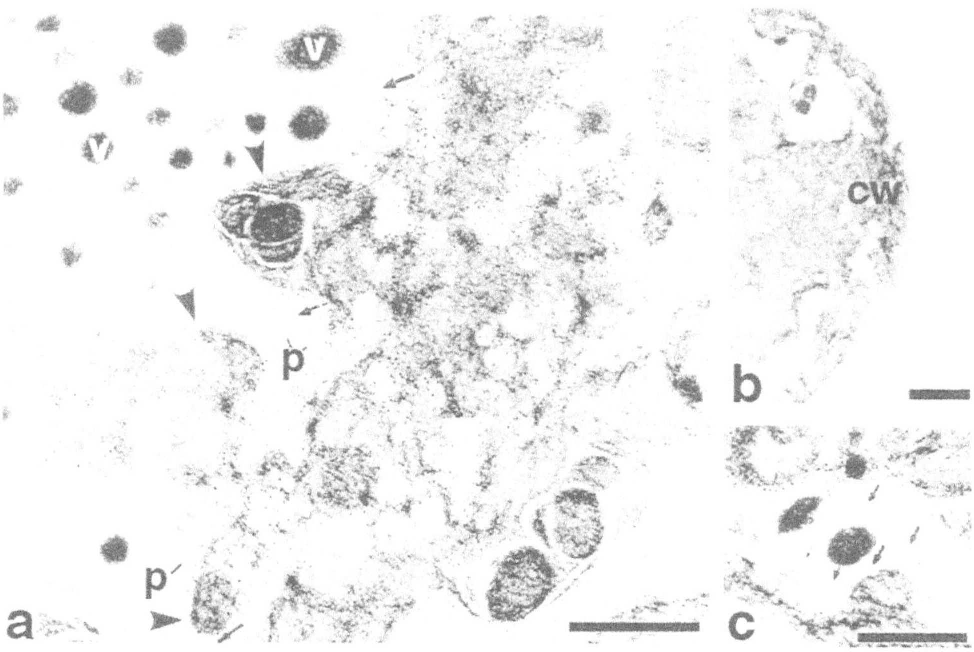

Figure 9. Immunolocalization of callose in (β-D-Glc)$_3$-treated pollen tubes. Polyclonal antibody for β (1→3) glucopyranose (callose) with 18 nm gold. (a) Labeling is limited to the areas of electron-translucent material. Gold particles surround the plasmalemma (arrows) except when fibrillar wall material lines the membrane (arrowheads). (b) Labeling at the pollen tube tip. The callose layer (translucent areas of wall) is not continuous. (c) Labeling is restricted to the outside of the osmiophilic material that surrounds the extruded vesicle content in the periplasm (arrows). cw, cell wall; p, plasmalemma; v, secretory vesicle. Bars = 0.5 μm. Reproduced with permission from Roy *et al* 1998, *Planta* **204**: 450-458, Figs 18–20, Springer-Verlag.

A model depicting the control and (β-D-Glc)$_3$-treated pollen tube tip region is shown in Fig 10. We propose that AGPs are normally involved in the proper alignment of pectins in the wall as the vesicles containing both types of macromolecules fuse with the plasma membrane (Fig 10a). When (β-D-Glc)$_3$ binds AGPs in the wall and plasma membrane of treated pollen tubes, the result is accumulation of the (β-D-Glc)$_3$ plus the newly secreted pectins and AGPs beneath the former pectin wall, along with callose synthesis induced at the plasma membrane near the sites of vesicle fusion (Fig 10b). The vesicle contents of esterified pectins do incorporate into the expanded periplasm and become de-esterified. This pectin incorporation leaves the AGPs, (β-D-Glc)$_3$ and the associated callose in isolated islands within the pectins. It is possible that the (β-D-Glc)$_3$ binds the AGPs in the newly incorporated plasma membrane at the site of vesicle fusion and induces callose synthase at the plasma membrane (Turner *et al* 1998). This hypothesis would explain the coincidence of callose, (β-D-Glc)$_3$ and AGPs in the expanded periplasm. Another possibility is that the (β-D-Glc)$_3$ is acting like an elicitor, and the response is analogous to formation of a papilla in cells challenged by elicitors such as chitin. This hypothesis also implies that elicitors may be binding AGPs in plant–pathogen interactions where papillae form. The ultrastructure of the expanded periplasm in (β-D-Glc)$_3$-treated lily pollen tubes resembles that of the papillae in French bean cell cultures, with islands of callose embedded in electron dense wall material (Brown *et al* 1998). If the tube tip is reacting to (β-D-Glc)$_3$ as these pathogen-challenged cells are reacting to elicitors, then crosslinking of wall materials could also be a component of the reaction. Such crosslinking would explain the quick response that leads to cessation of expansion for the existing pectin wall in pollen tubes arrested by (β-D-Glc)$_3$.

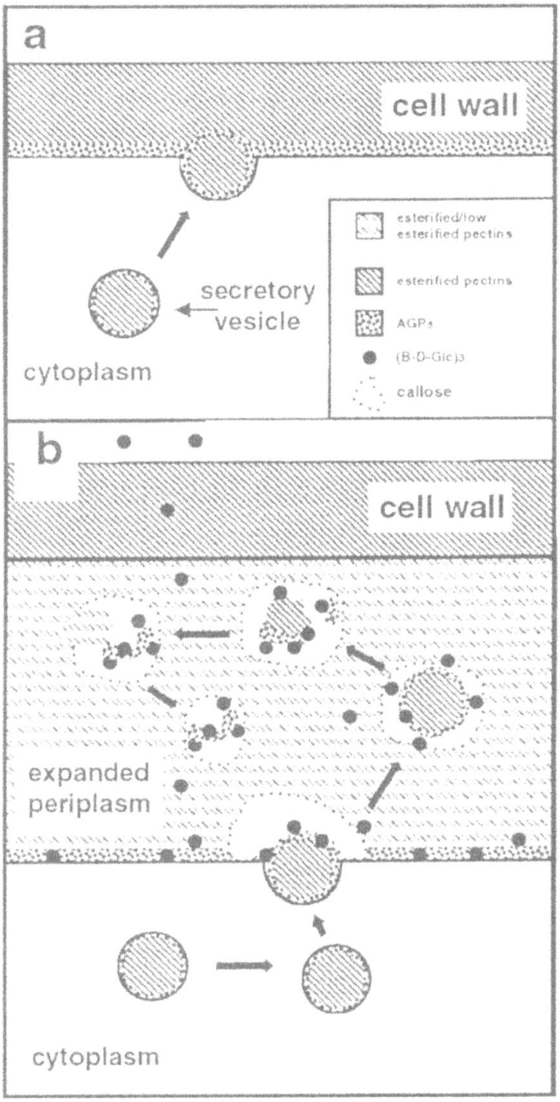

Figure 10. Model of effects of (β-D-Glc)$_3$ on cell wall assembly in lily pollen tube tip. (a) Control pollen tube. The content of the secretory vesicle immediately incorporates in the previously synthesized cell wall. (b) Pollen tube treated with (β-D-Glc)$_3$. As the secretory vesicle opens up, AGPs are exposed to (β-D-Glc)$_3$, which blocks the deposition of pectins in the original cell wall and may also induce callose synthase in the plasmalemma. The periplasm expands as a result of accumulation of wall components. The AGP/(β-D-Glc)$_3$ complexes and callose segregate as islands within the pectin fibrillar materials. Reproduced with permission from Roy *et al* 1998, *Planta* **204**: 450–458, Fig 21, Springer-Verlag.

The presence of AGP secretion at the pollen tube tip in lily coincides with the ability of (β-D-Glc)$_3$ to block tip growth (Jauh and Lord 1996). The (β-D-Glc)$_3$ has no effect on tube growth in tobacco pollen (*Nicotiana tabacum*, Table 1) where no AGPs can be detected at the tip of pollen tubes (Li *et al* 1995). This lack of AGPs at the tube tip and lack of (β-D-Glc)$_3$ effect is also true for pollen tubes of *Lycopersicum esculentum* and *Aquilegia eximia*. Pollen tubes of *Zea maize* and *Annona cherimoya* (a primative dicot in the *Annonaceae*) do secrete AGPs at their tips as evidenced by MAb JIM 13 staining and by (β-D-Glc)$_3$ arresting growth in these two species (Mollet *et al*, page 284, this volume) (Table 1).

Obviously, variation in this trait occurs, but the means of AGP deposition in the plasma membrane and walls of tobacco needs to be explained if the Golgi vesicles fusing with the tip are not the vehicles. Some earlier work, using a MAb to α-L-arabinofuranosyl residues (Harris *et al* 1987), shows localization to young *Nicotiana alata* pollen tube tips, as well as a banded pattern in the wall back from the tip. If this MAb recognizes AGPs, as further work has suggested (Ferguson *et al* 1998), then we have contradictory data on AGP localization in pollen tubes from the Solanaceae.

3. AN AGP–CALCIUM CONNECTION

The finding that (β-D-Glc)$_3$ inhibited lily pollen tube growth prompted us to investigate its effect on both the intracellular Ca^{2+} gradient and the extracellular Ca^{2+} flux. In previous studies, using a variety of different agents including BAPTA buffers, caffeine, elevated osmoticum, and mild thermal shock, we found that pollen tube elongation was inhibited with the concomitant dissipation of the intracellular gradient and the elimination of the extracellular current (Pierson *et al* 1994 1996). The generalization thus emerged that tube growth is tightly coupled to these unique expressions of Ca^{2+}, with the conclusion, in agreement with several other studies, that the intracellular Ca^{2+} gradient facilitates or biases exocytosis, which is essential for elongation in pollen tubes (Pierson *et al* 1994 1996).

At least one major feature distinguishes the mechanism of growth inhibition by (β-D-Glc)$_3$ from the mechanisms of the aforementioned agents, namely that while blocking growth, (β-D-Glc)$_3$ still permits a great amount of exocytosis, in which vesicles are secreted, but in which the contents fail to become properly incorporated into the cell wall. The (β-D-Glc)$_3$ thus uncouples the process of secretion from pollen tube elongation (Roy *et al* 1998).

Studies on pollen tubes loaded with the indicator dye fura-2-dextran show that application of (β-D-Glc)$_3$, rather than dissipating the Ca^{2+} gradient, causes the level of the ion to increase and to spread throughout the apical region of the pollen tube (Roy *et al* 1999). Although these results initially had not been expected, when considered together with the clear observation of extensive exocytosis, they made sense. From the relationship between Ca^{2+} elevation and growth inhibition it became evident that elongation was not abruptly halted, as with BAPTA buffers, caffeine, etc., but slowed, often taking 5-10 min to stop completely. However, elevation of Ca^{2+} can be detected within 1–2 min. The details of the Ca^{2+} elevation, shown in Fig 11, reveal that when treated with (β-D-Glc)$_3$ lily pollen tubes initially exhibit a normal tip focused gradient in which the [Ca^{2+}] is highest immediately adjacent to the plasma membrane at the tip of the tube but then drops off sharply, approaching basal levels within 20 μm back from the apex. However, by 120 s the elevated domain extends for 10 μm, before declining, and by 150 s the [Ca^{2+}] throughout 75 μm of the apical region has risen. Although the extreme tip still exhibits the highest level, now reaching 1 μM, the zone 75 μm from the tip has also increased markedly from 250 nM to 700 nM. During these first 2.5 min, growth still can be detected, but it has slowed considerably. By 4 min in (β-D-Glc)$_3$, growth has stopped completely, and the [Ca^{2+}] is above 800 nM throughout the apical 75 μm. Interestingly the extreme apical region now exhibits slightly lower levels of Ca^{2+} than more distal zones, generating a "reverse" gradient.

In this example the [Ca^{2+}] eventually climbs to 1.5 μM by 5 min; however, we know from studies of several cells over long periods of time that the elevation of [Ca^{2+}] does not continue indefinitely but plateaus, often below 1 μM. While growth has stopped, the clear zone is often extended in length, and cytoplasmic streaming, although still occurring, is considerably reduced in rate. A further noteworthy observation is the appearance of spot regions of elevated [Ca^{2+}]; in cells analyzed for an hour or longer in (β-D-Glc)$_3$, these appear to become the preferred loci for secretion. Taken together, these results provide the first example in which inhibition of pollen tube elongation is not coupled to a dissipation of the tip-focused Ca^{2+} gradient. At the same time these results provide strong support for the idea that spatial regions of elevated Ca^{2+} stimulate or bias the exocytosis of vesicles.

Studies on the extracellular influx of Ca^{2+} similarly show that growth inhibition, induced by application of the (β-D-Glc)$_3$, does not eliminate the current but rather fails to cause a significant change when compared to the control. By contrast, inhibition of tube growth with caffeine generates a significant reduction in the intracellular influx (Fig 12). Although we cannot as yet identify with certainty the source of the large increases in Ca^{2+}, the localization of highest Ca^{2+} to new loci of secretion indicates that Ca^{2+} enters the cytoplasm across the plasma membrane. An increased area of membrane in which Ca^{2+} influx channels are open or in which reduced efflux occurs via Ca^{2+}-ATPases could account for the elevated Ca^{2+} levels in the cytoplasm.

In summary, the results are clear and unambiguous; application of (β-D-Glc)$_3$, which inhibits elongation, causes pollen tubes to generate elevated levels of intracellular Ca^{2+} that extend for considerable distances back from the tip. In addition, treated cells maintain normal levels of inward Ca^{2+} current. Finally, the highest domains of intracellular Ca^{2+} correlate closely with the regions of preferred vesicle secretion. These observations thus strengthen the conclusion that the (β-D-Glc)$_3$ uncouples secretion from pollen tube elongation.

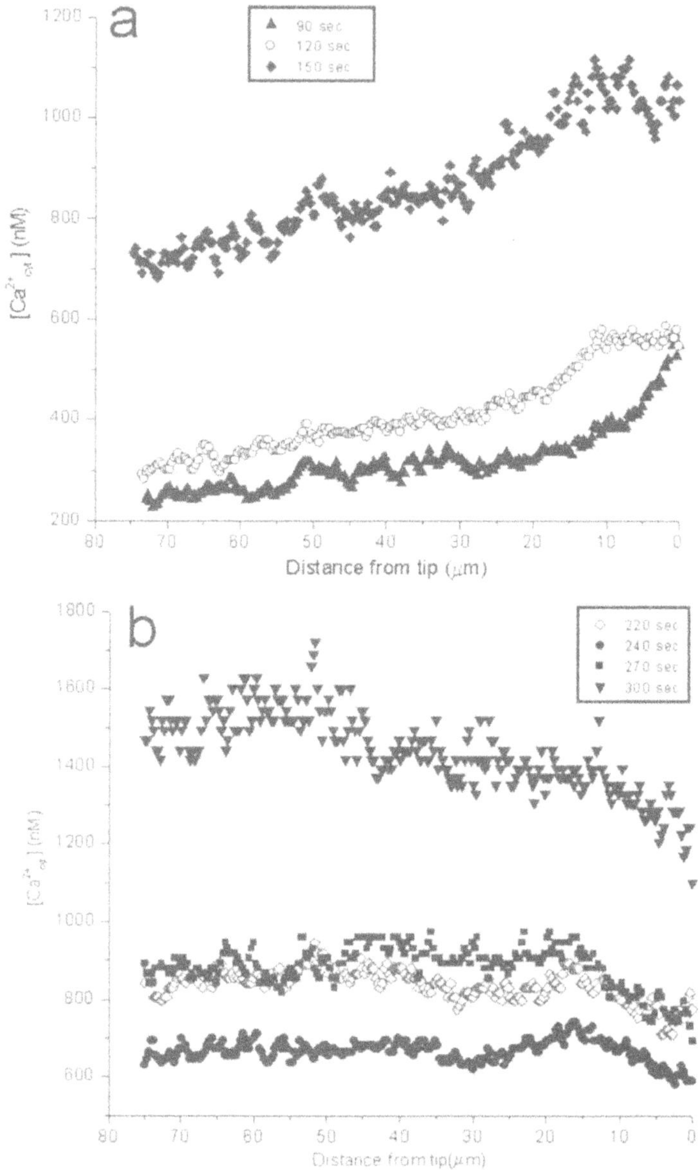

Figure 11. Line scan plots of cytoplasmic [Ca^{2+}] through the median part of the pollen tube tip. (a) At 90 s after (β-D-Glc)$_3$ treatment, there is a steep tip-focused [Ca^{2+}] gradient (▲). At 120 s after (β-D-Glc)$_3$ treatment the region of high [Ca^{2+}] has spread into the dome of the pollen tube tip, and [Ca^{2+}] back from the tip has increased slightly (○). At 150 s after (β-D-Glc)$_3$ treatment [Ca^{2+}] has increased throughout the pollen tube, the region of high [Ca^{2+}] has extended even further back from the tip, and there is a global increase of [Ca^{2+}] (◆). (b) Selected line scans showing fluctuations in [Ca^{2+}]. Profile of [Ca^{2+}] 220 s (◊), 240 s (●), 270 s (■) and 300 s (▼) after addition of (β-D-Glc)$_3$. Reproduced with permission from Roy *et al* 1999, Uncoupling secretion and tip growth in lily pollen tubes: evidence for the role of calcium in exocytosis, *Plant J.* **19**: 379-386, Fig 2, Blackwell Science.

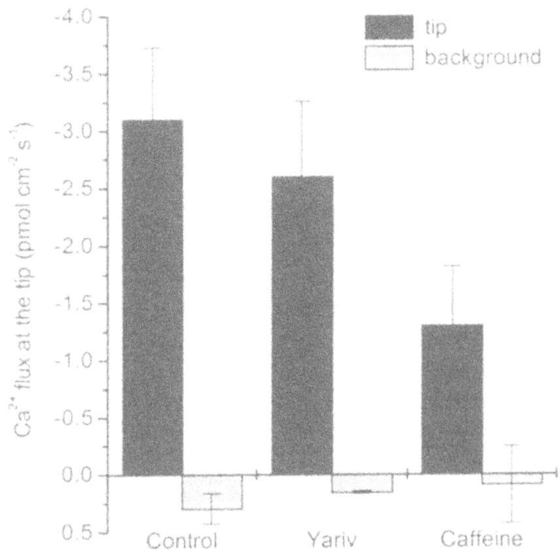

Figure 12. Extracellular Ca^{2+} influx at the tip of pollen tubes treated with 30 μM (β-D-Glc)$_3$, 3 mM caffeine or washed in growth medium (control). Pollen tubes treated with (β-D-Glc)$_3$ maintain Ca^{2+} influx at the tip not significantly different from the influx in control tubes, but significantly different from the extracellular Ca^{2+} flux in caffeine-treated tubes. Reproduced with permission from Roy *et al* 1999, *Plant J.* **19**: 379-386, Fig 4, Blackwell Science.

In attempting to decipher the underlying causes and explanation for the observations, what stands foremost in our minds is the ability of (β-D-Glc)$_3$ to bind and, by some mechanism, to prevent the normal activity of the AGPs that may be intimately involved with cell wall deposition, and with cell growth and development. Of particular importance, Serpe and Nothnagel (1994) have shown that the (β-D-Glc)$_3$ cross-links AGPs in the plasma membrane and decreases the mobility of probes associated with membrane proteins and glycoconjugates. It seems possible that reduced mobility of proteins associated with the plasma membrane might locally increase tension, which in turn could open stretch-activated ion channels. Influx through these channels would lead to the observed increase in intracellular [Ca^{2+}], which would then stimulate exocytosis. Although we do not observe an increase in extracellular Ca^{2+} influx in (β-D-Glc)$_3$-treated pollen tubes, if (β-D-Glc)$_3$ somehow interacts with the Ca^{2+} influx channels to increase the area of membrane through which the flux occurs, as indeed seems to be the case, this would result in more Ca^{2+} ions entering the cytoplasm than in control tubes. Further, we have calculated that the Ca^{2+} influx measured across the membrane at the growing pollen tube tip is 10- to 20-fold greater than that needed to drive the observed Ca^{2+} gradient (Holdaway-Clarke *et al* 1997), so even a doubling of the influx across the membrane may not result in an observable increase in total extracellular Ca^{2+} influx. Thus to explain the (β-D-Glc)$_3$-induced Ca^{2+} changes, it seems most likely that the fraction of AGPs associated with the plasma membrane will occupy a pivotal role. It is less clear if and how these local activities account for the failure of the vesicle contents to be incorporated into the cell wall proper (Roy *et al* 1998).

REFERENCES

Anderson, M. A., Harris, P. J., Bonig, I., and Clarke, A. E., 1987, Immuno-gold localization of α-L-arabinofuranosyl residues in pollen tubes of *Nicotiana alata* Link et Otto, *Planta* **171**: 438–442.

Brown, I., Trethowan, J., Kerry, M., Mansfield, J., and Bolwell, G. P., 1998, Localization of components of the oxidative cross-linking of glycoproteins and of callose synthesis in papillae formed during the interaction between non-pathogenic strains of *Xanthomonas campestris* and French bean mesophyll cells, *Plant J.* **15**: 333–343.

Clarke, A. E., Anderson, R. L., and Stone, B. A., 1979, Form and function of arabinogalactans and arabinogalactan-proteins, *Phytochemistry* **18**: 521–540.

Du, H., Simpson, R. J., Moritz, R. L., Clarke, A. E., and Bacic, A., 1994, Isolation of the protein backbone of an arabinogalactan-protein from the styles of *Nicotiana alata* and characterization of a corresponding cDNA, *Plant Cell* **6**: 1643–1653.

Ferguson, C., Teeri, T. T., Siika-aho, M., Read, S. M., and Bacic, A., 1998, Location of cellulose and callose in pollen tubes and grains of *Nicotiana tabacum, Planta* **206**: 452–460.

Fincher, G. B., Stone, B. A., and Clarke, A. E., 1983, Arabinogalactan-proteins; structure, biosynthesis, and function, *Annu. Rev. Plant Physiol.* **34**: 47–70.

Gleeson, P. A., and Clarke, A. E., 1979, Structural studies on the major component of *Gladiolus* style mucilage, an arabinogalactan-protein, *Biochem. J.* **181**: 607–621.

Gleeson, P. A., and Clarke, A. E., 1980, Arabinogalactans of sexual and somatic tissues of *Gladiolus* and *Lilium*, *Phytochemistry* **19**: 1777–1782.

Harris, P. J., Freed, K. Anderson, M. A., Weinhandl, J. A., and Clarke, A. E., 1987, An enzyme-linked immunosorbent assay (ELISA) for *in vitro* pollen growth based on binding of a monoclonal antibody to the pollen-tube surface, *Plant Physiol.* **84**: 851–855.

Holdaway-Clarke, T. L., Feijó, J, A., Hackett, G. R., and Hepler, P. K., 1997, Pollen tube growth and the intracellular cytosolic calcium gradient oscillate in phase while extracellular calcium influx is delayed. *Plant Cell* **9**: 1999–2010.

Jauh, G. Y., and Lord, E. M., 1996, Localization of pectins and arabinogalactan-proteins in lily (*Lilium longiflorum* L.) pollen tube and style, and their possible roles in pollination, *Planta* **199**: 251–261.

Knox, J.P., 1992, Molecular probes for the plant cell surface, *Protoplasma* **167**: 1–9.

Lennon, K. A., Roy, R., Hepler, P. K., and Lord, E. M., 1998, The structure of the transmitting tissue of *Arabidopsis thaliana* (L.) and the path of pollen tube growth, *Sex Plant Reprod.* **11**: 49–59.

Li, Y-Q., Bruun, L., Pierson, E., and Cresti, M., 1992, Periodic deposition of arabinogalactan epitopes in the cell wall of pollen tubes of *Nicotiana tabacum* L., *Planta* **188**: 532–538.

Li, Y-Q., Faleri, C., Geitmann, A., Zhang, H-Q., and Cresti, M., 1995, Immunogold localization of arabinogalactan proteins, unesterified and esterified pectins in pollen grains and pollen tubes of *Nicotiana tabacum* L. *Protoplasma* **189**: 26–36.

Lord, E. M., Walling, L. L., and Jauh, G. Y., 1996, Cell adhesion in plants and its role in pollination, In *Membranes: Specialized Functions in Plants* (M. Smallwood, J. P. Knox, and D. J. Bowles, eds.), Bios Scientific, Oxford, pp. 21–38.

Mau, S-L., Chen, C-G., Pu, Z-Y., Moritz, R. L., Simpson, R. J., Bacic, A., and Clarke, A. E., 1995, Molecular cloning of cDNAs encoding the protein backbones of arabinogalactan-proteins from the filtrate of suspension-cultured cells of *Pyrus communis* and *Nicotiana alata*, *Plant J.* **8**: 269–281.

Meikle, P. J., Bonig, I., Hoogenraad, N. J., Clarke, A. E., and Stone, B. A., 1991, The location of (1→3)-β-glucans in the walls of pollen tubes of *Nicotiana alata* using a (1→3)-β-glucan-specific monoclonal antibody, *Planta* **185**: 1–8.

Miki-Hirosige, H., Hoek, I. H. S., and Nakamura S., 1987, Secretions from the pistil of *Lilium longiflorum*, *Am. J. Bot.* **74**: 1709–1715.

Nothnagel, E. A., 1997, Proteoglycans and related components in plant cells, *Int. Rev. Cytol.* **174**: 195–291.

Park, S. Y., Jauh, G-Y., Mollet, J-C., Eckard, K. J., Nothnagel, E. A., Walling, L. L., and Lord, E. M., 2000, A lipid transfer-like protein is necessary for lily pollen tube adhesion to an in vitro stylar matrix, *Plant Cell* **12**: 151-163.

Pennell, R. I., Janniche, L., Kjellbom, P., Scofield, G. N., Peart, J. M., and Roberts, K., 1991, Developmental regulation of a plasma membrane arabinogalactan protein epitope in oilseed rape flowers, *Plant Cell* **3**: 1317–1326.

Pierson, E. S., Miller, D. D., Callaham, D. A., Shipley, A. M., Rivers, B. A., Cresti, M., and Hepler, P. K., 1994, Pollen tube growth is coupled to the extracellular calcium ion flux and the intracellular calcium gradient: Effect of BAPTA-type buffers and hypertonic media, *Plant Cell* **6**: 1815–1828.

Pierson, E. S., Miller, D. D., Callaham, D. A., van Aken, J., Hackett, G., and Hepler, P. K., 1996, Tip-localized calcium entry fluctuates during pollen tube growth, *Dev. Biol.* **174**: 160–173.

Roy, S., Eckard, K. J., Lancelle, S., Hepler, P. K., and Lord, E. M., 1997, High-pressure freezing improves the ultrastructural preservation of *in vivo* grown lily pollen tubes, *Protoplasma* **200**: 87–98.

Roy, S., Jauh, G. Y., Hepler, P. K., and Lord, E. M., 1998, Effects of Yariv phenylglycoside on cell wall assembly in the lily pollen tube, *Planta* **204**: 450–458.

Roy, S. J., Holdaway-Clarke, T. L., Hackett, G. R., Kunkel, J. G., Lord, E. M., and Hepler, P. K., 1999, Uncoupling secretion and tip growth in lily pollen tubes: evidence for the role of calcium in exocytosis. *Plant J.* **19**: 379-386.

Schultz, C., Gilson, P., Oxley, D., Youl, J., and Bacic, A., 1998, GPI-anchors on arabinogalactan-proteins: implications for signalling in plants, *Trends Plant Sci.* **3**: 426–431.

Serpe, M. D., and Nothnagel, E. A., 1994, Effects of Yariv phenylglycosides on *Rosa* cell suspensions: evidence for the involvement of arabinogalactan-proteins in cell proliferation, *Planta* **193**: 542–550.

Svetek, J., Yadav, M. P., and Nothnagel, E. A., 1999, Presence of a glycosylphosphatidylinositol lipid anchor on rose arabinogalactan-proteins, *J. Biol. Chem.* **274**: 14724–14733.

Turner, A., Bacic, A., Harris, P. J., and Read, S. M., 1998, Membrane fractionation and enrichment of callose synthase from pollen tubes of *Nicotiana alata* Link et Otto, *Planta* **205**: 380–388.

Van der Woude, W. J., Morré, D. J., and Bracker, C. E., 1971, Isolation and characterization of secretory vesicles in germinated pollen of *Lilium longiflorum*, *J. Cell Sci.* **8**: 331–351.

Yariv J., Rapport, M. M., and Graf, L., 1962, The interaction of glycosides and saccharides with antibody to the corresponding phenylazo glycosides, *Biochem J.* **85**: 383–388.

Youl, J. J., Bacic, A., and Oxley, D., 1998, Arabinogalactan-proteins from *Nicotiana alata* and *Pyrus communis* contain glycosylphosphatidylinositol membrane anchors, *Proc. Natl. Acad. Sci. USA* **95**: 7921–7926.

Chapter 15

Arabinogalactan-Proteins, Place-Dependent Suppression and Plant Morphogenesis

Dominick V. Basile, Margaret R. Basile, and Madeline M. Mignone
Department of Biological Sciences, Lehman College, City University of New York, Bronx, NY 10468

1. INTRODUCTION

Experiments in which several antagonists of the synthesis of Hyp-containing proteins (Hyp-proteins) induced profound changes in the morphogenetic patterns exhibited by five different species of leafy liverworts led to a hypothesis that some of the Hyp-proteins had a morphoregulatory function (Basile 1980, Basile and Basile 1982). The class to which the morphoregulatory Hyp-proteins belonged could not be known from the methods used. Two main classes of Hyp-proteins had been distinguished in plants by the time the series of experiments was completed. One class was referred to as "extensins" (Lamport 1963) and the other as arabinogalactan-proteins (AGPs) (Clarke *et al* 1979). Although there is no direct evidence to exclude "extensins", available evidence obtained in subsequent studies suggested that the type of Hyp-protein modified would be characterized as AGPs (Basile *et al* 1986). What follows are some examples of how by antagonizing Hyp-protein synthesis at different stages of plant ontogeny, three separate aspects of plant morphogenesis are modified. In each case the experimentally induced changes in plant morphogenesis could be correlated with quantitative and/or qualitative differences in AGPs.

2. TRANSITION FROM DIFFUSE TO APICAL GROWTH

The morphogenetic change that is influenced by Hyp-protein antagonists earliest in plant ontogeny is the transition from diffuse, non-polar growth to localized, polar (apical) growth. This pattern change is characteristic of the early ontogeny of almost all land plants (Fig 1). It is a pattern change that helps distinguish vegetative development of all land plants from that of most photoautotrophs. Depending on the kind of plant, the pattern change that results in the initiation of apical meristems can occur at different times after the initiation of development from a spore or fertilized egg. In many, if not most, mosses the transition from branching uniseriate filaments (protonemata) to

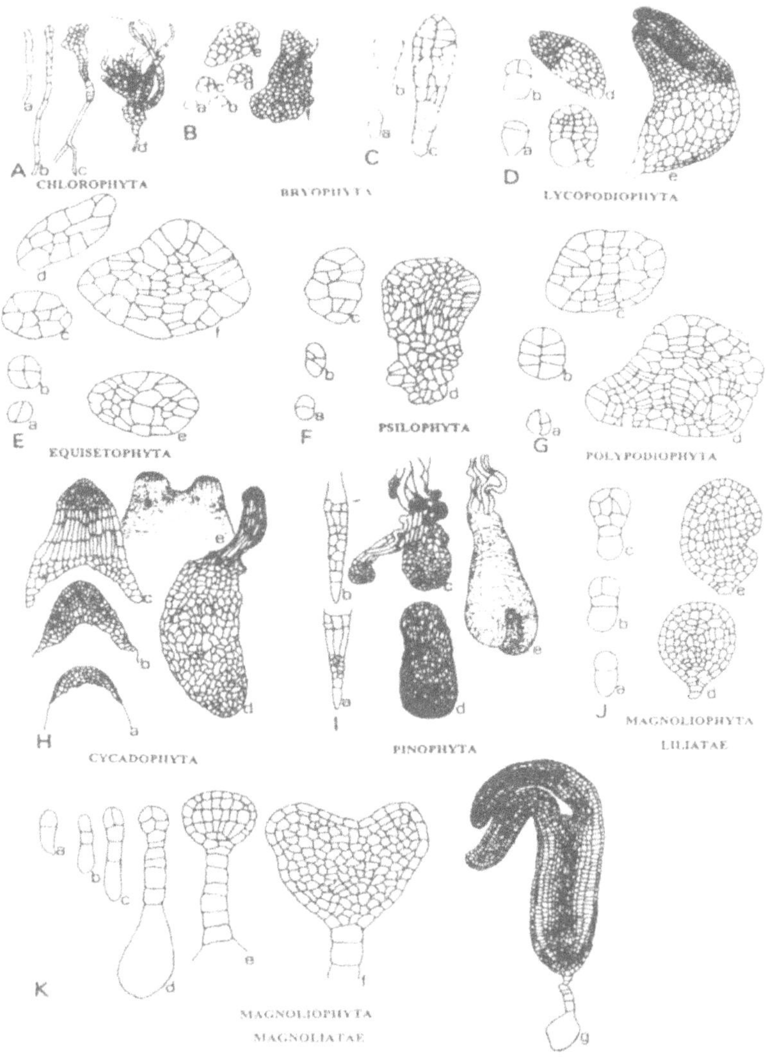

Figure 1. Illustrations showing the early stages of development of a chlorophyte (A) and representatives of all extant phyla of embryophytes (B–K). The degree to which development proceeds before the establishment of apical meristems differs in different phyla of embryophytes. Until then, all cells of the developing plants are essentially meristematic. A. *Fritschiella tuberosa,* like most chlorophytes, does not undergo the transition from non-apical to apical meristematic growth. B. *Scapania nemorosa*. C. *Conocephalum conicum*. D. *Lycpodium sp*. E. *Equisetum debile*. F. *Tmesipteris tannensis*. G. *Gymnogramme sulphurea*. H. a–c *Zamia sp*.; d–e *Macrozamia sp*. I. *Torreya californica*. J. *Luzula fosteri*. K. *Capsella bursa-pastoris*. A, C–K redrawn from Wardlaw (1955). B from Basile (1990). Drawings not to scale.

parenchymatous leafy shoots (Fig 2) is typically delayed. This delay was found to be significantly longer if any of three Hyp-protein antagonists were present in axenic cultures of *Physcomitrella patens* (Mignone and Basile [this volume]). Although this observation is neither evidence that a synthesis and/or deposition of a Hyp-protein was required to bring about this transition, nor evidence that the kind of morphoregulatory Hyp-protein was an AGP, such an explanation is consistent with the results of related studies. The results of two other studies provided evidence to support the idea that AGPs help mediate this phase change. In one study, the amount of AGPs extractable from the protonemal cells was approximately 10% of that from leafy shoots (Fig 3). In another study, the AGPs from leafy shoots had electrophoretic mobilities that differed from the mobilities of the AGPs from protonemal cells (Fig 4). Taken together, these observations suggest that certain AGPs must become and remain associated with protonemal cells to bring about the transition from diffuse to polar, apical growth. They further suggest that certain AGPs must remain in association with cells comprising the leafy shoots if the pattern of polar, apical growth is to be maintained.

Figure 2. Two stages in the vegetative development of a moss, *Physcomitrella patens* Hedwig, gametophyte. A. A portion of a protonema. B. A leafy shoot that developed from a protonema after the establishment of an apical meristem.

From research conducted in our laboratory, the only data to support a hypothesized role for AGPs in mediating the change from diffuse to apical growth was obtained from experiments conducted with bryophytes. Nevertheless, results from experiments conducted by others with flowering plants support our hypothesis. In a study in which a monoclonal antibody (JIM8) directed against an AGP epitope was used, Pennell *et al* (1992) found that the epitope it recognized was only expressed in cultured carrot cells that were capable of undergoing somatic embryogenesis. Because the epitope was not detected in the cells when they progressed beyond the proembryo stage, it did not appear essential to maintaining an embryogenic pattern once begun. In another study, Kreuger and van Holst (1993) reported that AGPs extracted from carrot seeds could cause non-embryogenic cell lines to become embryogenic. Subsequently, an epitope of an AGP immunoselected from the mixture of carrot seed AGPs by the monoclonal antibody ZUM18 was found capable of increasing embryonic potential in both carrot and cyclamen cell cultures (Kreuger and van Holst 1995, Kreuger *et al* 1995). A reasonable interpretation of these results is that the non-embryogenic cell lines manifest a diffuse, non-polar pattern of development because their cell surfaces are deficient in AGPs that possess certain epitopes. From this hypothesis it follows that the AGPs derived from embryogenic cell lines are serving as an exogenous replacement for a failed endogenous source of these epitopes. Whether this interpretation is correct or reasonable remains to be determined.

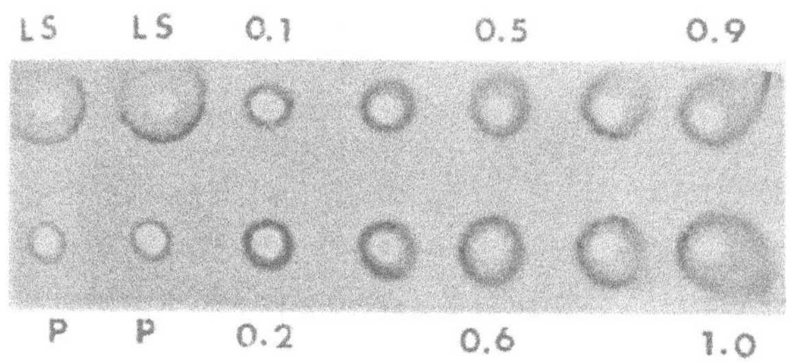

Figure 3. A comparison of the AGP content of 1 g freeze-dried weight of protonema (P) versus leafy shoots (LS) of *P. patens* by means of a radial diffusion assay (van Holst and Clarke 1985). Numbered wells correspond to concentrations of gum arabic in mg/ml. Leafy shoots contain approximately 10-fold more AGPs than do protonemata.

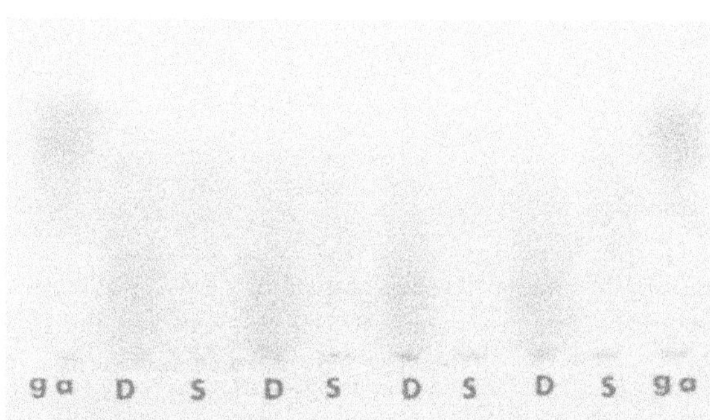

Figure 4. Results of comparative electrophoresis (Basile *et al* 1989) of AGPs extracted from protonemata (D) versus leafy shoots (S) of a moss, *P. patens*. Equal amounts of AGPs were loaded in each well. The AGPs from the different sources differ with respect to both the intensity of the staining with ß-glucosyl Yariv reagent and the relative migration of the more intensely stained portions. ga, gum arabic.

3. LOCAL SUPPRESSION/DESUPPRESSION OF CELL PROLIFERATION AND CELL ENLARGEMENT

Experiments with Hyp-protein antagonists indicate that AGPs may mediate two other aspects of plant morphogenesis/organogenesis. They first became apparent when plants were treated at a stage of ontogeny closely following the onset of organ initiation (Basile 1967). Furthermore, they only became apparent in plants that were deemed to be the products of "reductive" or "regressive" evolution. In other words, the form of the experimental plants was considered to be "derived" because of an "evolutionarily reduction or loss" in the kind and/or number of their parts (Stebbins and Basile 1986).

The five species of leafy liverworts with which our original experiments were conducted were chosen because they met these criteria. Each of the species was believed to have lost the capacity to form leaves in a ventral position. In addition, these same five species were presumed to have lost the capacity to initiate branches from all but one of the possible 11 sites of origin described by Crandall (1969). All five species, when treated with Hyp-protein antagonists, developed ventral leaves and more than one of the 11 branch types, and these much more frequently (Basile and Basile 1982). Clearly, the morphogenetic capacity to form leaves and/or branches in specific positions had not been "lost" by these plants, just "suppressed".

We think that the most parsimonious explanation for these results is that some of the Hyp-proteins that were antagonized were involved in locally suppressing organ development. Assuming that this explanation is correct, the question becomes: What Hyp-protein? Since the time one of us (DVB) read the early papers authored by Adrienne Clarke and coworkers (Clarke *et al* 1978 1979, Fincher *et al* 1983 *inter alia*), research in our lab has focused on AGPs as being the Hyp-proteins most likely to be morphoregulatory (mAGPs). Specifically, the aspects of morphoregulation we think are mediated by mAGPs are the suppression/desuppression of cell proliferation and cell enlargement in localized cell populations.

Thus far, the best evidence to implicate mAGPs in the suppression/desuppression of localized leaf and branch development comes from research conducted with a leafy liverwort, *Gymnocolea inflata*. Because of its response to the presence of ammonium ion in its culture medium, *G. inflata* was chosen for an investigation of whether changes in AGPs could be correlated with changes in the pattern of morphogenesis (Basile and Basile 1980). Of the five species of liverworts used in our experiments, only *G. inflata* responds to ammonium ion by exhibiting the same pattern changes as with Hyp-protein antagonists. Figures 5–8 provide examples of *G. inflata* cultured in the absence and presence of ammonium ion. The advantage of using the ammonium ion is that the induced metabolic and concomitant morphogenetic effects last much longer than with the Hyp-protein antagonists. Consequently, it has been possible to compare the AGPs synthesized by *G. inflata* plants exhibiting the morphology illustrated in Fig 5 with AGPs synthesized by plants exhibiting the morphology shown in Figs 7 and 8. For convenience, the plants exhibiting the morphology shown in Fig 5 are referred to as being suppressed and those shown in Figs 7 and 8 as desuppressed. Figure 6 shows a plant that had undergone a transition from suppressed to desuppressed. It illustrates the form change that is brought about by replacing an ammonium-free culture medium with one that contains ammonium ion part way through the incubation period.

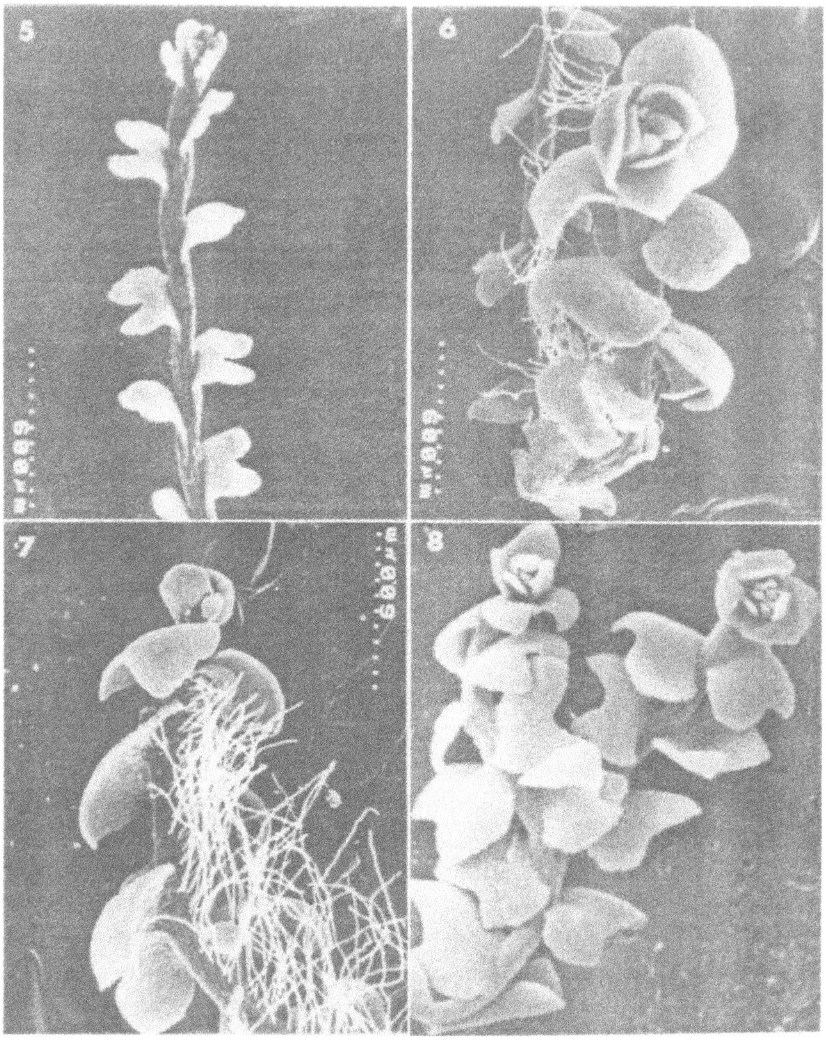

Figures 5–8. SEMs showing some of the differences in the size, shape and distribution of leaves of *G. inflata* plants cultured in the absence or presence of ammonium ion. 5. Portion of plant cultured on ammonium-free medium and exhibiting suppressed leaf and branch development. 6. A plant showing an abrupt transition from suppressed to desuppressed leaf development correlated with an exchange of an ammonium-free with an ammonium ion-containing culture medium. 7. Portion of plant cultured solely on ammonium ion-containing culture medium showing desuppressed leaf development. 8. Portion of plant cultured solely on ammonium ion-containing culture medium showing desuppressed branch and leaf development.

When AGPs released into the culture medium by suppressed and desuppressed plants were compared by electrophoresis (Basile *et al* 1989), their electrophoretic patterns were distinctly different (Fig 9). These results were interpreted to be support for the idea that ammonium ion was somehow able to influence AGP synthesis in this species and that the altered synthesis is somehow related to the observed differences in the forms that developed. Because the AGPs do not migrate as distinct bands, comparative electrophoresis did not provide a means of identifying AGPs that might be mediating the observed differences in morphogenetic pattern. Another method used to compare AGPs, cesium chloride gradient analysis, proved to be somewhat more informative.

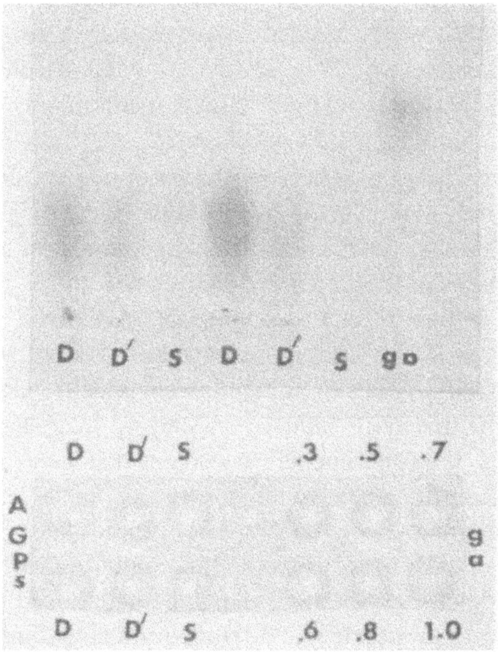

Figure 9. The upper portion shows results of a comparative electrophoresis of AGPs extracted from media in which *G. inflata* plants were cultured solely on ammonium-free medium (S), solely on ammonium-containing medium (D) or after the ammonium-free medium was replaced with ammonium-containing medium (D′). The lower portion shows the results of the radial diffusion assay (van Holst and Clarke 1985) used to determine the relative concentration of AGPs being compared. Numbered wells correspond to concentrations of gum arabic (ga) in mg/ml. Note that although the concentrations of AGPs being compared were similar, the capacity to bind ß-glucosyl Yariv reagent after electrophoresis was distinctly different, as was the relative migration of the more darkly staining regions. These differences are similar to those observed when AGPs from moss protonemata and leafy shoots were compared (Fig 4).

When AGPs extracted from suppressed plants were compared with those from desuppressed plants by CsCl density gradient analysis, the range in buoyant densities (BDs) was distinctly different (Basile and Basile 1993). Specifically, there was a wider range of BDs of the AGPs in the extracts from suppressed plants (Fig 10). Just the opposite was observed when AGPs extracted from the culture medium of the two morphotypes were compared (Fig 10). Instead of the wider range in BD of AGPs being associated with the suppressed plants, it was associated with those that were desuppressed. Our interpretation of these data is that when suppressed organ development was desuppressed, mAGPs that act to suppress organogenesis locally if they remained *in situ* were somehow prevented from doing so and consequently released into the culture medium. In an attempt to identify specific AGPs that mediated suppression in *G. inflata*, we undertook to develop antibodies that were monospecific for AGPs that are released into the culture medium when culture conditions were changed from those that supported suppressed organogenesis to those that promoted desuppressed organogenesis. With the collaboration of Keith Roberts and coworkers at The John Innes Institute, we succeeded in developing a panel of five monoclonal antibodies (MAbs) (Basile *et al* 1999). The MAbs were selected on the basis of their capacity to bind AGPs that appeared in the culture medium only after culture conditions were changed to induce desuppressed development of organs in plants in which they were normally suppressed (Fig 6). These MAbs are designated John Innes Monoclonals (JIMs) 101–105. A sixth monoclonal, JIM 106, detected AGPs in the culture media from both suppressed and desuppressed plants. An example of a dot-blot assay used to compare binding affinities of this panel of MAbs is given in Fig 11. These MAbs were determined to belong to two different isotypes: JIMs 101, 103, 105 and106 are IgMs, and JIMs 102 and 104 are IgGs. Development of this panel of MAbs provided the first clear evidence that specific AGPs are altered when specific changes in the pattern of morphogenesis is experimentally induced in one of the plant species used in our research. Unfortunately, the experimental procedures that were used should result in the altered synthesis, function and "behavior" of any or all AGPs. It is not surprising, therefore, that our attempts to use JIMs 101–105 to desuppress leaf and branch development by adding each of them to *G. inflata* culture medium were unsuccessful. Ventral leaves and branch development remained suppressed. It is possible that the plant cell surfaces were impermeable to the immunoglobins. Perhaps the use of fragments (light chains) of the MAbs and/or the AGPs immunoselected by them will yield more positive results. At this time, however, there is no direct evidence that any one of the AGPs specified by our panel of MAbs acts to suppress locally any aspect of plant growth and development. Fortunately, it is not necessary for us to have isolated the AGPs responsible for mediating growth suppression for us to adhere to our hypothesis that certain mAGPs play a pivotal role in the temporal and spatial regulation of growth suppression. We can think of no other explanation for the fact that whenever we do something to prevent growth suppression from occurring at a specific time or place during the course of vegetative development, plants released AGPs with different physical/chemical characteristics into their culture medium. Because the methods that have been employed can result in the altered synthesis of any AGP and do not target specific AGPs, it has proven difficult to distinguish the altered morphoregulatory AGPs from the rest. What seems to be required is a way of getting those AGPs that function in growth suppression to be released without impairing their structure and function. Antibodies that specify these unimpaired AGPs should prove to be potent investigative aids. Their development would help us to proceed much further in the quest to identify and characterize the mAGP(s) suspected of regulating aspects of plant morphogenesis by virtue of its/their capacity to suppress locally cell proliferation and/or cell enlargement.

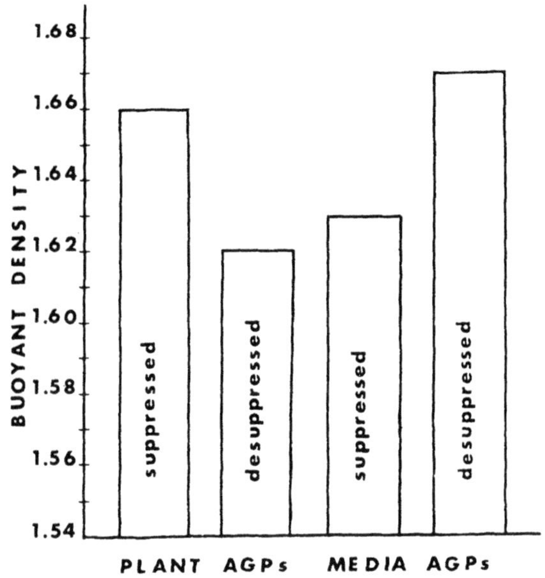

Figure 10. A comparison of the mean buoyant densities of AGPs extracted from *G. inflata* plants and the media in which they were cultured axenically. An analysis of variance based on six experiments indicated that the differences in AGP buoyant densities between suppressed and desuppressed plants and between the media in which they were growing was significant at 95% probability.

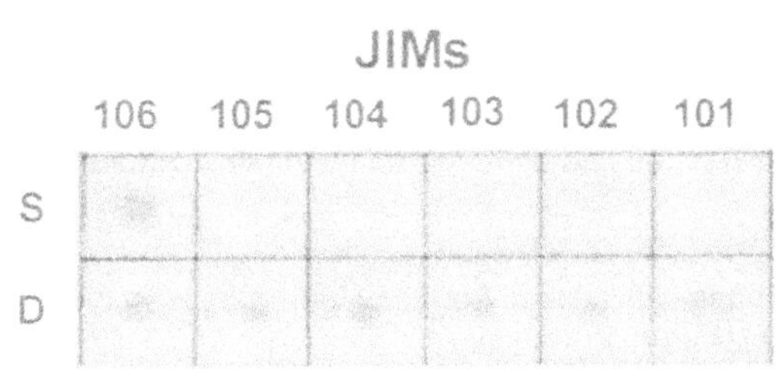

Figure 11. Dot-blot of monoclonal antibodies raised against AGPs that are released by *G. inflata* plants into their culture medium in correlation with an experimentally induced change from suppressed to desuppressed leaf and branch development. The AGPs spotted in row S were extracted from pooled culture media that supported a pattern of leaf and branch development shown in Fig 5. The AGPs spotted in row D were extracted from pooled media that induced desuppressed leaf and branch development as shown in Figs 6–8. Squares were spotted with equal amounts of AGPs based on a radial diffusion assay (van Holst and Clarke 1985). The JIMs 101–105 only reacted with epitopes of AGPs released by desuppressed plants, while JIM 106 reacted with an epitope common to AGPs released by both suppressed and desuppressed plants.

REFERENCES

Basile, D. V., 1967, The influence of hydroxy-L-proline on ontogeny and morphogenesis of the liverwort *Scapania nemorosa*, *Am. J. Bot.* **54**: 977–983.

Basile, D. V., 1980, A possible mode of action for morphoregulatory hydroxyproline-proteins, *Bull. Torrey Bot. Club* **107**: 325–338.

Basile, D. V., 1990, Morphoregulatory role of hydroxyproline-containing proteins in liverworts, In *Bryophyte Development: Physiology and Biochemistry* (R. N. Chopra and S. C. Bhatla, eds.), CRC Press, Boca Raton, FL, pp. 225–243.

Basile, D. V., and Basile, M. R., 1980, Ammonium ion-induced changes in form and hydroxyproline content of wall protein in the liverwort *Gymnocolea inflata*, *Am. J. Bot.* **67**: 500–507.

Basile, D. V., and Basile, M. R., 1982, Evidence for a regulatory role of cell surface hydroxyproline-containing proteins in liverwort morphogenesis, *J. Hattori Bot. Lab.* **53**: 221–227.

Basile, D. V., and Basile, M. R., 1993, The role and control of place-dependent suppression of cell division in plant morphogenesis and phylogeny, *Mem. Torrey Bot. Club* **25**: 63–84.

Basile, D. V., Basile, M. R, and Varner, J. E., 1986, The occurrence of arabinogalactan protein in *Plagiochila arctica* Bryhn and Kaal. (Hepaticae), *Am. J. Bot.* **73**: 603–604.

Basile, D. V., Kushner, B. K., and Basile, M. R, 1989, A new method for separating and comparing arabinogalactan proteins for the chemosystematics of the Hepaticae, *Bryologist* **92**: 164–169.

Basile, D. V., Basile, M. R., Salama, N., Peart, J., and Roberts, K., 1999, Monoclonal antibodies to arabinogalactan-proteins (AGPs) released by *Gymnocolea inflata* when leaf and branch development is desuppressed, *Bryologist* **102**: 304-308.

Clarke, A. E., Gleeson, P. E., Jermyn, M. A. and Knox, R. B., 1978, Characterization and localization of ß-lectins in lower and higher plants, *Aust. J. Plant Physiol.* **5**: 707–722.

Clarke, A. E., Anderson, R. L., and Stone, B. A., 1979, Form and function of arabinogalactans and arabinogalactan-proteins, *Phytochemistry* **18**: 521–540.

Crandall, B., 1969, Morphology and development of branches in the leafy Hepaticae, *Nova Hedwigia* **30**: 1–261.

Fincher, G. B., Stone, B. A., and Clarke, A. E., 1983, Arabinogalactan proteins: structure, biosynthesis and function, *Annu. Rev. Plant Physiol.* **34**: 47–70.

Kreuger, M., and van Holst, G-J., 1993, Arabinogalactan proteins are essential in somatic embryogenesis of *Daucus carota* L., *Planta* **189**: 243–248.

Kreuger, M., and van Holst, G-J., 1995, Arabinogalactan protein epitopes in somatic embryogenesis of *Daucus carota* L., *Planta* **197**: 135–141.

Kreuger, M., Postma, E., Brouwer, Y., and van Holst, G-J., 1995, Somatic embryogenesis of *Cyclamen persicum* in liquid medium, *Physiol. Plant.* **94**: 605–612.

Lamport, D. T. A., 1963, O_2 fixation into hydroxyproline of plant cell wall protein, *J. Biol. Chem.* **238**: 1438–1440.

Pennell, R. I., Janniche, L., Scofield, G. N., Booij, H., de Vries, S. C., and Roberts, K., 1992, Identification of a transitional cell state in the developmental pathway to carrot somatic embryogenesis, *J. Cell Biol.* **119**: 1371–1380.

Stebbins, G. L., and Basile, D. V., 1986, Phyletic phenocopies: A useful technique for probing the genetic and developmental basis of evolutionary change, *Evolution* **40**: 422–425.

van Holst, G., and Clarke, A. E., 1985, Quantification of arabinogalactan-protein in plant extracts by single radial diffusion, *Anal. Biochem.* **148**: 446–450.

Wardlaw, C. W., 1955, *Embryogenesis in Plants*, Methuen, London.

Chapter 16

Xylem-Specific Expression of Arabinogalactan-Protein-Like Genes

C.A. Loopstra, E.-G. No, H. Wang and J. Puryear
Department of Forest Science and Crop Biotechnology Center, Texas A&M University, College Station, TX 77843-2135

1. INTRODUCTION

Woody plants are biologically different from their herbaceous counterparts because of their perennial habit and ability to form wood, an extremely ecologically and economically important trait. Wood is by far the most abundant component of the terrestrial biomass and therefore has a significant influence upon the planetary carbon cycle and global climate. Wood is also a leading industrial raw material and an important component of the global economy.

Wood cell development is characterized by five phases: cell division, cell enlargement, cell wall thickening, lignification, and death (Thomas 1991). Three of these phases involve deposition or modification of the cell wall. During the cell enlargement phase, a thin primary cell wall encases the protoplasm. A multilayered secondary cell wall is later deposited inside the primary cell wall. The amount of thickening depends upon the cell type and time of year. Lignification of the middle lamella and primary cell wall takes place early and rapidly. Lignification of the secondary cell wall is more gradual and lags just behind wall formation (Thomas 1991). In a cross-section of a woody stem, files of cells along a developmental continuum from the vascular cambium to fully differentiated xylem cells can be observed. This pattern, along with the availability of large quantities of relatively pure xylem, makes woody plants valuable model systems for studies of xylogenesis.

The major chemical components of wood are cellulose (45–50%), hemicellulose (20–25%), lignin (20–30%) and extractives (0–10%), these four components accounting for greater than 99% of the mass. Until recently, proteins have not been considered to be important components of wood. Proteins, however, play important structural roles in cell walls and are thought also to have key roles in signaling processes. Perhaps as many as a few hundred different proteins are specific to cell walls.

A class of proteins that may play important roles in the development of xylem or wood are the arabinogalactan-proteins (AGPs). The AGPs (recently reviewed by Nothnagel [1997]) have been found in almost all plant species examined, including angiosperms, gymnosperms, and lower plants such as bryophytes. These highly glycosylated proteins are frequently expressed in a tissue-specific manner and have been localized on the plasma membrane at the plasma membrane–cell wall interface, in the cell wall, and in some cytoplasmic organelles. The AGPs have also been observed to be secreted into tissue culture media or into special tissues or organs. Many AGPs contain a hydrophobic putative transmembrane domain at the C-terminus. The presence of glycosylphosphatidylinositol (GPI) anchors may indicate that these AGPs are transiently found on the plasma membrane and are released through processing of the C-terminal domain. Processing of the C-terminal of two AGPs from *Nicotiana alata* and *Pyrus communis* has been detected, along with evidence of GPI anchors (Youl *et al* 1998).

Although AGPs have been characterized for decades, they have fairly recently received increased attention because of mounting evidence linking them to various aspects of plant development (Kreuger and van Holst 1996). Some potential roles of AGPs are particularly interesting with respect to wood development. As suggested by Kieliszewski and Lamport (1994), AGPs may play a role in secondary wall initiation by intercalating phenolics for subsequent orderly polymerization. If so, AGPs may play an important role in lignin structure. Alternatively, AGPs and other non-crosslinked extensins may be crystal growth inhibitors or nucleation promoters and may thus act as matrices for the orderly addition of nascent wall precursors to the growing wall matrix. Kreuger and van Holst (1996) hypothesize that despite their large size, AGPs or their epitopes may pass through cell walls from one cell type to another and thereby alter cell wall composition and impact differentiation. The possibility that AGPs could play a role in the developmental progression during xylogenesis is very intriguing. Various AGPs have been detected in all conifer tissues tested and presumably play roles in conifer development. Bobalek and Johnson (1983) reported significant differences in the amino acid compositions of AGPs from loblolly pine and Douglas fir dry seed, stratified seed, cotyledon stage seedlings, 2-month-old seedlings, 2-year-old saplings and callus. Egertsdotter and von Arnold (1995) have demonstrated that AGPs play a role in the development of Norway spruce somatic embryos. Cell lines capable of embryo maturation were shown to contain differing amounts and compositions of AGPs than lines incapable of maturation. The addition of seed AGPs to non-maturing cell lines caused them to gain characteristics of the maturing cell lines. Several genes encoding AGPs or AGP-like proteins have been cloned from differentiating xylem of loblolly pine (Loopstra and Sederoff 1995, Allona *et al* 1998, Zhang *et al* 1998). Gaining an understanding of the roles of AGPs in wood formation would provide insight on an interesting biological question and may have great economic value.

2. AGPS IN XYLEM OF LOBLOLLY PINE

The first two pine AGP-like genes to be cloned, *PtX3H6* and *PtX14A9*, were obtained by differential screening of a xylem cDNA library using xylem and megagametophyte cDNA probes (Loopstra and Sederoff 1995). Both of these genes are abundantly and preferentially expressed in differentiating xylem. As the two most extensively characterized pine genes encoding AGPs, *PtX3H6* and *PtX14A9* will be the focus of this paper. Additional AGP-like genes were identified as part of a loblolly pine expressed sequence tag (EST) sequencing project (Allona *et al* 1998, Zhang *et al* 1998). An AGP clone not previously determined to be AGP-like was identified in the EST collection following purification of a xylem AGP and N-terminal sequencing (Loopstra *et al* 2000). Most, if not all, of these genes also have strong preferential expression in xylem.

2.1 PtX3H6 and PtX14A9 — Structure

The polypeptide predicted to be encoded by *PtX3H6* has 168 amino acids and contains many features of classical AGPs. It includes a hydrophobic putative signal peptide followed by a Thr-rich region, a Pro/Hyp-rich region, and a hydrophobic carboxy-terminus hypothesized to contain a transmembrane domain (Fig 1). Repeats of Pro-Pro-Pro-Val-X-X are similar to repeats found in Pro-rich cell wall proteins. The PtX3H6 contains no tyrosine residues, however, and is only 1.4% lysine. The amino acid composition of PtX3H6 is more similar to that of an AGP since it is rich in Pro/Hyp, Thr, Ala and Ser (Table 1). Like the extensins, AGPs are usually rich in Pro/Hyp and Ser, although some AGPs, which are low in Hyp, have been identified. The AGPs are also frequently rich in Ala, Thr, and sometimes Gly. The amino acid composition of PtX3H6 is similar to that of AGP*Pc*1 from pear suspension cultures (Chen *et al* 1994). The polypeptide predicted to be encoded by *PtX14A9* has 264 amino acids and also contains putative signal peptide and transmembrane domains. The PtX14A9 amino acid composition is less like the classical AGPs than that of PtX3H6 (Table 1). The most abundant amino acids in PtX14A9 are Ala, Ser, and Pro, however, and the polypeptide contains the sequence Ala-Pro-Ala-Pro-Ser-Pro-Ala-Ser found near the N-termini of several angiosperm AGPs. Both PtX3H6 and PtX14A9 are hypothesized to contain possible GPI membrane anchor sites, although experimental results indicate that mature PtX3H6 may not be cleaved at this site.

2.2 PtX3H6 and PtX14A9 — Homologs

Searches of the GenBank NR (non-redundant) and dBEST (non-redundant EST divisions) databases did not reveal any non-loblolly pine proteins with a high level of similarity to PtX3H6. A putative Pro-rich cell wall protein, H6, specifically expressed in cotton fibers contains three Pro-Pro-Pro-Val- X_1-X_2 repeats and many more repeats of Pro-Pro-Pro- X_1-X_2, where X_1 is usually Ala and X_2 is Ser or Thr (John and Keller 1995). The PtX3H6 contains similar motifs but has fewer repeats than H6, and they are less ordered. The cotton H6 protein is also hypothesized to be an AGP and contains putative signal peptide and a hydrophobic C-terminus. The protein is posttranslational

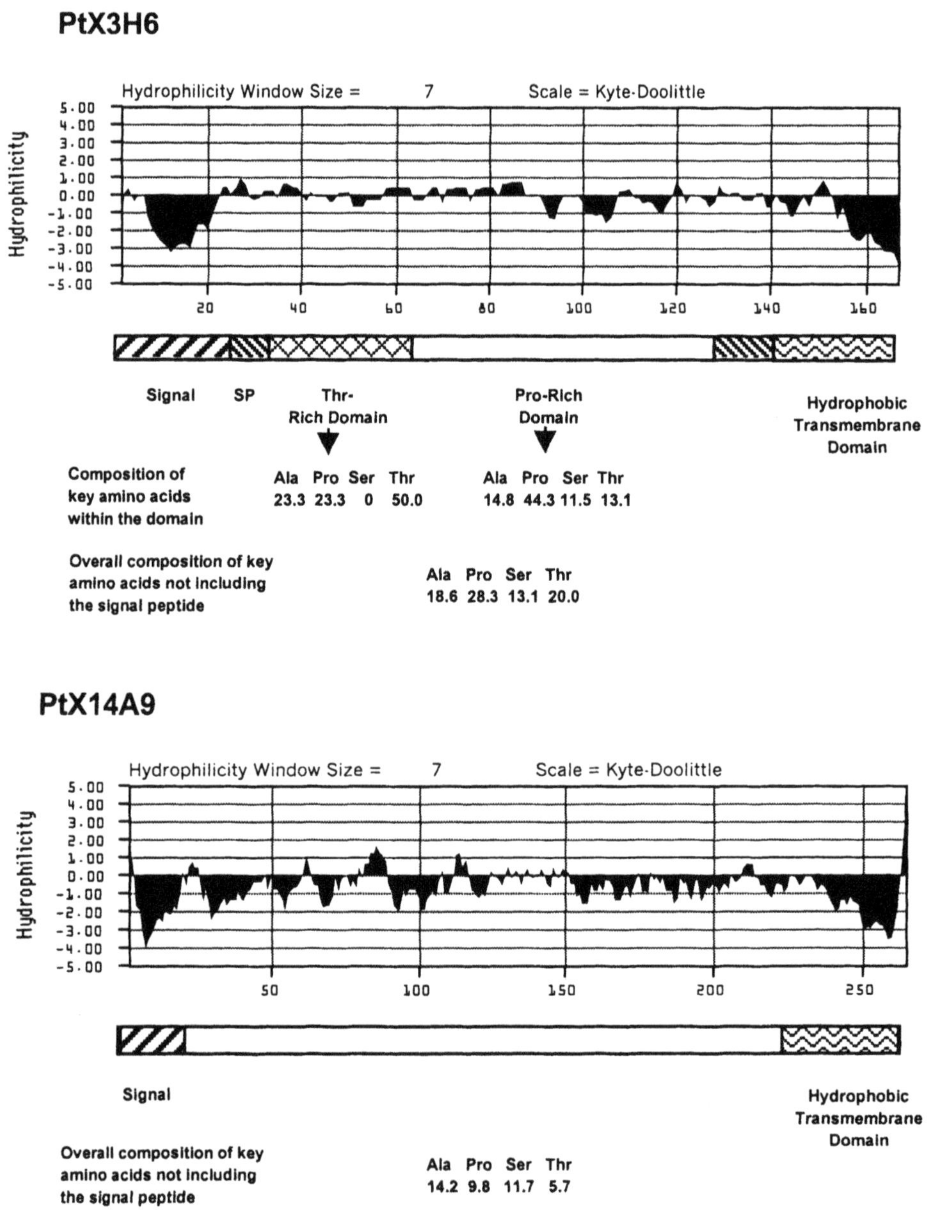

Figure 1. PtX3H6 and PtX14A9 polypeptide structures. The PtX3H6 and PtX14A9 hydrophilicity plots show profiles similar to those of other AGPs. Line drawings show distinguishable features, including putative signal peptides (signal), Ser-Pro repeats (SP), a Thr-rich domain, a Pro-rich domain, and hydrophobic putative transmembrane domains. The percentages of Ala, Pro, Ser, and Thr are given for the entire PtX3H6 and PtX14A9 proteins, excluding the signal peptide, and for the Thr-rich and Pro-rich domains of PtX3H6.

ly modified and is rich in Pro, Thr, Ala, and Ser (73%). It has been suggested that H6 may be an integral part of the plasmalemma, taking part in the development and architecture of the secondary wall of cotton fiber. An *Arabidopsis* genomic sequence and several *Arabidopsis* cDNA clones also encode proteins containing motifs similar to those in PtX3H6. Both PtX3H6 and the *Arabidopsis* clones contain the motif X_1-X_2-Pro_3-Z, where X_1 and X_2 are usually Ala, Ser, or Thr, and Z is Val or Ala. Both the cotton and *Arabidopsis* genomic sequences encode proteins larger than PtX3H6, however, and none contain a region as rich in Thr as the one in PtX3H6. It is possible that many AGPs exist with motifs similar to those in PtX3H6, but no close counterparts seem to exist in angiosperms.

Table 1. Amino acid compositions, per 100 residues, excluding putative signal peptides. Hyp and Pro contents are combined under Pro.

Amino acid	PtX3H6[a]	PtX14A9[a]	PtXAGP3[b]	AGP*Pc*1[c]
Ala	18.6	14.2	14.7	21.3
Arg	0	2.0	2.2	0.8
Asx	0.7	5.3	4.4	1.6
Cys	0	1.2	2.2	0
Glx	1.4	7.3	8.1	0.8
Gly	1.4	7.3	2.9	4.1
His	0	0.8	2.2	0
Ile	1.4	2.8	3.7	3.2
Leu	3.5	7.7	8.8	1.6
Lys	1.4	2.8	3.7	2.4
Met	0	1.2	0	0
Phe	1.4	4.9	6.6	1.6
Pro/Hyp	28.3	9.8	8.8	24.5
Ser	13.1	11.7	17.6	19.6
Thr	20.0	5.7	2.2	13.1
Trp	0	1.6	2.2	0
Tyr	0	2.0	4.4	0
Val	9.0	8.5	5.1	4.9

[a]*PtX3H6* and *PtX14A9*, xylem-specific genes cloned from loblolly pine (Loopstra and Sederoff 1995).
[b]*PtXAGP3*, AGP clone from differentiating xylem of loblolly pine (Loopstra *et al* 2000).
[c]*AGPPc1*, AGP clone from pear suspension culture (Chen *et al* 1994).

Homologs of *PtX14A9* have been found and are much more similar to the pine gene than are the homologs of *PtX3H6*. A homolog from a poplar xylem cDNA library was isolated in our laboratory by differential screening using xylem, phloem, and leaf cDNAs to probe for xylem-specific genes (H. Wang, unpublished data). The pine and poplar proteins are 52% identical (68% positives) within a 197-amino acid region not including the first 35 and last 32 amino acids of PtX14A9. With use of the pine and poplar sequences, we searched the dBEST database. Homologs were identified in several species, including *Arabidopsis*, soybean and rice. Additional poplar clones have recently appeared in the dBEST database as a result of a poplar xylem EST sequencing project (Sterky *et al* 1998). A cotton fiber homolog has also been identified

(John 1995). The cotton gene is primarily expressed in developing cotton fibers, with low expression in leaves and no detectable expression in roots, ovules, and petals. Because PtX14A9 homologs are found in gymnosperms (loblolly pine), woody angiosperms (poplar), herbaceous dicots (*Arabidopsis* and soybean), and a monocot (rice), we feel it is probably expressed in most, if not all, higher plants and is a protein critical during xylem formation. We also find it very interesting that although homologs have been identified in herbaceous plants through EST sequencing, their potential roles in xylem development are only apparent because of the pattern of expression (xylem specific) in woody species such as pine and poplar. The poplar and *Arabidopsis* proteins are 54.8% identical over their entire length and 66.5% identical within a 182 amino acid region not including the first 19 and last 40 amino acids of the poplar protein. This homology may be the highest level found so far between AGP sequences from different genera. Previously, Gerster *et al* (1996) reported the *Brassica napus* AGP Sta39-4 and the pear AGP*Pc*1 to be 36.8% identical.

2.3 *PtX3H6* and *PtX14A9* Encode AGPs

Although *PtX3H6* and, to a lesser extent, *PtX14A9* contain characteristics of other AGPs as described above, it is not possible to determine conclusively from just their predicted amino acid sequences that they do indeed encode AGPs. Only a few known AGPs have been cloned and sequenced, and sequences are generally not thought to be conserved between AGPs. We felt it was important to ascertain whether *PtX3H6* and *PtX14A9* encode AGPs to complement our work on gene function. Two different strategies were used.

The first approach was to obtain amino acid sequence data from the predominant loblolly pine xylem AGP and compare it to the sequences of PtX3H6 and PtX14A9. The AGPs from differentiating loblolly pine xylem were purified using a low salt–detergent extraction and precipitation with Yariv phenylglycoside, followed by size exclusion and anion exchange chromotography. Following deglycosylation with anhydrous hydrofluoric acid, one major and one minor band were visible by PAGE. The major band was transferred to polyvinylidene difluoride membrane, and 20 amino acids of N-terminal sequence were obtained. Although the sequence obtained contained a five amino acid sequence identical to one in PtX14A9, the purified protein was not encoded by *PtX3H6* or *PtX14A9*. The 20 amino acid sequence was searched in the loblolly pine EST database and was found in clone *5n2d*. Clone *5n2d* had not previously been identified as an AGP. The protein encoded by the cloned gene contains a hydrophobic putative signal peptide followed by a hydrophilic peptide that is also cleaved. The N-terminal sequence derived from the purified protein starts at amino acid number 52 of the inferred protein. A 68 amino acid AGP-like domain contains 29.4% Ser, 25% Ala, and 17.6% Pro/Hyp. A hydrophobic putative transmembrane domain is found at the C-terminus. The protein, which we call PtXAGP3, contains a putative GPI anchor site. It is therefore possible that this protein is cleaved three times: once to remove the signal peptide, once to remove the hydrophilic domain, and once to remove the transmembrane domain, resulting in a small Ser, Ala, Pro/Hyp-rich protein core (Loopstra *et al* 2000).

A second approach used to determine whether *PtX3H6* and/or *PtX14A9* encode AGPs was to express the pine genes in a heterologous system and to analyze the transgenic plants for the presence of a new AGP. We produced transgenic tobacco and poplar containing *PtX3H6-c-myc* and *PtX14A9-c-myc*, in which a 9 amino acid segment of the *c-myc* gene was inserted into the pine genes. In each case, the genes were under the control of their own promoters and terminators. Proteins were extracted from transgenic tobacco using a low salt buffer (20 mM NaCl, 0.1% Triton-X 100), followed by a high salt buffer containing 4 M guanidine hydrochloride. The remaining pellet was then treated with cellulase, and the released proteins were extracted. The extracted proteins were analyzed using dot-blots and the *c-myc* antibody. The PtX3H6-c-myc was not extracted by low salt but was extracted by high salt. Additional protein was extractable following cellulase treatment. The PtX14A9-c-myc was extracted with the low salt buffer. Additional protein was obtained in the high salt extraction. Very little additional protein was extracted following the cellulase treatment.

To determine if PtX3H6-c-myc and PtX14A9-c-myc are AGPs, the proteins extracted with the low salt and high salt buffers were further purified. Many of the non-AGPs were precipitated from each fraction using trichloroacetic acid (TCA) . The AGPs were precipitated using ethanol followed by a precipitation with Yariv phenylglycoside. Following disassociation of the AGPs and Yariv phenylglycoside with sodium hydrosulfite, the Yariv-AGP pellet and supernatant fractions were separated by size exclusion chromatography. The fractions eluted from the column were analyzed for AGPs by radial diffusion assays and reacted with the *c-myc* antibody to identify fractions containing PtX3H6-c-myc and PtX14A9-c-myc. In the PtX3H6 study, the c-myc antibody reacted only with proteins extracted with high salt, not precipitated by TCA, and precipitated by both ethanol and Yariv phenylglycoside. The intensity of the antibody labeling was highly correlated with the amount of AGPs in each fraction. In the PtX14A9 study, the antibody reacted with proteins extracted with low or high salt, not precipitated by TCA, and precipitated by both ethanol and Yariv phenylglycoside. Again, the column fractions reacting with the c-myc antibody were the same as those reacting with Yariv phenylglycoside in the radial diffusion assay. These results, along with the previously mentioned similarities to AGPs, provide very strong evidence that PtX3H6 and PtX14A9 are AGPs as hypothesized.

2.4 *PtX3H6* and *PtX14A9* — Expression and Regulation

Both *PtX3H6* and *PtX14A9* are expressed at very high levels in differentiating xylem, at much lower levels in needles and embryos, and at nondetectable levels in megagametophytes (Loopstra and Sederoff 1995). Within differentiating xylem, expression of *PtX3H6* is greater than that of *PtX14A9*, and both are expressed at many times the levels of phenylalanine ammonia lyase and cinnamyl alcohol dehydrogenase, two enzymes in the lignin biosynthetic pathway. Transcript levels at different stages of pine tree development were compared by Northern blot analyses. The RNA was isolated from the stems of 6-week-old seedlings and from xylem scraped from the stems of 1, 2, or 10-year-old trees. Transcript levels were slightly lower in the seedlings but were high and similar in the other samples (Fig 2). Numerous morphological differences have been observed in xylem laid down early or late in the growing season (Thomas 1991). Earlywood has larger cells, thinner cell walls, lower density and a lighter color than latewood. The transition from earlywood to latewood is dramatic and may be related to environmental regulation of auxin production (Zobel and van Buijtenen 1989). Northern blot hybridizations failed to reveal any differences in the levels of *PtX3H6* and *PtX14A9* transcripts in RNA isolated from earlywood and latewood samples.

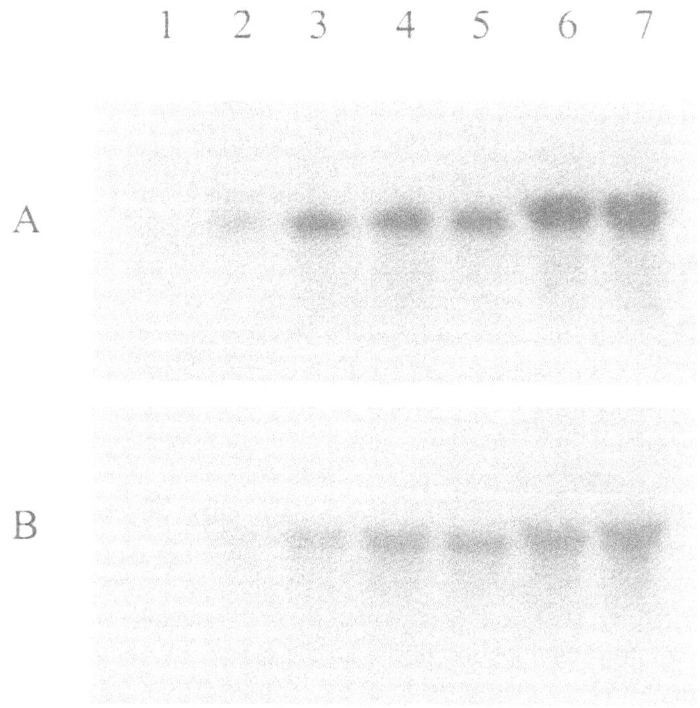

Figure 2. Northern blot analyses comparing levels of *PtX3H6* (panel A) and *PtX14A9* (panel B) transcripts over development. Lane 1, embryo. Lane 2, 6 week old seedlings. Lanes 3, 4, and 5, xylem scraped from the stems of 1, 2, and 10 year old trees. Lanes 6 and 7, earlywood and latewood.

Wounded xylem tissue was obtained by stripping the xylem off bolts at different times after peeling away the bark and cambium. Samples were taken 0, 30, 60, and 120 min after the wounding treatment. Megagametophytes were either frozen immediately after dissection or wounded with a scalpel and allowed to incubate overnight. Following Northern blot analyses, no hybridization was observed to RNAs from either wounded or unwounded megagametophytes. Strong hybridization was observed to all xylem RNA samples, and with neither gene were differences observed between transcript levels in wounded and unwounded tissues.

To examine the roles of plant growth regulators in the expression of *PtX3H6* and *PtX14A9*, loblolly pine seedlings were treated with TIBA, an auxin transport inhibitor; uniconazole, a gibberellin inhibitor; and silver nitrate, an ethylene inhibitor (No and Loopstra 2000). Transcript levels in untreated seedlings were compared to those in seedlings treated with one of the inhibitors. The *PtX3H6* transcripts decreased following treatment with TIBA or silver nitrate and were unaffected by uniconazole. The *PtX14A9* transcripts decreased moderately in response to TIBA and greatly following treatment with silver nitrate and uniconazole. Because *PtX3H6* and *PtX14A9* exhibited similar patterns of expression in various tissues, over development and in response to stresses, we had previously thought that the regulation of these two genes was very similar. These experiments show that although both genes appear to be regulated by these growth regulators, they are affected differently by the various inhibitors. Hormone inhibitor experiments indicate that *PtX3H6* transcript levels decrease following application of auxin and ethylene inhibitors, whereas *PtX14A9* transcript levels decrease following application of ethylene and gibberellin inhibitors.

Both *PtX3H6* and *PtX14A9* were originally cloned as part of a project to isolate xylem-specific promoters for use in genetic engineering of wood properties. Promoters were obtained from genomic clones and fused to the reporter gene *uid*A, encoding β-glucuronidase (GUS) (5' constructs). When tested by microprojectile bombardment into loblolly pine xylem, embryos, and megagametophytes, neither promoter conferred xylem-specific expression. In transgenic tobacco, the PtX14A9 promoter resulted in little expression, and the *PtX3H6* promoter-*uid*A fusion resulted in GUS expression in nonvascular as well as vascular tissues. In transgenic poplar, the *PtX14A9* construct was more effective than in tobacco, and most, but not all, expression was in vascular tissues. The *PtX3H6* promoter again resulted in GUS expression in most tissues. These results caused us to hypothesize that elements downstream of the transcription start site are involved in the regulation of PtX3H6 and *PtX14A9*. Constructs were made fusing the promoters, *uid*A, the 3' untranslated regions, and approximately 1 kb of 3' flanking sequence (5'3' constructs). Transgenic tobacco and poplar were produced. We found that an element at the 3' end of *PtX3H6* reduced overall GUS expression and greatly increased xylem specificity (E.-G. No, unpublished data). The 3' end when used with the CaMV35S promoter was insufficient to cause this increase in specificity. Therefore, elements at both the 5' and 3' ends of *PtX3H6* were required for xylem specificity. An element at the 3' end of *PtX14A9* increased expression in vascular tissues. This element had this enhancing effect even when used with the CaMV35S promoter. The 5' and 5'3' constructs are now being tested in loblolly pine.

3. DISCUSSION

Monoclonal antibodies have been used for localizing AGPs in vascular tissues (reviewed by Nothnagel 1997). In particular, the JIM4 and JIM13 antibodies have been used for immunolocalizations in roots of carrot, radish, pea and onion with varying results (Casero *et al* 1998). Labeling patterns in carrot had previously been characterized (Knox *et al* 1989 1991). In carrot, JIM4 labels the pericycle cells in front of the xylem poles and the protoxylem elements. In contrast, JIM4 labels some, but not all, of the pericycle cells in front of the xylem and phloem in radish, and it does not label cells in pea or onion roots. The JIM13 antibody labels developing xylem cells in carrot and radish roots and labels some of the pericycle cells in pea roots. In onion roots, the developing xylem cells are not labeled with JIM13, but cells in the region of the phloem are. In *Arabidopsis*, the JIM13 antibody labels an epitope associated with xylem development (Dolan and Roberts 1995). A discrete ring of cells in the periderm is labeled during xylem vessel element differentiation, and then the epitope disappears as lignification proceeds.

The JIM13 antibody has been used to detect an AGP appearing during zinnia tracheid culture (Stacey *et al* 1995). The JIM13 epitope appeared in the primary wall of some cells 72 hours after initiation and later in the secondary thickenings in walls of mature tracheids. Both JIM13 and JIM14 have been used to label cells in maize coleoptiles, where they recognize epitopes only in tissues of the vascular bundles. The JIM13 epitope is localized in the plasma membrane, plasma membrane invaginations and multivesicular bodies within the vacuole of future sclerenchyma cells with obvious protoplasmic disintegration, and in future tracheids. Exhibiting a different pattern, the JIM14 epitope is detectable in the innermost layer of sclerenchyma cell walls of cells undergoing protoplasmic degeneration. Schindler *et al* (1995) hypothesized that the JIM13 and JIM14 epitopes may identify cells subject to programmed cell death. The stems of woody plants are an excellent example of cells that have undergone programmed cell death. Both PtX3H6 and PtX14A9 are expressed in cells destined to undergo programmed cell death and may play a role in autolysis. However, the cells undergo other developmental steps first, and PtX3H6 and PtX14A9 may play their roles in xylogenesis earlier in development. It would be interesting to determine whether JIM13 and/or JIM14 bind to PtX3H6-c-myc or PtX14A9-c-myc, which can be extracted from transgenic plants. If so, it would then be valuable to examine JIM13 and JIM14 epitope expression along the developmental continuum found in a cross-section of a woody stem.

We have used expression of two pine AGP-like genes in transgenic tobacco to prove that both genes encode AGPs by tracking the expression of the c-myc epitope during AGP purifications. These transgenic plants have also provided insight into the extractabilities of PtX3H6 and PtX14A9. Like the JIM13 epitope, PtX14A9-c-myc can be extracted using a low salt–detergent buffer. In maize coleoptiles, the JIM13 epitope is localized in the plasma membrane, plasma membrane invaginations and in multivesicular bodies within the vacuole (Schindler *et al* 1995). A PtX14A9 *Glycine max* homolog identified in the dBEST database was isolated by screening an expression library using antibodies to proteins purified from the plasma membrane (Shi *et al* 1995). Therefore, it is likely that PtX14A9 and its homologs are found in the plasma membrane and possibly other locations.

Like PtX3H6-c-myc, the JIM14 epitope is not extracted using low salt but can be extracted with 4 M guanidine hydrochloride (Schindler *et al* 1995). In maize coleoptiles, the JIM14 epitope is detected only in the innermost layer of cell walls. Additional PtX3H6-c-myc can be extracted from transgenic tobacco following treatment with cellulase. Some AGPs released by cellulase digestion have previously been found in cabbage leaves (Kido *et al* 1996). The cabbage AGPs and PtX3H6 may be associated with cellulose microfibrils and pectic substances. In both cases, the cellulase used may not have been pure and may have contained hemicellulases as well as cellulase.

Several genes encoding AGPs or AGP-like proteins have been cloned from differentiating xylem of loblolly pine. These genes have high levels of expression in differentiating xylem and low or nondetectable expression in other tissues. Pine EST sequencing is still in the early stages, with 1,097 single-pass sequences reported (Allona *et al* 1998). In addition, at least one previously sequenced EST not identified as an AGP has been shown to be an AGP by protein purification and sequencing. Therefore, it is likely that other pine xylem AGPs remain to be cloned. With a relatively large number of AGPs preferentially expressed in differentiating pine xylem, it is likely they play important roles in the development of wood ranging from controlling the progression of development to lignification to programmed cell death. The presence of PtX14A9 homologs in many plant species, with amino acid identities between genera much higher than previously reported for other AGPs, leads us to believe that it plays an important role in xylem development in most higher plants.

REFERENCES

Allona, I., Quinn, M., Shoop, E., Swope, K., St. Cyr, S., Carlis, J., Riedl, J., Retzel, E., Campbell, M. M., Sederoff, R., and Whetten, R. W., 1998, Analysis of xylem formation in pine by cDNA sequencing, *Proc. Natl. Acad. Sci. USA* **95**: 9693–9698.

Bobalek, J. F., and Johnson, M. A., 1983, Arabinogalactan-proteins from Douglas fir and loblolly pine, *Phytochemistry* **22**: 1500–1503.

Casero, P. J, Casimiro, I., and Knox, J. P., 1998, Occurrence of cell surface arabinogalactan-protein and extensin epitopes in relation to pericycle and vascular tissue development in the root apex of four species, *Planta* **204**: 252–259.

Chen, C.-G., Pu, Z.-Y., Moritz, R. L., Simpson, R. J., Bacic, A., Clarke, A. E., and Mau, S.-L., 1994, Molecular cloning of a gene encoding an arabinogalactan-protein from pear (*Pyrus communis*) cell suspension culture, *Proc. Natl. Acad. Sci. USA* **91**: 10305–10309.

Dolan, L., and Roberts, K., 1995, Secondary thickening in roots of *Arabidopsis thaliana*: anatomy and cell surface changes. *New Phytol.* **131**: 121–128.

Egertsdotter, U., and von Arnold, S., 1995, Importance of arabinogalactan proteins for the development of somatic embryos of Norway spruce (*Picea abies*), *Physiol. Plant.* **93**: 334–345.

Gerster, J., Allard, S., and Robert, L. S., 1996, Molecular characterization of two *Brassica napus* pollen-expressed genes encoding putative arabinogalactan proteins, *Plant Physiol.* **110**: 1231–1237.

John, M. E., 1995, Characterization of a cotton (*Gossypium hirsutum* L.) fiber mRNA, *Plant Physiol.* **107**: 1477–1478.

John, M. E., and Keller, G., 1995, Characterization of mRNA for a proline-rich protein of cotton fiber, *Plant Physiol.* **108**: 669–676.

Kido, S., Yasufuku, H., and Azuma, J., 1996, Isolation and characterization of arabinogalactan proteins released by cellulase digestion of cabbage leaves, *J. Agric. Food Chem.* **44**: 3432–3436.

Kieliszewski, M. J., and Lamport, D. T. A., 1994, Extensin: repetitive motifs, functional sites, post-translational codes, and phylogeny, *Plant J.* **5**: 157–172.

Knox, J. P., Day, S., Roberts, K., 1989, A set of cell surface glycoproteins forms an early marker of cell position, but not cell type, in the root apical meristem of *Daucus carota* L., *Development* **106**: 47–56.

Knox, J. P., Linstead, P. J., Peart, J., Cooper, C., and Roberts, K., 1991, Developmentally regulated epitopes of cell surface arabinogalactan proteins and their relation to root tissue pattern formation, *Plant J.* **1**: 317–326.

Kreuger, M., and van Holst, G.-J., 1996, Arabinogalactan proteins and plant differentiation, *Plant Mol. Biol.* **30**: 1077–1086.

Loopstra, C. A., and Sederoff, R. R., 1995, Xylem-specific gene expression in loblolly pine, *Plant Mol. Biol.* **27**: 277–291.

Loopstra, C. A., Puryear, J. D., and No, E.-G., 2000, Purification and cloning of an arabinogalactan-protein from xylem of loblolly pine, *Planta* **210**: 686-689.

No, E.-G., and Loopstra, C. A., 2000, Hormonal and developmental regulation of two arabinogalactan-proteins in xylem of loblolly pine (*Pinus taeda*), *Physiol. Plant.* (in press).

Nothnagel, E. A.,1997, Proteoglycans and related components in plant cells, *Int. Rev. Cytol.* **174**: 195–291.

Schindler, T., Bergfeld, R., and Schopfer, P., 1995, Arabinogalactan proteins in maize coleoptiles: developmental relationship to cell death during xylem differentiation but not to extension growth, *Plant J.* **7**: 25–36.

Shi, J., Dixon, R. A., Gonzales, R. A., Kjellbom, P., and Bhattacharyya, M. K., 1995, Identification of cDNA clones encoding valosin-containing protein and other plant plasma membrane associated proteins by general immunoscreening strategy, *Proc. Natl. Acad. Sci. USA* **92**: 4457–4461.

Stacey, N. J., Roberts, K., Carpita, N. C., Wells, B., and McCann, M. C., 1995, Dynamic changes in cell surface molecules are very early events in the differentiation of mesophyll cells from *Zinnia elegans* into tracheary elements, *Plant J.* **8**: 891–906.

Sterky, F., Regan, S., Karlsson, J., Hertzberg, M., Rohdo, A., Holmberg, A., Amini, B., Bhalerao, R., Larsson, M., Villarroel, R., Van Montagu, M., Sandberg, G., Olsson, O., Terri, T. T., Boerfan, W., Gustafsson, P., Uhlen, M., Sundberg, B., and Lundeberg, J., 1998, Gene discovery in the wood-forming tissues of poplar: analysis of 5,692 expressed sequence tags, *Proc. Natl. Acad. Sci. USA* **95**: 13330–13335.

Thomas, R. J., 1991, Wood: formation and morphology, In *Wood Structure and Composition*, (M. Lewin and I. S. Golstein, eds.), Marcel Dekker, Inc., New York, pp.7–47.

Youl, J. J., Bacic, A., and Oxley, D., 1998, Arabinogalactan-proteins from *Nicotiana alata* and *Pyrus communis* contain glycosylphosphatidylinositol membrane anchors, *Proc. Natl. Acad. Sci. USA* **95**: 7921–7926.

Zhang, Y., Allona, I., Kieliszewski, M., and Sederoff, R., 1998, Isolation and characterization of putative cell wall proteins in loblolly pine, In *Proceedings of the American Society of Plant Physiologists Annual Meeting*. Madison, WI June 27–July 1, 1998. Abstract 1002.

Zobel, B. J., and van Buijtenen, H., 1989, *Wood Variation, Its Causes and Control*, Springer-Verlag, New York.

Chapter 17

Induction of Phyletic Phenocopies in *Streptocarpus* (Gesneriaceae) by Three Antagonists of Hydroxyproline-Protein Synthesis

Richard A. Rauh and Dominick V. Basile
The Graduate School and University Center, City University of New York, Lehman College, 250 Bedford Park Boulevard West, Bronx, New York 10468

1. INTRODUCTION

Plants in the family Gesneriaceae display a wide range of vegetative morphologies. The family separates into two subfamilies on the basis of the relative size of the cotyledons of their respective plants (Burtt 1963). In the subfamily Gesnerioideae, cotyledons remain equal, a phenomenon that is a basal synapomorphy of dicotyledons. In the subfamily Cyrtandroideae, cotyledons develop unequally, a condition known as anisocotyly. Burtt (1970) suggests that the evolutionary advantage of the untypical and unusual seedling morphology be tied to environmental factors. He couples anisocotyly with numerous, tiny seeds lacking endosperm. Microspermy, which in other forest plants is ameliorated by symbiosis or saphrophytism, is in this case helped by the immediate production of photosynthetic tissue because of the precocious enlargement of one of the cotyledons. In some species of the genus *Streptocarpus*, as well as other genera scattered throughout the Cyrtandroideae, the single cotyledon continues development to become the entire vegetative structure of the mature plant. This reaches an extreme in the genus *Monophyllaea*, which completely lacks a shoot apical meristem (Cronk and Moeller 1997). The "fuzzy morphology" (Rutishauser 1995) of this self-contained organism (half leaf-like, half stem-like) has been termed a phyllomorph, and the axis, a combination stem/petiole, is known as a petiolode (Jong 1970 1973).

It is possible to arrange the morphotypes of the genus *Streptocarpus* into a series. At one extreme is this highly derived unifoliate form where the accrescent, larger cotyledon continues to grow. The subgenus *Streptocarpus* defines those species that display this morphology and those with closely related acaulescent organization. At the other extreme is the more familiar caulescent, node–internode type shoot organization that occurs after anisocotyly, in the subgenus *Streptocarpella*. In a study by Moeller and Cronk (1997), designed to clarify the phylogenetic relationships of the genus and the related genus *Saintpaulia*, evidence was obtained that *Streptocarpus* is not monophyletic. This hypothesis was supported by a further analysis by Smith *et al* (1998), whose results indicate a monophyletic *Saintpaulia* nested within a paraphyletic *Streptocarpus*. The subgenera *Streptocarpus* and *Streptocarpella* are revealed as separate clades, with *Streptocarpella* being further divided geographically (Fig 26).

The morphological development of the genus *Streptocarpus* thus appears extremely flexible, expressed in the variety of vegetative forms exhibited by the various species (Hilliard and Burtt 1971). Occasionally it will manifest itself in a single species. *S. nobilis* C. B. Clarke, usually a caulescent variety belonging to the subgenus *Streptocarpella*, under adverse conditions will act as a facultative unifoliate.

In between the extremes of caulescent and monocarpic unifoliate patterns in this series are a number of intermediate forms. Certain species in the subgenus *Streptocarpus* form rosette-like groupings of phyllomorphs, and are called "rosulates". Jong (1978) breaks these down into two further classifications. One is "centric rosulate", in which the additional phyllomorphs are grouped about a vertical central axis; the other is "excentric rosulates", in which the prostrate petiolode is not unlike a rhizome. The additional phyllomorphs cluster adventitiously along the dorsal side. A third form, "plurifoliates", act as unifoliates for most of their life span but initiate additional phyllomorphs at the time of flowering.

The species used in these experiments, *S. prolixus* C. B. Clarke, is classed as a plurifoliate. It was the primary species used in experiments by Rosenblum (1981) and Rosenblum and Basile (1984) for ease and speed of growing, as well as for its facultative unifoliate habit. As may be seen from Figs 1–4, although its normal growth begins with isocotyly, it soon becomes anisocotyledonous, with one cotyledon growing larger 3 or 4 days after germination. The smaller microcotyledon ceases enlargement, is gradually displaced by growth of the petiolode between the two cotyledons, and eventually falls off (see also Fig 1, Rosenblum and Basile 1984). Rosenblum (1981) used this plant to study the morphogenetic effects of various hormones, but it also served as a model for induction of "phyletic phenocopies". The term "phyletic

Figures 1–11. (See next page.) Normal and phenovariant seedlings of plurifoliate *Streptocarpus prolixus*. Co, macrocotyledon; co, microcotyledon; sp, secondary phyllomorph; sp2, additional phyllomorph; pl, plumule; lp, leaf primordium. 1–4. Stages in the normal development of seedlings; note suppression of the plumule. Bar = 2 mm. 1. At 6 days after germination (DAG), note equal cotyledons. 2. At 20 DAG, the upper cotyledon has begun accrescence. 3. At 30 DAG, a distinct size difference is visible between the two cotyledons. 4. At 60 DAG; note expansion of petiolode, separating the two cotyledons (arrow). 5–11. *S. prolixus* seedlings from pot cultures treated with antimetabolites. 5–8. Phenovariation #3, the appearance of a plumule with anisocotyly. 5. At 23 DAG showing plumule arising from apparently equal cotyledons and two leaf primordia. 6. Frontal view 37 DAG showing developing leaves and accrescent macrocotyledon. 7. Side view of same stage of growth clearly showing anisocotyly. Bar = 2 mm.

phenocopies" was proposed by Stebbins and Basile (1986) to define changes in form that mimic the typical form of a related phenotype, particularly one characteristic of a different taxon. The authors suggest that by inducing plants to produce these phenocopies, investigators can obtain clues to the developmental basis of evolutionary change.

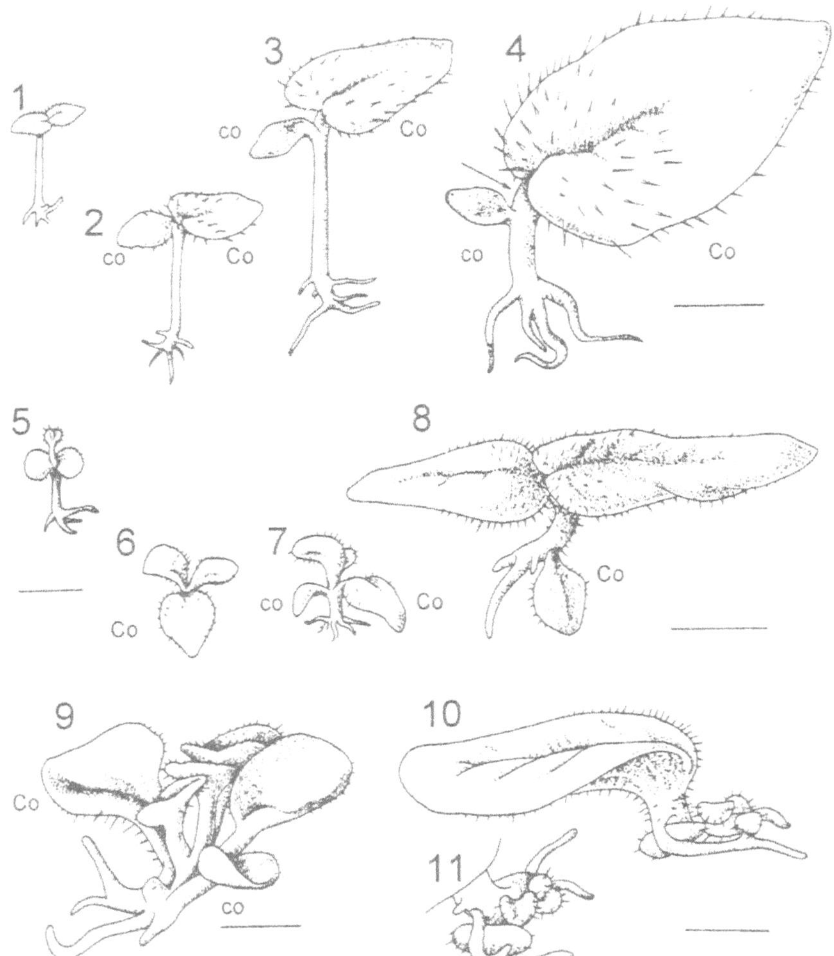

Figures 1–11 continued. 8. Later stage (71 DAG). The two leaves continue to enlarge, but there is no indication of further shoot development. In the next month the leaf on the left withered and eventually fell off, and the plant resumed its unifoliate pattern. The visible cotyledon is the macrocotyledon; the microcotyledon had fallen off by this stage. Bar = 5 mm. 9. Side view of multiple phyllomorphs (also shown in Figs 17 and 25) phenocopying a centric rosulate habit. Bar = 2 mm. 10–11. Seedling showing similarity to the excentric rosulate pattern. Bar = 2 mm. 10. Side view of phyllomorph showing horizontal rhizome-like petiolode, with a number of roots and four secondary phyllomorphs. 11. Detail view, looking down the petiolode. Within the next two weeks, all of the secondary phyllomorphs withered, and the plant resumed its traditional plurifoliate habit.

Basile (1967 1969 1970 1973a 1979, Basile and Basile 1980a), in a series of experiments, induced phyletic phenocopies in species belonging to five distinct families of leafy liverworts. In each case exogenously supplied antagonists of Hyp-protein synthesis produced phenotypic changes that phenocopied characteristics of more primitive species of different genera and families of leafy liverworts (see Basile 1990, for review). The most consistent phenovariation in Basile's series of experiments was the reappearance of a suppressed ventral leaf and the shift in organization of the plant axis from a dorsi-ventral to a radial symmetry. In other examples, leaf size and shape changes and more primitive branching patterns were induced (Basile and Basile 1980b 1982).

Rosenblum (1981, Rosenblum and Basile 1984), working with *S. prolixus*, succeeded in inducing phenovariants by primarily using applications of gibberellic acid (GA_3). Although experiments with GA_3 showed that the capacity to express a node–internode type plant body was not "lost or lacking" in *S. prolixus,* the authors did not test what Basile and Basile (1984 1993) hypothesized to be the more pivotal role of Hyp-proteins, presumably arabinogalactan-proteins (AGPs), in place-dependent suppression.

To test this hypothesis, three distinctly different-acting antagonists of the normal synthesis of Hyp-proteins are employed here, as they had previously been in the leafy liverwort experiments. The antagonist 3,4-dehydro-L-proline (3,4-dhp) is a specific inhibitor of prolyl hydroxylase, the enzyme that converts Pro to Hyp (Nolan *et al* 1978, Cooper and Varner 1983). The antagonist 2,2'-dipyridyl (2,2'-dp) is a chelator of ferrous iron that is an essential cofactor for the enzymatic conversion of Pro to Hyp (De Kock and Vaughan 1975). The antagonist *trans*-4-hydroxy-L-proline (Hyp) is a structural analog of Pro that can be inserted in the position normally occupied by Pro, thereby disrupting the usual sequence of amino acids (Basile *et al* 1987). The objective of the experiments reported here was to determine whether the exogenous application of micromolar concentrations of the three Hyp-protein-antimetabolites would induce phyletic phenocopies in *S. prolixus* analogous to those induced in leafy liverworts (Basile and Basile 1984).

2. MATERIALS AND METHODS

2.1 Plant materials

Seeds of *Streptocarpus prolixus* C. B. Clarke used in this study were from the seed fund of the American Gloxinia-Gesneriad Society, and from plants grown to maturity from these seeds and hand-pollinated by the authors. Seeds were stored at 5–6°C until needed.

2.2 Culture conditions

Two different procedures were used to germinate seeds and treat seedlings. In one, folded 110-mm diameter discs of Whatman #4 filter paper formed platforms approximately 4 × 4 × 4 cm on which the seeds/seedlings were supported within "baby food" jars covered with Magenta Caps™ (Sigma, St. Louis, MO). After heat-sterilization of the filters and jars, the test solutions were added in the amount of 25–30 ml, which saturated the filter paper platforms but remained well below the level of the top of the platform. Treatments consisted of solutions of 3,4-dhp, 2,2'-dp, and Hyp (Sigma) supplied in 10-, 50-, and 100-μM concentrations. After germination, solutions of Knop's medium plus "metals 49" (Basile and Basile 1988) mixed with the same antimetabolite concentrations replaced the original solutions. In a repeat series of experiments, the seeds were surface sterilized for 5 min in 2% Clorox (5.25% sodium hypochlorite) using the "washing machine" technique developed by Basile (1973b). The sterilized seeds were placed on filter paper platforms in autoclaved baby food jars containing 27-ml aliquots of Knop's medium plus "metals 49". Filter-sterilized antimetabolites were added at the same concentrations as in the first experiments, before seeding the platforms.

In the second procedure, a commercial soil-less mix (Grace Sierra Metromix 360™, E. C. Geiger, Inc., Harleysville, PA) in 5.5 × 4.7-cm plastic pots supported the seeds. Twenty to thirty seeds were placed in each pot. The pot cultures were drenched with water or micromolar solutions of antimetabolites at seeding time and every 7 days thereafter for 2 months. Solutions of Hyp were supplied at higher, millimolar concentrations in a continuance of the pot culture experiments. The pot cultures were incubated in an aquarium tank on water-saturated capillary matting, with the tank covered with clear acrylic.

All cultures, both platform and pot, were maintained in a Percival (Boone, IA) model WE-106 plant growth chamber at 20°C, 80% humidity. Light with a photoperiod of 14 hours was provided from a mixture of incandescent and cool white fluorescent lamps at 4520 lux for pot culture and 3875 lux for platform culture.

2.3 Photography

A number of the seedlings measuring 1–3 mm were fixed, dehydrated, critical point dried in a Tousimis (Rockville, MD) Samdri-790, and sputter coated with gold with use of a Technics Hummer II (Technics Corp., Alexandria, VA). The seedlings were then photographed at low magnification (25 kV) on a Hitachi S 2700 scanning electron microscope with a Polaroid camera attachment. Photographs of the pot-grown plants were taken with an Olympus OM-2 camera mounted on a Zeiss dissecting microscope using Kodak TP-2415 black-and-white panchromatic and Kodak Kodacolor 100 color print films.

2.4 Illustrations

Drawings were made from living plants by referring to details observed through a Zeiss dissecting microscope. The specimens were set in Petri dishes above a sheet of 1-mm lined graph paper for accurate measurement. Diagrammatic interpretations were made by referring to SEM micrographs.

3. RESULTS

In all cases, the exogenous application of the antagonists of Hyp-protein synthesis "desuppressed" the normally suppressed shoot apical meristem of *S. prolixus*. This desuppression resulted in a variety of morphotypes that did not appear to be correlated with the type of inhibitor. The desuppression of the shoot apical meristem, however, apparently triggered suppression of the growth rate of other parts of the seedlings. The size of the phenovariants remained in the 1–3-mm range, whereas the control seedlings reached lengths of 13–20 mm during the same time. The pattern of development without antimetabolite inhibition was typical for *S. prolixus* (Figs 1–4; #1 in Table 1).

Table 1. Effect of Hyp-protein antimetabolites on morphology of *S. prolixus* seedlings as observed at 80 days after the onset of germination on filter paper platforms. Relative numbers of plants are shown under columnar headings designating the five morphotypes exhibited: 1. Anisocotyly with continuous growth of the macrocotyledon and suppression of the shoot apical meristem. 2. Anisocotyly with secondary phyllomorph displaced to the base of the macrocotyledon. 3. Anisocotyly with phyllomorph and/or plumular growth at the axis of the two cotyledons. 4. Multiple phyllomorphy. 5. Isocotyly with plumular growth. Plants in column 1 exhibited the normal phenotype for *S. prolixus*. Plants in columns 2–5 are phenovariations (PV) of the normal growth pattern. The last column is the overall frequency of (% of plants exhibiting) PV.

	1	2	3	4	5	PV	%PV[a]
Control	91	6	4	0	2	12	11.7
2,2'-dp[b]	38	10	8	1	1	20	34.5
3,4-dhp	59	7	12	0	1	20	25.3
Hyp	96	11	27	7	12	57	37.3

[a] Significant at G = 23.8, $P<0.0001$.
[b] 2,2'-dp = 2,2'-dipyridyl; 3,4-dhp = 3,4-dehydro-L-proline; Hyp = *trans*-4-hydroxy-L-proline.

In experiments conducted with the pot-cultured plants, micromolar concentrations of the antimetabolites induced no discernible variation from the controls, whereas millimolar concentrations of Hyp induced pronounced phenovariation. The pot-cultured phenovariants displayed a variety of forms. Some phenovariants phenocopied features characteristic of species of subgenus *Streptocarpella* (Figs 5–8), whereas other phenovariants phenocopied features characteristic of centric (Fig 9) and excentric (Figs 10, 11) rosulates of the subgenus *Streptocarpus*. The plant shown in Fig 10, after a time of displaying typical cotyledonous accrescence, suddenly produced a number of small phyllomorphs simultaneously aligned along the prostrate rhizome-like petiolode. This development very closely approximates the growth pattern of the mature morphotype of *S. primulifolius* Gand., an excentric rosulate (Hilliard and Burtt 1971).

However, the rhizome-like habit in this latter species is a condition that results from many seasons of continuous development. The small, additional phyllomorphs of the induced variant only lasted about 2 weeks, withering and disappearing as suddenly as they appeared, even though the antimetabolite Hyp was regularly applied at 10 mM over an extended period. Another pot-grown phenovariant (Fig 9) strongly resembled *S. gardenii* Hook., a typical centric rosulate species (Jong 1978). The seedlings were small (total measurement across the entire plant was 17–18 mm) and very young (54 days after the onset of germination), yet they temporarily mimicked the structure of plants of several seasons' growth.

The results obtained when seeds and seedlings were continuously exposed to antimetabolites via filter paper platforms, however, were decidedly different from when they were periodically drenched in soil mix. In experiments conducted with the platform cultures, micromolar concentrations of all three of the antagonists were sufficient to induce phenovariation. Although some phenovariation occurs naturally (10–15%, according to Rosenblum 1981), the antimetabolites induced noticeably higher percentages (see Table 1). A 2 × 4 contingency table analysis of Table 1 showed a highly significant relationship between the presence of phenovariation and the type of experimental treatment. A total of 11.7% of controls produced phenovariants as compared with 34.5% of 2,2'-dp-treated seedlings, 25.3% of 3,4-dhp-treated seedlings, and 37.3% of Hyp-treated seedlings ($G = 23.8$, 3 df, $P < 0.0001$).

The observed morphotypes exhibited five different patterns:

1. Normal (for *S. prolixus*) anisocotyledonous growth coupled with suppression of the shoot apical meristem and continuous growth of the macrocotyledon (Fig 4).
2. Anisocotyledonous, with a second phyllomorph displaced to the base of the macrocotyledon. This morphotype is a very early and untypical manifestation of the plurifoliate habit of *S. prolixus* (Figs 12, 18, 19). A continued growth of second and third phyllomorphs arising from the base of the preceding phyllomorph mimicked the pattern of mature plants in the species *S. fanniniae* Harvey ex C. B. Clarke and *S. davyii* S. Moore (Figs 13, 20).
3. Some accrescence, with a phyllomorph or plumular growth appearing in the axis of both cotyledons, which phenocopies the pattern of species in the subgenus *Streptocarpella* (Figs 14, 21).
4. Multiple phyllomorphy, characteristic of the rosulate species of the subgenus *Streptocarpus* and the closely related genus *Saintpaulia* (Figs 17, 25).
5. Isocotyly with plumular growth, which reproduces the growth pattern of members of the subfamily Gesnerioideae (Figs 15, 16, 22–24).

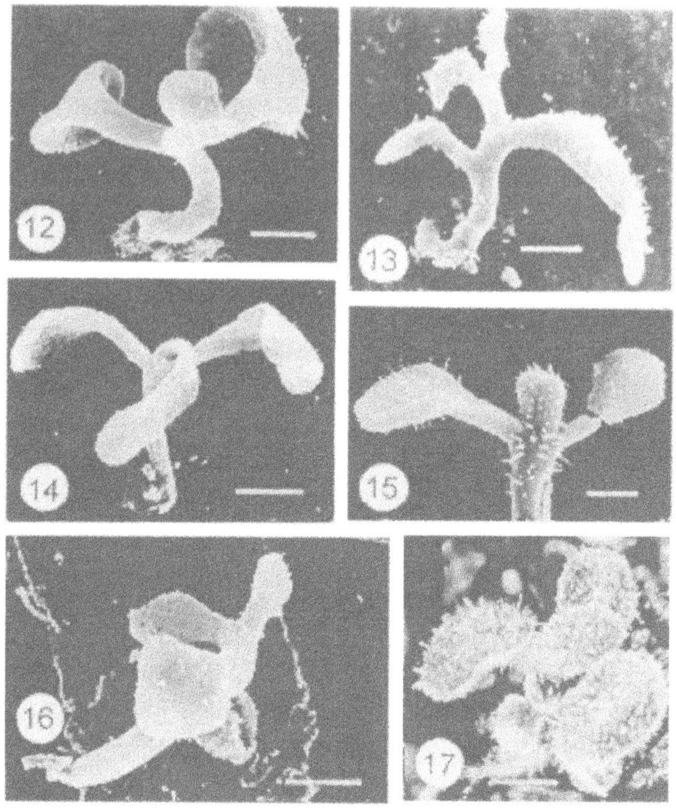

Figures 12–17. Phenovariations in *Streptocarpus* seedling morphology. 12–16 (SEMs). *S. prolixus* from platform culture, treated with antimetabolites. Bar = 0.5 mm. 12. Anisocotyly, with a secondary phyllomorph appearing at the base of the macrocotyledon (#2 in Table 1). 13. Further development of this variation; the secondary phyllomorph shows an additional phyllomorph at the base of its lamina, phenocopying the habits of *S. fanniniae* and *S. davyii.* 14. Some accrescence, with secondary phyllomorph in the axis of the two cotyledons (#3 in Table 1). 15. Isocotyly with plumule (#5 in Table 1); note leaf primordia. 16. Further development of isocotyly with plumular growth (#5 in Table 1): leaves show opposite decussate arrangement, typical of the subfamily Gesnerioideae. 17. (light micrograph). Pot-grown seedling showing multiple phyllomorphs (#4 in Table 1).

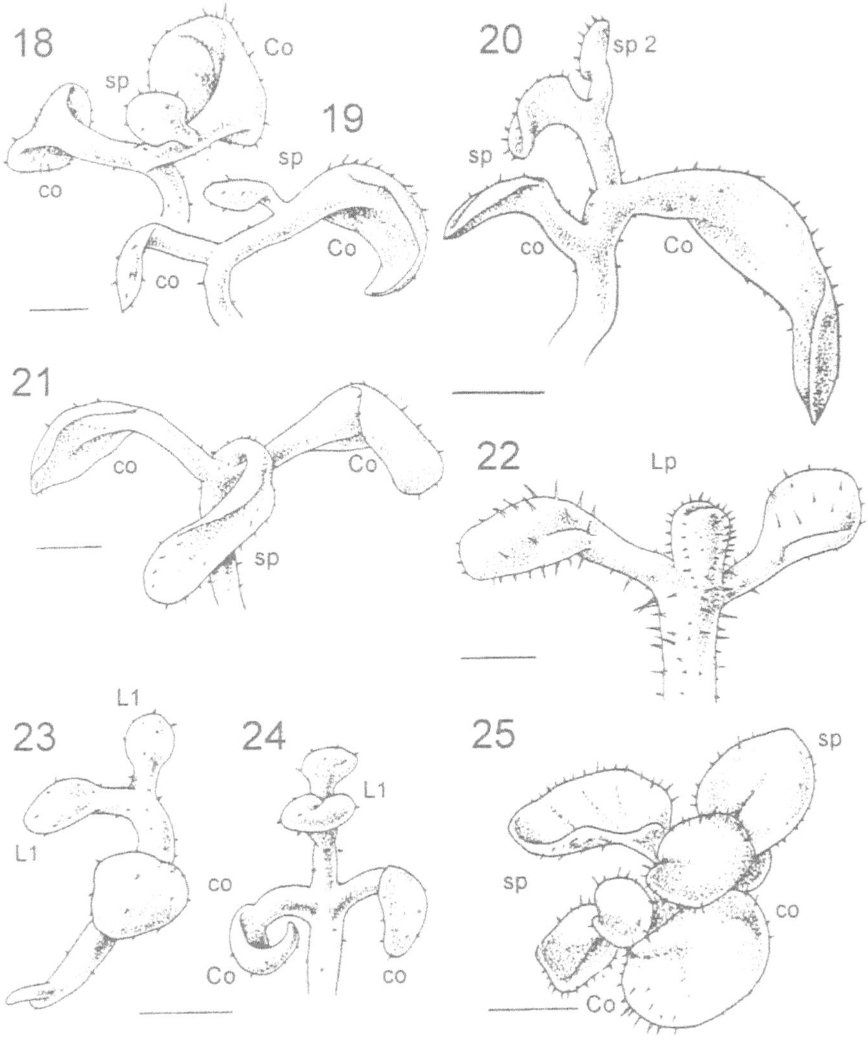

Figures 18–25. Diagrammatic interpretations of Figs 12–17. Bar = 0.5 mm. 18. Top view of Fig 12. Note position of the secondary phyllomorph. 19. A constructed side view of the same seedling, assisted by reference to micrographs of this angle of other seedlings showing similar morphology. 20. Phenovariation of Fig 13, with phyllomorphs arising from the base of preceding phyllomorphs. 21. Rendering of Fig 14. The appearance of size difference between the two cotyledons is slight because of a somewhat distorted view of the right cotyledon, which was actually 0.5 mm longer than the left cotyledon. 22. Diagram of Fig 15, ignoring damage to the specimen during mounting. Note the two leaf primordia. 23–24. Drawings of Fig 16 from slightly different-angled SEM micrographs, making the opposite decussate leaf arrangement more apparent. 25. Top view of Fig 17. Five secondary phyllomorphs are shown. The macrocotyledon is the larger leaf form, on the lower right. The microcotyledon is the small dark shape immediately above. Side view of the seedling is shown in Fig 9. Abbreviations as in Figs 1–11.

4. DISCUSSION

As stated earlier, the objective of this study was to determine whether exogenously supplied antagonists of Hyp-protein synthesis would induce phenovariation in a flowering plant family as had been previously observed in leafy liverworts. The results obtained show that they can. The Hyp-protein(s) responsible for the observed effects is not known. Evidence from studies conducted with bryophytes (Basile and Basile 1990 1993) suggests that one or more AGPs are the likely candidates. The AGPs are Hyp-containing glycoproteins and proteoglycans associated with the plasma membranes and cell walls of all classes of land plants (Clarke *et al* 1978, Basile and Basile 1987 1993). Prior to this study, both experimental and inferential evidence had been advanced to indicate that various AGPs may function in cell recognition, cell–cell communication and morphoregulation (Clarke and Knox 1978, Clarke *et al* 1978, Fincher *et al* 1983, van Holst and Clarke 1986). Basile and Basile (1990 1993) have found changes in the physical properties of some AGPs correlated with Hyp-protein antagonist-induced phenovariation in liverworts. Serpe and Nothnagel (1994), using an agent to perturb the functions of AGPs in living cells, suppressed cell proliferation of cell-suspension cultures of rose. Cell proliferation resumed (was "desuppressed") when the AGP-perturbing agent was removed. Thompson and Knox (1998), using the same reagent, suppressed the development of shoots in embryonic carrot cultures. Investigations to determine the nature of the Hyp-protein(s) altered by the antagonists that induced phenovariation in *S. prolixus*, however, constitute the basis for a separate study. In this study, emphasis is on the phenotypic consequences of perturbing Hyp-protein synthesis.

By the exogenous application of three Hyp-protein antagonists, a single species of *Streptocarpus* was induced to express morphological features extending not only through both subgenera of *Streptocarpus* but also through two subfamilies of the Gesneriaceae. The phenovariants induced in *S. prolixus* can be divided into four basic forms: anisocotyly (Cyrtandroideae) with the premature introduction of a second phyllomorph at the base of the macrocotyledon (*Streptocarpus*, plurifoliate form), anisocotyly with plumular growth (*Streptocarpella*), anisocotyly with multiple phyllomorphs (*Streptocarpus* rosulate form), and isocotyly with plumular growth (Gesnerioideae). All the morphotypes induced, except for number five, are represented in the various clades shown in the molecular analysis of *Streptocarpus* and *Saintpaulia* by Moeller and Cronk (1997). The two rosulate types are not clearly defined in this tree but appear randomly in the rosulate clade (Fig 26). Isocotyledonous phenovariants (number five in the table) display a character trait that appears even earlier on the evolutionary time line than is shown in Moeller and Cronk's (1997) analysis.

It is probable that the varied morphologies of the genus *Streptocarpus*, as in the liverworts previously chosen for experimentation, occur because of "differences with respect to the time and place cell proliferation and/or enlargement is suppressed" (Basile and Basile 1993). (The unifoliate species of *Streptocarpus* and the several species of liverworts selected for experimentation were considered to be evolutionarily derived via reduction in the number and/or the kind of organs they could

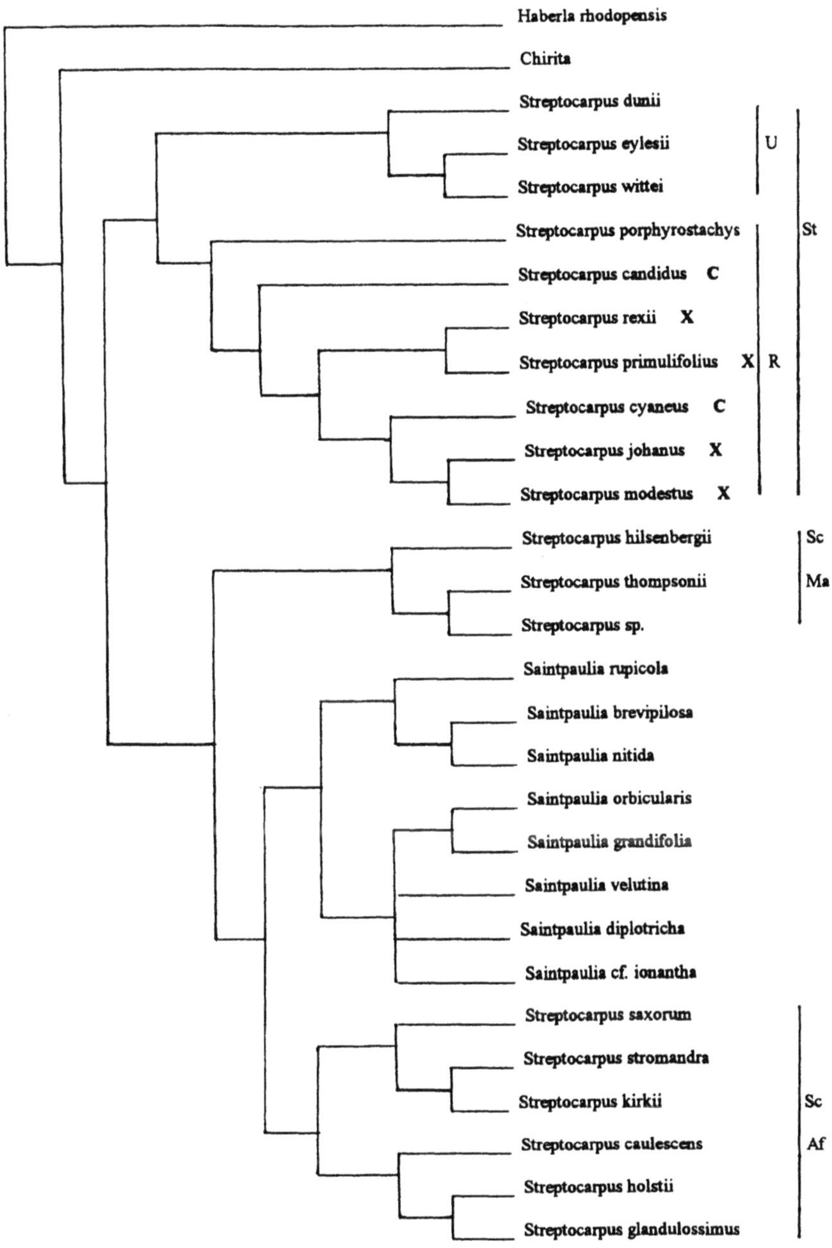

Figure 26. Most parsimonious tree of 419 steps (CI = 0.79, RI = 0.90) based on analysis of nuclear ribosomal DNA internal transcribed spacer sequences, including aligned gaps. Subgenera of *Streptocarpus* are abbreviated as follows: St subg, *Streptocarpus*; Sc subg, *Streptocarpella*; Ma, Madagascan species: Af, African species; Sections within St subg are indicated and abbreviated as U = unifoliate, R = rosulate, C = centric, X = excentric. Adapted from Smith *et al* (1998). All species in this tree, including the outgroups, are anisocotyledonous. Of the rosulates, according to Jong (1978), *S. candidus* and *S. cyaneus* are centric. *S. rexii, S. primufolius, S. johannis* and *S. modestus* are excentric.

develop.) In the studies using liverworts, antagonism of Hyp-protein synthesis resulted in desuppressed leaf and branch development. In the present study, antagonists of Hyp-protein synthesis led to desuppression of the primary shoot apical meristem and microcotyledon of *S. prolixus*. It appears then, that certain Hyp-proteins have essentially the same role in mediating place-dependent suppression in plants as diverse as liverwort gametophytes and flowering plant sporophytes. The wide evolutionary divergence of these two groups suggests a molecular homology that may be the same for all land plants, and provides additional support for the relevance of the "suppression hypothesis" (Basile and Basile 1993) in general, and the role of Hyp-proteins in place-dependent suppression in particular.

ACKNOWLEDGEMENTS

The authors wish to acknowledge a PSC/CUNY award # 6-67180 from the Research Foundation of the City of New York for support on this research, and Dr. Dwight Kincaid, for assistance with the statistical analysis of the data.

REFERENCES

Basile, D. V., 1967, The influence of hydroxy-L-proline on ontogeny and morphogenesis of the liverwort, *Scapania nemerosa*, *Am. J. Bot.* **54**: 977–983.

Basile, D. V., 1969, Toward an experimental approach to the systematics and phylogeny of leafy liverworts, In *Current Topics in Plant Science* (J. E. Gunekel, ed.), Academic Press Inc., New York, pp. 120–133.

Basile, D. V., 1970, Hydroxy-L-proline and 2,2'-dipyridyl-induced phenovariations in the liverwort, *Nowellia currifolia*, *Science* **170**: 1218–1220.

Basile, D. V., 1973a, Hydroxy-L-proline and 2,2'-dipyridyl-induced phenovariations in the liverwort, *Jungermannia lanceolata*, *Bull. Torrey Bot. Club* **100**: 350–352.

Basile, D. V., 1973b, A simple method of initiating axenic cultures of pteridophytes from spores, *Am. Fern J.* **63**: 1147–1151.

Basile, D. V., 1979, Hydroxyproline-induced changes in form, apical development and cell wall protein in the liverwort *Plagiochila arctica*, *Am. J. Bot.* **66**: 776–783.

Basile, D. V., 1990, Morphoregulatory role of hydroxyproline-containing proteins in liverworts, In *Bryophyte Development: Physiology and Biochemistry* (R. N. Chopra and S. C. Bhatla, eds.), CRC Press, Boca Raton, FL, pp. 226–243.

Basile, D. V., and Basile. M. R., 1980a, Ammonium ion-induced changes in form and hydroxyproline content of wall protein in the liverwort *Gymnocolea inflata*, *Am. J. Bot.* **67**: 500–507.

Basile, D. V., and Basile. M. R., 1980b, True dichotomous branching in the liverwort *Jungermannia lanceolata*, *Bot. Soc. Am. Misc. Pub.* **158**: 10.

Basile, D. V., and Basile, M. R., 1982, Evidence for a regulatory role of cell surface hydroxyproline-containing proteins in liverwort morphogenesis. *J. Hattori Bot. Lab.* **53**: 221–227.

Basile, D. V., and Basile. M. R., 1984, Probing the evolutionary history of bryophytes experimentally, *J. Hattori Bot. Lab.* **55**: 173–185.

Basile, D. V., and Basile. M. R., 1987, The occurrence of cell wall-associated arabinogalactan proteins in the Hepaticae, *Bryologist* **90**: 401–404.

Basile, D. V., and Basile. M. R., 1988, Procedures used for the axenic culture and experimental treatment of bryophytes, In *Methods in Bryology: Proceedings of the Bryology Methods Workshop, Mainz* (J. M. Glime, ed.), Hattori Botanical Laboratory, Nichinan, pp. 1–16.

Basile, D. V., and Basile. M. R., 1990, Hydroxyproline-containing glycoproteins in leafy liverworts, In *Bryophytes: Their Chemistry and Chemical Taxonomy* (H. D. Zinsmeister and R. Mues, eds.), Clarendon Press, Oxford, pp. 275–288.

Basile, D. V., and Basile. M. R., 1993, The role and control of the place-dependent suppression of cell division in plant morphogenesis and phylogeny, *Mem. Torrey Bot. Club* **25**: 63–84.

Basile, D. V., Lin, J. J., and Varner, J. E., 1987, The metabolism of exogenous hydroxyproline by gametophytes of *Plagiochila arctica* Bryhn et Kaal. (Hepaticae), *Planta* **175**: 539–545.

Burtt, B. L., 1963, Studies in the Gesneriaceae of the Old World, XXIV: Tentative keys to the tribes and genera, *R. Bot. Gard. Edinb. Notes* **24**: 205–220.

Burtt, B. L., 1970, Studies in the Gesneriaceae of the Old World, XXXI: Some aspects of functional evolution, *R. Bot. Gard. Edinb. Notes* **30**: 1–10.

Clarke, A. E., and Knox, R. B., 1978, Cell recognition in flowering plants, *Quart. Rev. Biol.* **53**: 3–28.

Clarke, A. E., Gleeson, P. A., Jermyn, M. A., and Knox, R. B., 1978, Characterization and localization of β-lectins in lower and higher plants, *Aust. J. Plant Physiol.* **5**: 707–722.

Cooper, J. B., and Varner, J. E., 1983, Selective inhibition of proline hydroxylation by 3,4-dehydroproline, *Plant Physiol.* **73**: 324–328.

Cronk, Q., and Moeller, M., 1997, Strange morphogenesis — organ determination in *Monophyllaea, Trends Plant Sci.* **2**: 327–328.

DeKock, P. C., and Vaughn, D., 1975, Effects of some chelating and phenolic substances on the growth of excised pea root segments, *Planta* (Berlin) **26**: 187–195.

Fincher, G. B., Stone, B. A., and Clarke, A. E., 1983, Arabinogalactan proteins: structure, biosynthesis and function, *Annu. Rev. Plant Physiol.* **34**: 47–70.

Hilliard, O. M., and Burtt, B. L., 1971, *Streptocarpus: An African Plant Study,* University of Natal Press, Pietermaritzburg.

Jong, K., 1970, Developmental aspects of vegetative morphology in *Streptocarpus*, Ph.D. Thesis, University of Edinburgh, Scotland.

Jong, K., 1973, *Streptocarpus* (Gesneriaceae) and the phyllomorph concept, *Acta Bot. Neerl.* **23**: 244–245.

Jong, K., 1978, Phyllomorphic organization in rosulate *Streptocarpus, R. Bot. Gard. Edinb. Notes* **36**: 369–396.

Moeller, M., and Cronk, Q. C. B., 1997, Origin and relationships of *Saintpaulia* (Gesneriaceae) based on ribosomal DNA internal transcribed spacer (ITS) sequences, *Am. J. Bot.* **71**: 52–64.

Nolan, J. C., Ridge, S., Oronsky, A. L., and Kerwar, S. S., 1978, Studies on the mechanism of reduction of prolyl hydroxylase activity by 3,4-dehydroproline, *Arch. Biochem. Biophys.* **189**: 448–453.

Rosenblum, I. M., 1981, An approach toward understanding some of the morphogenic bases of phylogeny of *Streptocarpus* (Gesneriaceae), Ph.D. Thesis, City University of New York.

Rosenblum, I. M., and Basile, D. V., 1984, Hormonal regulation of morphogenesis in *Streptocarpus* and its relevance to evolutionary history of the Gesneriaceae, *Am. J. Bot.* **71**: 52–64.

Rutishauser, R., 1995, Developmental patterns of leaves in Podostemataceae compared with more typical flowering plants — saltational evolution and fuzzy morphology, *Can. J. Bot.* **73**: 1305–1317.

Serpe, M. D., and Nothnagel, E. A., 1994, Effects of Yariv phenylglycosides on *Rosa* cell suspensions: evidence for the involvement of arabinogalactan-proteins in cell proliferation, *Planta* **193**: 542–550.

Smith, J. F., Kresge, M. E., Moeller, M., and Cronk, Q. C. B., 1998, A cladistic analysis of ndhF sequences from representative species of *Saintpaulia* and *Streptocarpus,* subgenera *Streptocarpus* and *Streptocarpella* (Gesneriaceae), *Edinb. J. Bot.* **55**: 1–11.

Stebbins, G. L., and Basile, D. V., 1986, Phyletic phenocopies: a useful technique for probing the genetic and developmental basis of evolutionary change, *Evolution* **40**: 422–425.

Thompson, H. M. J., and Knox, J. P., 1998, Stage-specific responses to embryogenic carrot cell cultures to arabinogalactan-protein-binding β-glucosyl Yariv reagent, *Planta* **205**: 32–38.

van Holst, G. J., and Clarke, A. E., 1986, Organ-specific arabinogalactan-proteins of *Lycopersicon peruvianum* (Mill.) demonstrated by crossed electrophoresis, *Plant Physiol.* **80**: 786–789.

Chapter 18

Evidence for the Interrelated Actions of Auxin, Ethylene, and Arabinogalactan-Proteins on the Transition from Non-Apical to Apical Growth of *Physcomitrella patens* Hedw. (Funariaceae)

Madeline M. Mignone[1] and Dominick V. Basile
[1]*Department of Biological Sciences, Lehman College, City University of New York, Bronx, New York 10468; and Graduate School, City University of New York, New York, New York*

1. INTRODUCTION

We are investigating the role and control of the spacial-temporal suppression of cell proliferation and/or cell enlargement in plant morphogenesis. A prime example of this phenomenon is the suppressed development of every third leaf primordium in a large proportion of the species of leafy liverworts (Jungermanniales, Hepatcae). Through a series of experiments, it was found that the correlative inhibition of the development of these leaves past a primordial stage was regulated by the interrelated actions of auxin, ethylene, and certain arabinogalactan-proteins (AGPs) (see Basile 1990, Basile and Basile 1990 1993 1994, for reviews). In addition, it became evident from experiments with species of *Streptocarpus*, a flowering plant (Rosenblum and Basile 1984), that this correlative control system was not restricted to determining whether or not leaves developed in liverworts at a specific place and time. The results of more than 25 years of research in this area indicated that the auxin-ethylene-AGP system played a critical role in three important aspects of pattern formation and pattern change in leafy liverworts, and probably all plants (Basile and Basile 1993). One aspect, already mentioned, is to stop (suppress) the further development of specific organ primordia as soon as they are established. The failure of primordia to develop into mature organs is the basis for the recurrent trend in leafy liverwort families toward "reductive" or "recessive" evolution via "loss" in the numbers (i.e., leaves) and/or kinds (i.e., branching-types) of parts. A second aspect is the modulation in the size and shapes of leaves. The third aspect is the setting apart of the apical meristematic regions during the transition from a diffuse (e.g., protonemal, prothallial, or embryogenic) growth pattern to one that is strictly apical/polar. This transition from diffuse to polar/apical growth is characteristic of the early stages of development of both the gametophytes and sporophytes of all land plants. It is this third hypothesized role for suppression that the experiments reported here were designed to test.

For these experiments, a moss, *Physcomitrella patens* (Funariaceae), was chosen. During its relatively short life cycle, this small, easily cultured plant undergoes a series of distinct pattern changes. These pattern changes have proven to be amenable to both physiological and genetic manipulation and analysis (Engel 1968, Ashton and Cove 1977, Cove *et al* 1978, Cove and Ashton 1984 1988, Ashton *et al* 1988 1993, Cove 1992).

Vegetative development of *P. patens* normally proceeds through four distinct stages (Cove *et al* 1980, Bhatla 1994). The first, following spore germination, is termed the "primary chloronema stage" (Fig 1). The cells of the branching uniseriate filaments that comprise the plant body at this stage are distinguished by their high chloroplast content and by their perpendicular cross walls. The second stage is termed the "caulonemal stage". This stage represents a somewhat more suppressed development in that the cells of the filaments formed during this stage are narrower than those of the chloronema, have noticeably fewer chloroplasts per cell, and are separated by oblique cross walls (Fig 2). The third or "secondary chloronema stage" is distinguished from the primary chloronema by virtue of their origin as branches from caulonemal cells and, more importantly, by their greater tendency to give rise to the gametophore buds that mark the beginning of the fourth or "leafy gametophore stage" (Fig 3). According to our hypothesis, gametophore initiation is brought about by the suppression of protonemal growth or, more specifically, by the suppression of protonemal growth that is brought about by the interrelated actions of auxin, ethylene and a Hyp-containing protein, probably an AGP.

Mutants resistant to the effects of both auxins and cytokinins, termed Category I mutants, were isolated by Ashton *et al* (1979b). These mutants grow to form a plant that is more or less spherical in shape and is made up entirely of chloronemata (Ashton *et al* 1979b 1993, Cove *et al* 1980). If mutants with this phenotype lack the ability to respond to auxin, then they are named NAR mutants (Ashton *et al* 1979a). Ashton *et al* (1993) found evidence that wild-type *P. patens,* when grown on medium with naphthylphthalamic acid or 2,3,5 triiodobenzoic acid (TIBA), produces plants with morphology similar to that of the NAR mutants.

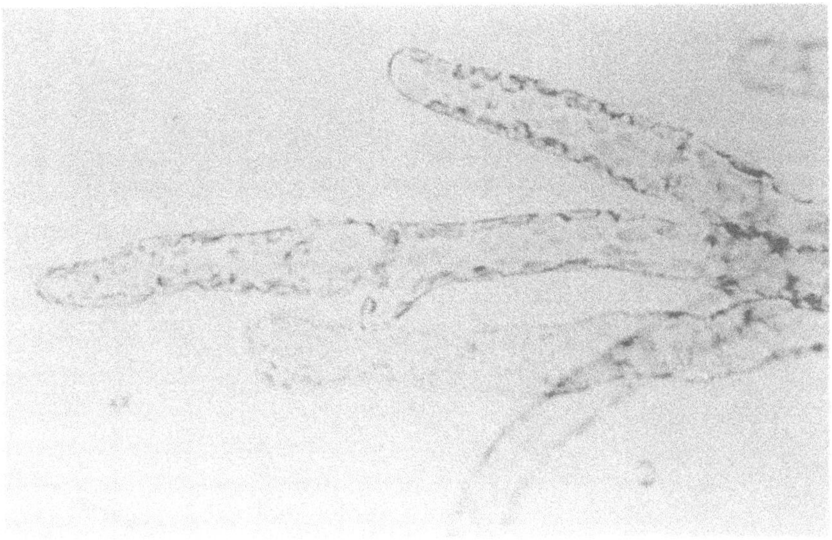

Figure 1. The primary chloronema stage. Note the perpendicular cross walls and the high content of chloroplasts. (1160x)

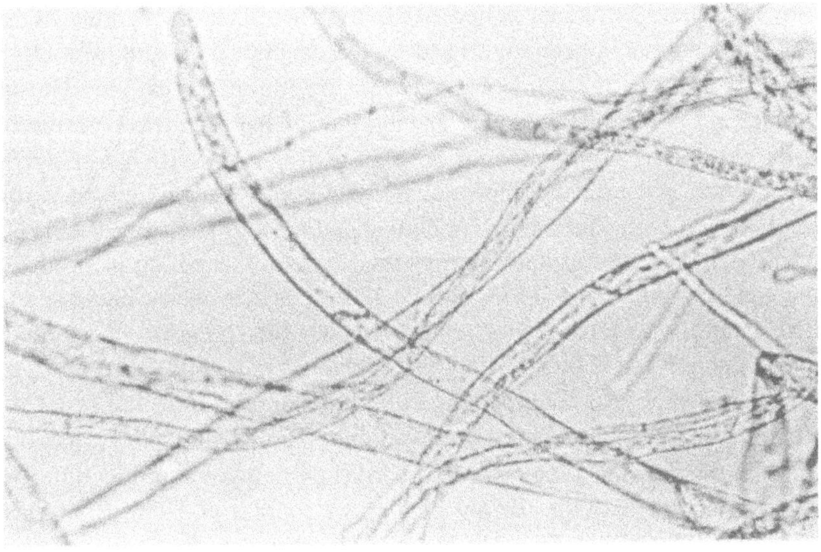

Figure 2. The caulonema stage. Note the oblique cross walls and noticeable lack of chloroplasts. (1070x)

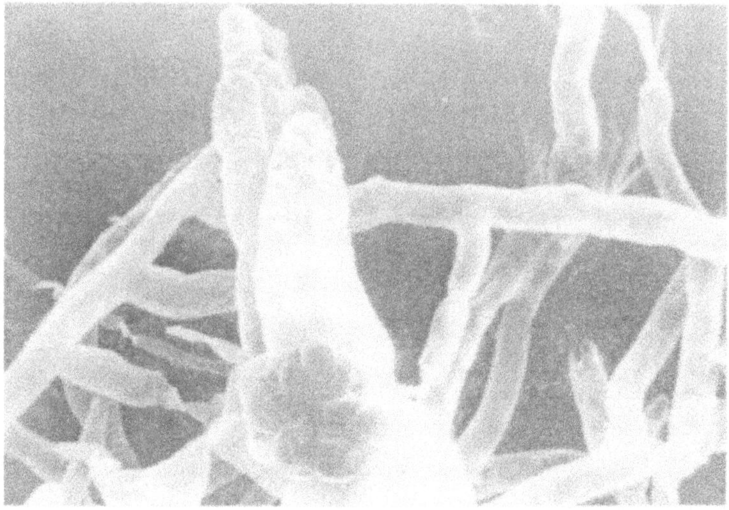

Figure 3. SEM of the leafy gametophore stage. This stage marks the transition from a diffusely branching filamentous body to an apically growing parenchymatous body. (352x)

Ashton *et al* (1979a) also isolated several phenotypically distinct groups of gametophore-overproducing mutants called OVE mutants. The OVE mutants have slightly abnormal gametophores, some with a stem that is short or absent. The OVE leaves are smaller than those of the wild type cultured on minimal medium. The gametophyte growth rate in OVE mutants is slower than that of the wild type on minimal medium (Ashton *et al* 1979a), and its overall development is similar to that of wild-type strains cultured in the presence of high level cytokinins. These mutants initiate several gametophores on each caulonema (Ashton *et al* 1979a 1993, Cove *et al* 1980).

Ample evidence shows that auxin and ethylene are naturally produced by moss protonema (Johri and Desai 1973, Johri 1978). It is also well established that auxin is required before development of chloronemata can progress to gametophore bud initiation (Ashton and Cove 1977 1990, Cove *et al* 1980, Cove and Ashton 1984, Ashton *et al* 1985 1990). Comparable data indicating an essential role for ethylene in gametophore bud induction is not yet available. Nevertheless, an exogenous auxin and 1-aminocyclopropane-1-carboxylic acid can trigger ethylene production in at least one moss, as they can in a wide variety of plants (Bopp *et al* 1986, Bhatla and Dhingra-Babbar 1990). The AGPs, the only class of Hyp-proteins that appear to be ubiquitous components of plants, were detected at the cell surfaces of representatives of every family of mosses (J. Mankiewicz and D.V. Basile, unpublished data). In early studies, the presence of exogenous Hyp in culture media was reported to dramatically prolong (i.e., prevent the timely suppression of) protonemal growth of several species of moss (Burkholder 1959) and one leafy liverwort (Basile 1967). It was subsequently determined that exogenous Hyp can interfere with the normal synthesis and function of Hyp-proteins in general (Basile 1980, Basile *et al* 1987) and AGPs in particular (Basile and Basile 1987 1990). This literature, although scattered and seemingly unrelated, nevertheless provides the earliest indication that all three of the components of our proposed correlative control system are naturally occurring in mosses and may be involved in mediating the transition from protonemal to leafy gametophore stages in mosses. We now present the results of a series of experiments that support our proposition that the transition from diffuse growth to polar, apical growth is correlated with the suppression of protonemal growth and that this suppression is regulated by the interrelated actions of auxin, ethylene and a Hyp-protein, most probably an AGP.

2. MATERIALS AND METHODS

2.1 Plant Material

2.1.1 Wild-type species

Axenic cultures of wild-type *P. patens* (Hedw.) Br. Eur., derived from cultures kindly provided by John C. Wallace (Bucknell University, Lewisburg, PA), were the source of inoculum for all experiments. Inoculum was prepared by homogenizing stock cultures of *P. patens* in a heat sterilized microblender head at 60 V for 25 sec. Ten drops (~1.0 ml) of the homogenized tissue was inoculated into culture vessels using sterile disposable transfer pipettes (Samco Sci. Inc, San Fernando, CA).

2.1.2 Mutant Species (OVE and NAR)

Cultures kindly provided by Neil Ashton (University of Regina, Saskatchewan, Canada) were subcultured on a regular basis.

2.2 Culture Conditions

The basal aqueous nutrient medium used for both stock and experimental cultures consisted of Knop's macronutients supplemented with Hutner's "metals 49" micronutrients, and 1% w/v glucose. Formulae for the macro- and micronutrient mixtures are given in Basile (1978). The pH of the medium was adjusted to 4.6, and 30 ml aliquots were delivered into culture ("baby food") jars with Magenta Caps™ (Sigma, St. Louis, MO) before being heat sterilized for 20 min at 15 psi. Cultures were incubated on shelves in rooms controlled to maintain 18°C and continuous illumination of 700–1000 lux from cool white fluorescent bulbs. The incubation period for each of the experiments employing antagonists was 6 weeks.

Mutant species (OVE and NAR) were grown in moss medium formulated by Ashton and Cove (1977) and incubated as above.

2.3 Antagonists

Five antagonists were individually tested for their capacity to prevent subapical growth suppression, thereby preventing apical meristem initiation. Two, TIBA (ICN Biochemicals, Costa Mesa, CA) and 2-[(1-napthalenylamino)carbonyl] benzoic acid (ALANAP) (a gift from the Uniroyal Co. Inc., Naugatuck, CT), were auxin transport antagonists; one, aminoethoxyvinylglycine (AVG) (Sigma), was a specific antagonist of ethylene synthesis; and two, 3,4-dehydro-L-proline (3,4-dhp) (Sigma) and *trans*-4-hydroxy-L-proline (Hyp) (Calbiochem, San Diego, CA), were antagonists of normal Hyp-protein synthesis. All stock solutions of antagonists were cold sterilized through 0.2-μm filters and added to the sterile basal medium prior to inoculation. As a result of pilot experiments, the concentrations of antagonists that were found to be effective were ALANAP, 8 μg /ml; TIBA, 11 μg/ml; AVG, 0.90 μg/ml; 3,4-dhp, 8 μg/ml; and Hyp 12 μg/ml.

2.4 Culture in Knop's and Knudsen's Media

After the results of the physiological experiments with antagonists were confirmed, a quicker mode of producing protonemal tissues was tested. Source tissue was briefly homogenized as above and then placed either in Knop's medium, which promoted gametophore growth, or in Knudsen's medium, which promoted protonemal growth. Basile and Basile (1980) observed that *Gymnocolea inflata* maintained its normal "suppressed" pattern of leaf and branch development when cultured on ammonium free nutrient medium (Knop's), and a "desuppressed" pattern on ammonium ion-containing medium. This same effect was observed with *P. patens*, so the tissue was grown for 14 days in Knudsen's and then harvested. The tissue proved to be as luxuriant in growth using Knudsen's as it was in the antagonists. The cultures grown on Knop's medium were harvested after 6 weeks. These tissues were freeze dried and put aside for electrophoretic studies.

2.5 Microscopy

The tissue from each culture vessel was harvested and carefully examined for the presence of gametophore bud formation. Some of this tissue was fixed for 1 hour in 2% (v/v) gluteraldehyde in fresh basal nutrient media. The fixed tissue was then rinsed three times with cacodylate buffer. Some of the fixed tissue was placed in 50% (v/v) aqueous glycerol overnight and then transferred to 100% glycerol solution. Semipermanent slides were made of representative material from each experimental treatment and labeled for subsequent light microscopy studies. Photomicrographs were recorded on Kodachrome film with an Olympus OM4 camera fitted to a Zeiss Axiovert microscope. Other aldehyde-fixed, buffer-washed tissue was dehydrated in a graded alcohol series, critical-point dried, mounted on copper studs, and gold coated (Sicko and Jensen 1973). SEM micrographs were taken with an Amray 1830 at 20 kV.

2.6 Histological Assays

The distribution of AGPs during different stages of development was determined by light microscopy. Whole samples of both suppressed and desuppressed tissue were cleared in 80% ethanol and then stained with Yariv's reagent, a known and universally accepted indicator of AGPs (Yariv *et al* 1962 1967).

2.7 Quantitative AGP Assays

Wild-type Species: The AGPs were extracted from equal amounts of lyophilized samples (1.0 g) of both suppressed and desuppressed tissue of *P. patens*. The tissue samples were pulverized after freezing with liquid nitrogen, transferred to a centrifuge tube with 10 ml of Barbital Buffer (Sigma Diagnostics #B6632, modified according to D. V. Basile, unpublished data), and 15 drops of 1% 3-[(3-cholamidopropyl)dimethylammonio]-1-propanesulfonate were added. The extraction mixture was vortexed (Multi-Tube Vortexer, Scientific Manufacturing Industries, Emoryville, CA) for 30 min, and centrifuged at 15,000 *g* for 30 min. The supernatant was decanted into another centrifuge tube, microwaved to the boiling point, and cooled. The pellet was then resuspended in buffer and detergent, and the above extraction process was repeated. The microwaved supernatants were once again centrifuged at 1500 *g* and transferred to Centriprep 30 tubes (Millipore, Bedford, MA) where they were spun at 3,500 *g* until concentrated to 0.5 ml. The concentrates were then filter-centrifuged through nylon membranes to remove any remaining particulates. The concentrations of all final extracts were determined by the radial diffusion assay of van Holst and Clarke (1985).

To obtain a concentration of 1.0 mg/ml of AGPs from the desuppressed tissue, the above protein extraction was performed three times. The resulting extracts were then pooled, and concentrated to approximately 0.5 ml.

Mutant Species of* P. patens *(OVE Bud and NAR): Tissue (1.0 mg) from OVE bud and NAR specimens was extracted, and the above protocol was followed.

2.8 Electrophoresis

A 1.6% (w/v) agarose gel was prepared using a glyoxyl agarose (NuFix™, FMC Bioproducts, Rockland, ME) with a framed template as per Basile *et al* (1989). The pooled protein extracts of the suppressed (Knop's-grown), desuppressed (Knudsen's-grown), and mutant (OVE bud and NAR) plants were diluted and assayed so that equivalent amounts were loaded on the gel. A total of 20 μl of the pooled protein samples were placed in the wells, as were 1.0 mg/ml gum arabic (Sigma) samples used as standards.

The gel was run at 500 volts, 15 milliamps, for approximately 5 hours or until the dye front reached 6 cm from the line of origin.

The gel was immersed in cyanoborohydride (0.08 M) for at least 2 hours to effect cross-linking of the AGPs to the gels. Excess cyanoborohyride reagent was removed by washing in running tap water for 2 hours. The gel was then placed in a heat-sealable plastic food bag containing 50 ml of Yariv's ß-glucosyl reagent (30 μl/ml) in 50 ml of 0.15 M NaCl solution (Basile *et al* 1989). The bags were sealed and the gels were incubated in the stain for approximately 3–4 days.

3. RESULTS

In the specified culture conditions, gametophore bud formation by *P. patens* was well under way in control cultures at the fifth week of incubation. Leafy gametophores were evident in all control cultures by the end of the 6-week incubation period (Fig 4). In sharp contrast and without exception, cultures in which the basal nutrient medium was supplemented with any one of the antagonists remained in a purely protonemal pattern of development. In fact, antagonist-treated cultures were never observed to proceeded beyond the primary chloronemal stage during the experimental incubation period. Because of the very small dry-weight mass of the tissues produced in individual cultures in 6 weeks, no attempt was made at a quantitative comparison of growth among and/or between the replicate cultures of each of the several experiments. At a qualitative level, however, it was clear that none of the antagonists caused any obvious inhibition of protonemal growth. If anything, growth of the protonemal tufts seemed more luxuriant in the presence of the antagonists than in the controls. It seems reasonable to conclude that each of the molecules being "antagonized" has some important role in bringing about the transition from diffusely proliferating, branching filaments to locally (apically) proliferating, unbranched leafy shoots. Examples from the various antagonist treatments are shown in Figs 5–7.

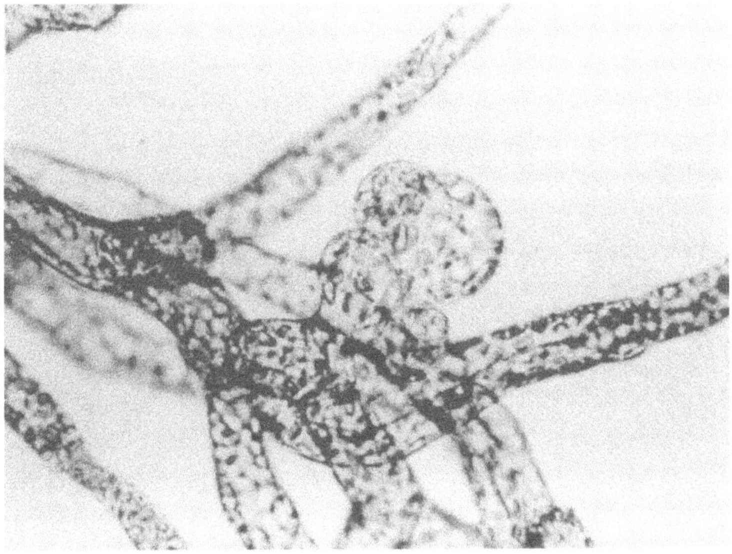

Figure 4. Gametophore bud rising from the apical meristem of 4-week-old control cultures. Note the transition from the diffuse filamentous protonemal growth to the suppressed state of apical growth. (781x)

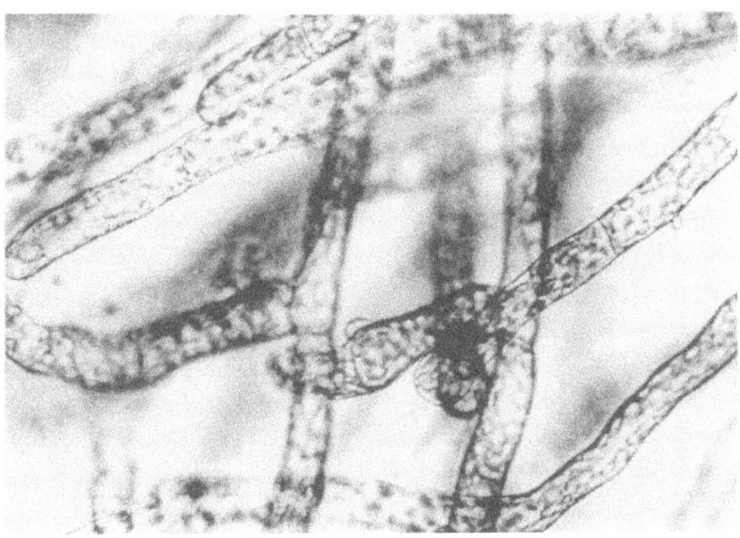

Figure 5. Six-week-old cultures treated with TIBA. Note that the development has remained at the most desuppressed stage of chloronemata growth. (773x).

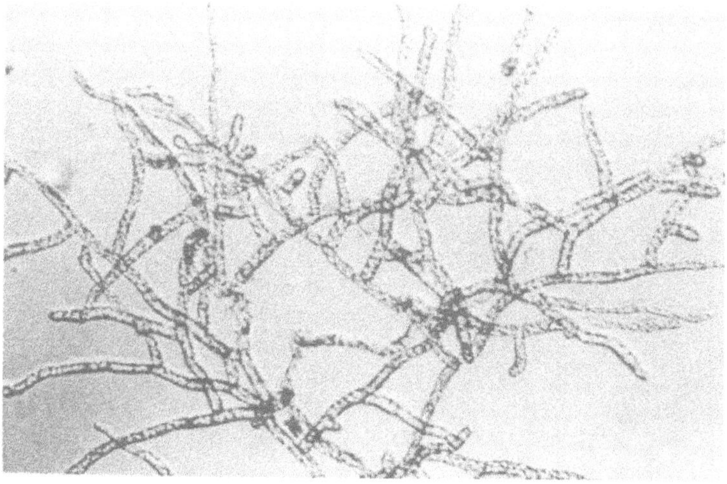

Figure 6. Six-week-old cultures treated with AVG. (489x)

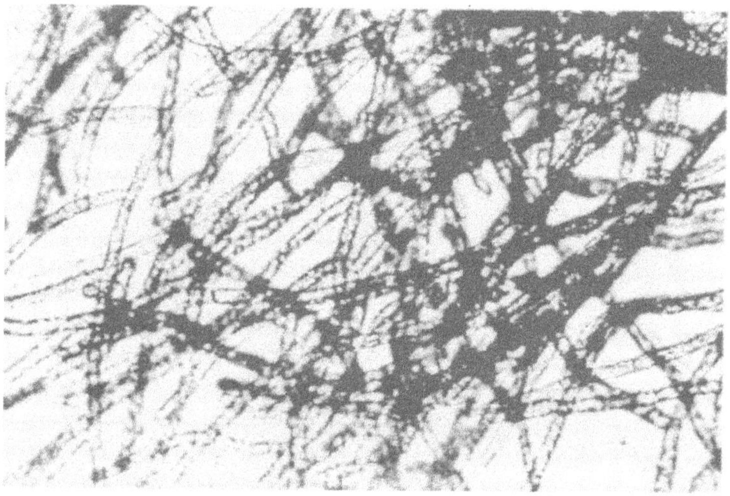

Figure 7. Six-week-old cultures treated with 3,4-dhp. Note the lush growth of the protonema (510x).

As can be seen in Fig 8, the AGPs extracted from plants cultured on the two different types of media (Knop's producing gametophores or suppressed tissue, and Knudsen's producing protonema or desuppressed tissue) produced significantly different radial diffusion assays. An estimated 15-fold more AGPs were produced by the suppressed leafy gametophores than by the desuppressed protonema. Histochemical staining of the cleared tissue of wild-type *P. patens* at the bud forming stage also demonstrated this marked difference in AGP distribution (Fig 9).

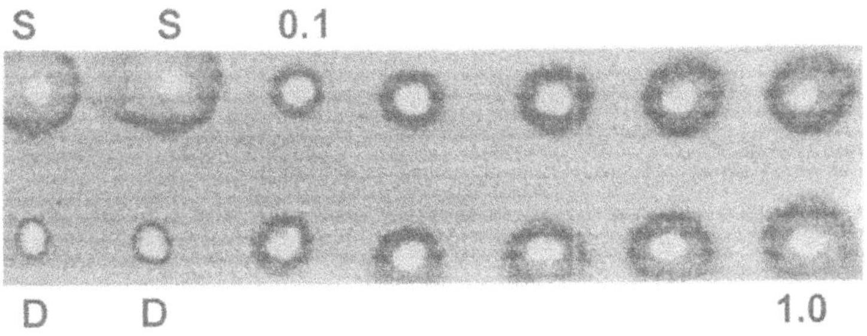

Figure 8. Radial diffusion assay of suppressed (S) (Knop's-grown) tissue vs. desuppressed (D) (Knudsen's-grown) tissue. Numbered (mg/ml) wells are gum arabic standards.

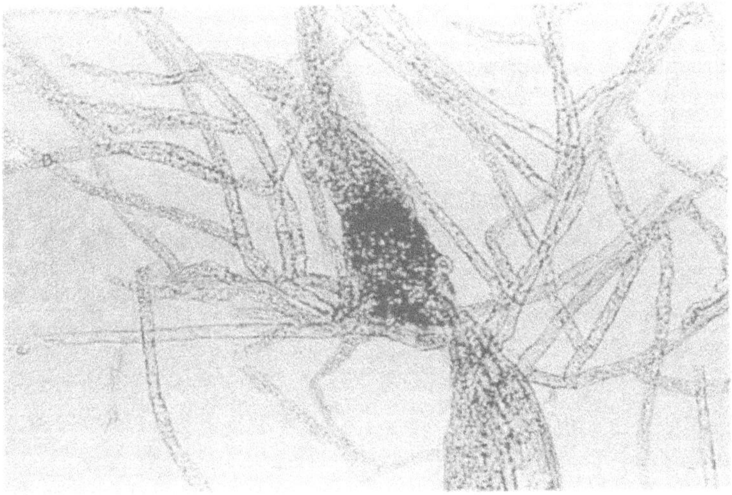

Figure 9. Gametophore bud primordium growing from *P. patens* protonema stained with Yariv's reagent. Note how the leaf primordium (S = suppressed) and surrounding protonemal area are strongly stained as opposed to the protonema (D = desuppressed) located distally from the meristematic region. (212x)

The AGPs extracted from both suppressed and desuppressed *P. patens* tissue were compared by agarose gel electrophoresis, in which a clear difference in the electrophoretic patterns was produced by the two samples (Fig 10).

Figure 11 shows further comparisons of the electrophoretic patterns of the suppressed wild-type (S), desuppressed wild-type (D), OVE mutant (O) and NAR mutant (N). Note that the OVE mutant (O), which is a transitional tissue charact-erized by high quantities of both protonemal and gametophore bud tissues, has an electrophoretic pattern that demonstrates the presence of both suppressed and de-suppressed types of AGPs. The NAR mutant (N), which is characterized by only chloronemal growth, shows a pattern similar to that of the desuppressed (D) tissue that was grown on Knudsen's medium.

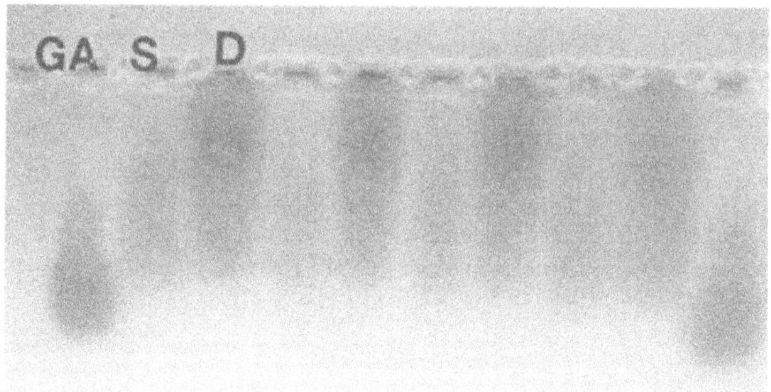

Figure 10. Electrophoresis of suppressed (S) (Knop's-grown) tissue vs. desuppressed (D) (Knudsen's-grown) tissue. Note the significant differences in the electrophoretic patterns between the tissue types. GA = gum arabic standard (1.0 mg/ml)

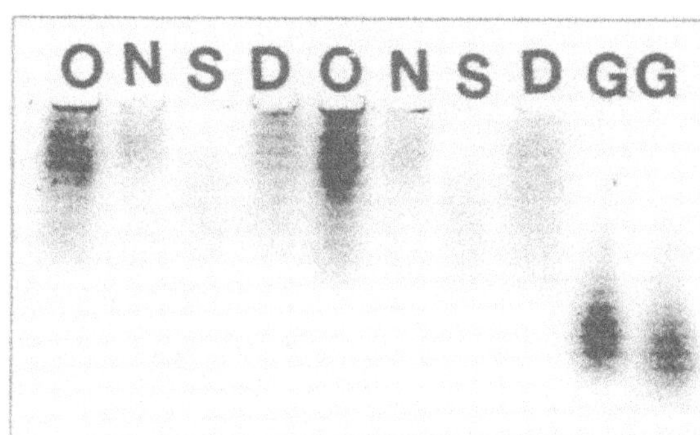

Figure 11. Comparison of electrophoretic patterns of the suppressed wild-type (S), desuppressed wild-type (D), OVE mutant (O) and NAR mutant (N). G = gum arabic standard.

4. DISCUSSION

How auxin, ethylene and Hyp-proteins individually function to bring about this growth pattern transition from a diffuse, branching filamentous body to an apically growing parenchymatous body is still to be determined. It seems clear, however, that their combined actions, if not antagonized, would be to suppress further division and enlargement of most of the cells comprising a protonema. By extrapolation, the establishment of a mechanism to suppress growth throughout most of the developing plant body is prerequisite to establishing highly localized growth centers (apical meristems).

If any one component of this correlative system is blocked and cannot function (Fig 12), then suppressed development is arrested to the most desuppressed level — the chloronematal level. Thus, through these series of experiments, using *P. patens* as a model system, we have given evidence that a correlative control system comprised of auxin, ethylene, and Hyp-proteins regulates plant development by place-dependent suppression of cell proliferation. The correlative control system brings about the transition from diffuse growth to strictly polar apical growth by suppressing growth in all cells except the population of cells comprising the apical meristem.

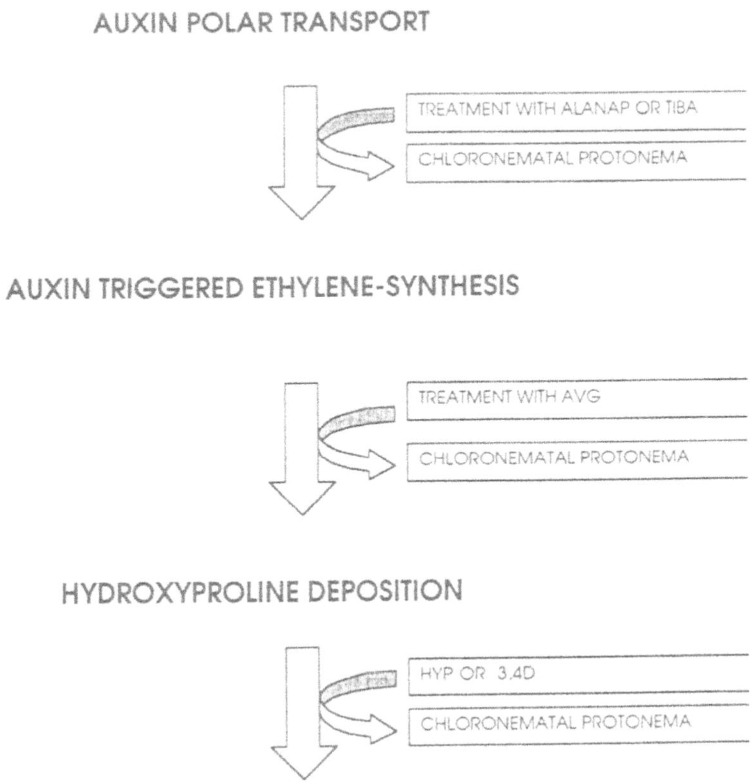

Figure 12. Model of correlative control system for transition from non-apical to apical growth of *P. patens*. If any one component of this correlative system is blocked and cannot function, then suppressed development is arrested to the most desuppressed level — the chloronematal level. 3,4D = 3,4-dehydro-L-proline.

REFERENCES

Ashton, N. W., and Cove, D. J., 1977, The isolation and preliminary characterization of auxotrophic and analogue resistant mutants in the moss, *Physcomitrella patens*, *Mol. Gen. Genet.* **154**: 87–95.

Ashton, N. W., and Cove, D. J., 1990, Mutants as tools for the analytical dissection of cell differentiation in *Physcomitrella patens* gametophytes, In *Bryophyte Development, Physiology, and Biochemistry* (R. N. Chopra and S. C. Bhatla, eds.), CRC Press, Boca Raton, pp. 17–31.

Ashton, N. W., Cove, D. J., and Featherstone, D. R., 1979a, The isolation and physiological analysis of mutants of the moss, *Physcomitrella patens*, which overproduce gametophores, *Planta* **144**: 437–442.

Ashton, N. W., Grimsley, N. H., and Cove, D. J., 1979b, Analysis of gametophytic development in the moss *Phycomitrella patens*, using auxin and cytokinin resistant mutants, *Planta* **144**: 427–435.

Ashton, N. W., Schulze, A., Hall, P., and Bandurski, R. S., 1985, Estimation of indole-3-acetic acid in gametophytes of the moss, *Physcomitrella patens, Planta* **164:** 142-144.

Ashton, N. W., Boyd, P. J., Cove, D. J., and Knight, C. D., 1988, Genetic analysis in *Physcomitrella patens*, In *Methods in Bryology: Proceedings of the Bryology Methods Workshop, Mainz* (J. M. Glime, ed.), Hattori Botanical Laboratory, Nichinan, pp. 59–72.

Ashton, N. W., Cove, D. J., Wang, T. L., and Saunders, M. J., 1990, Developmental studies of *Physcomitrella patens*, using auxin and cytokinin sensitivity mutants, In *Plant Growth Substances 1988* (P. Phares and S. B. Rood, eds.), Springer-Verlag, Berlin, pp. 57–64.

Ashton, N. W., Zacharias, R. A., and Fracchia, D. F., 1993, Genetic analysis of phytohormone action and morphogenesis in *Physcomitrella patens*, *Mem. Torrey Bot. Club* **25**: 11–20.

Basile, D. V., 1967, The influence of hydroxy–L-proline on the ontogeny and morphogenesis of the liverwort, *Scapania nemorosa*, *Am. J. Bot.* **54**: 977–983.

Basile, D. V., 1978, Culture media for bryophytes, In *CRC Handbook Series for Nutrition and Food*, Section G., Vol III (M. Rechcigl, Jr., ed.), CRC Press, Cleveland, OH, pp. 557–568.

Basile, D. V., 1980, A possible mode of action for morphoregulatory hydroxyproline-proteins, *Bull. Torrey Bot. Club* **107**: 325–338.

Basile, D. V., 1990, Morphoregulatory role of hydroxyproline-containing proteins in liverworts, In *Bryophyte Development: Physiology and Biochemistry* (R. N. Chopra and S. C. Bhatla, eds.), CRC Press, Boca Raton, FL, pp. 225–243.

Basile, D. V., and Basile, M. R., 1980, Ammonium ion-induced changes in form and hydroxyproline content of wall protein in the liverwort, *Gymnocolea inflata*, *Am. J. Bot.* **67**: 500–517.

Basile, D. V., and Basile, M. R., 1987, The occurrence of cell-wall associated arabinogalactan proteins in the hepaticae, *Bryologist* **90**: 401–404.

Basile, D. V., and Basile, M. R., 1990. Hydroxyproline metabolism and hydroxyproline-containing glycoproteins in leafy liverworts, In *Bryophytes: Their Chemistry and Chemical Taxonomy* (H. D. Zinsmeister and R. Mues, eds.), Clarendon Press, Oxford, pp. 275–288.

Basile, D. V., and Basile, M. R., 1993, The role and control of the place-dependent suppression of cell division in plant morphogenesis and phylogeny, *Mem. Torrey Bot. Club* **25**: 63–84.

Basile, M. R., and Basile, D. V., 1994, The role of growth suppression in leafy liverwort morphogenesis and phylogeny, *J. Hatt. Bot. Lab.* **76**: 75–85.

Basile, M. R., Basile, D. V., and Varner, J. E., 1987, Lycorene, ascorbate, prolyl-hydroxylase, and hydroxyproline proteins in relation to cellular suppression, desuppression, *Ann. NY Acad. Sci.* **494**: 175–177.

Basile, D. V., Kushner, B. K., and Basile, M. R., 1989, A new method for separating and comparing arabinogalactan proteins for the chemosystematics of the Hepaticae, *Bryologist* **92**: 164–169.

Bhatla, S. C., 1994, *Moss Protonema Differentiation*, John Wiley and Sons, Inc., New York.

Bhatla, S., and Dhingra-Babbar, S., 1990, Growth regulating substances in mosses, In *Bryophyte Development: Physiology, and Biochemistry* (R. N. Chopra and S. C. Bhatla, eds.), CRC Press, Boca Raton, FL, pp. 79–101.

Bopp, M., Gerhauser, D., and Kessler, U., 1986, On the hormonal system of mosses, In *Plant Growth Substances* (M. Bopp, ed.), Springer-Verlag, Berlin, pp. 263–269.

Burkholder, P. R., 1959, Organic nutrition of some mosses growing in pure culture, *Bryologist* **62**: 5–15.

Cove, D. J., 1992, Regulation of development in the moss *Physcomitrella patens*, In *Development: The Molecular Genetic Approach* (V. E. A. Russo, S. Brody, D. Cove, and S. Ottolenghi, eds.), Springer-Verlag, Berlin, pp. 179-193.

Cove, D. J., and Ashton, N. W., 1984, The hormonal regulation of gametophytic development in bryophytes, In *The Experimental Biology of Bryophytes* (A. F. Dyer and J. G. Ducket, eds.), Academic Press, Orlando, FL, pp. 177-202.

Cove, D. J., and Ashton, N. W., 1988, Growth regulation and development in *Physcomitrella patens*: an insight into growth regulation and development of Bryophytes. *Bot. J. Linn. Soc.* **98**: 247–252.

Cove, D. J., Schild, A., Ashton, N. W., and Hartmann, E., 1978, Genetic and physiological studies of the effect of light on the development of the moss *Physcomitrella patens*, *Phytochem. Photobiol.* **27**: 249–254.

Cove, D. J., Ashton, N. W., Featherstone, D. R., and Wang, T. L., 1980, The use of mutant strains, in the study of hormone action and metabolism in the moss *Physcomitrella patens*, In *Proceedings of the Fourth John Innes Symposium*, 1979 (D. R. Davies, and D. A. Hopwood, eds.), John Innes Charity, Norwich, England, pp. 231–241.

Engel, P., 1968, The induction of biochemical and morphological mutants in the moss *Physcomitrella patens*, *Am. J. Bot.* **55**: 438–446.

Johri, M. M., 1978, Regulation of cell differentiation and morphogenesis in lower plants, In *Frontiers of Plant Tissue Culture* (T. A. Thorpe, ed.), International Association for Plant Tissue Culture, University of Calgary Offset Printing Services, Calgary, Canada, pp. 27–36.

Johri, M. M., and Desai, S., 1973, Auxin regulation of caulonema formation in moss protonema, *Nat. New Biol.* **245**: 223–224.

Rosenblum, I., and Basile, D., 1984, Hormonal regulation of morphogenesis in *Streptocarpus* and its relevance to evolutionary history of the Gesneriaceae, *Am. J. Bot.* **77**: 52–64.

Sicko, L. M., and Jensen, T. E., 1973, Preservation of plant cell surfaces for scanning electron microscopy, In *31st Ann. Proc. Electron Microscopy Society of America* (C. J. Arceneaux, ed.), Claitor's Publishing Division, Baton Rouge, LA, pp. 418–419.

van Holst, G., and Clarke, A. E., 1985, Quantification of arabinogalactan protein in plant extracts by single radial gel diffusion, *Anal. Biochem.* **148**: 786–789.

Yariv, J. Rapport, M. M, and Graf, L., 1962, The interaction of glycosides and saccharides with antibody to the corresponding phenylazo and glycosides, *Biochem. J.* **85**: 383–388.

Yariv, J., Lis, H., and Katchalski, E., 1967, Precipitation of arabic acid and some seed polysaccharides by glycosylphenylazo dyes, *Biochem. J.* **105**: 1c.

Chapter 19

Bioactive Arabinogalactan-Proteins and Related Pectic Polysaccharides in Sino-Japanese Herbal Medicines

Haruki Yamada
Oriental Medicine Research Center, The Kitasato Institute, 5-9-1, Shirokane, Tokyo 108, Japan

1. INTRODUCTION

Many plants have long been used as traditional herbal medicines. Because chronic disorders of an endogenous nature and pressing medical problems such as non-specific, constitutional or psychosomatic diseases have increased during the second half of the 20th century, traditional herbal medicines have begun to attract worldwide attention (Yamada 1994a). Herbal extracts used in Sino-Japanese traditional herbal medicines contain both low- and high-molecular-weight substances. The former include alkaloids, terpenoids, saponins and flavonoids, whereas the latter include proteins, tannins and polysaccharides in the hot water extract. Although biologically active substances with low molecular weight in medicinal herbs have been well studied and characterized, they can not account for all of the clinical effects achieved. Among the high-molecular-weight components of medicinal herbs, polysaccharides have been shown to possess a variety of pharmacological activities. Most especially, pectic polysaccharides containing arabinogalactan have been shown to have immunostimulating, antiulcer, antimetastasis, hypoglycemic or other activities. (Yamada 1994b 1996, Yamada and Kiyohara 1999). These observations suggest that pectic polysaccharides are involved in the efficacy of traditional herbal medicines.

The present paper deals with our recent studies on bioactive AGPs and related pectic polysaccharides in Sino-Japanese traditional herbal medicines.

2. EXAMPLES OF BIOACTIVE POLYSACCHARIDES

2.1 Complement-Activating Arabinogalactan and Pectins from the Roots of *Angelica acutiloba* Kitagawa

The complement system is known to play an important role in humoral immunity, and it includes two activation cascades, the "classical" and "alternative" pathways. As the initial complement component of the classical pathway, C1 is activated by the immune complex, whereas the alternative pathway is directly activated from C3 by microorganisms, protozoa, or other activators working through an antibody-independent mechanism (Yamada and Kiyohara 1999). The activation of complement appears to be intrinsically associated with several immune reactions such as the activation of macrophages and lymphocytes, cellular cooperation, immunopotentiation and the regulation of cyclical antibody production (Yamada and Kiyohara 1999). Therefore, complement-activating substances might modulate the immune system.

The roots of *A. acutiloba* Kitagawa have been used for the treatment of gynecological diseases in the practice of Sino-Japanese herbal medicine. These roots contain several pectic polysaccharides with potent complement-activating capacity that arises from an acidic arabinogalactan (AGIIb-1) having an arabino-3,6-galactan (N-I) side chain (Yamada *et al* 1987, Yamada and Kiyohara 1999). The active N-I chain consists of a β(1→3) galactan backbone to which are attached 6-linked galactosyl side chains that in turn carry arabinofuranosyl or β-galactosyl side chains on some residues (Fig 1; Yamada and Kiyohara 1999). The 6-linked β-galactosyl side chains are responsible for the expression of the complement-activating activity of N-I, and the attachment of these active side chains to the β-D-(1→3)-galactan backbone is necessary to express potent activity (Kiyohara *et al* 1997). Although native N-I activated just the classical complement pathway, after N-I was digested with exo-α-L-arabinofuranosidase the resulting AF-N-I activated both the classical and alternative pathways (Fig 1). This observation suggests that arabinofuranosyl side chains prevent the activation of the alternative pathway by the galactan moiety.

Other complement-activating pectins were also isolated from the same herb (Kiyohara *et al* 1988). These pectins consist of galacturonan regions and ramified regions, which have a rhamnogalacturonan core carrying neutral sugar side chains. The ramified regions are largely responsible for the activity (Yamada 1994b, Yamada and Kiyohara 1999). To analyze whether the neutral sugar chains in the ramified region contribute to expression of complement-activating activity, the neutral sugar chains were released by base-catalyzed β-elimination, and their activity was tested (Yamada and Kiyohara 1999). Long galactosyl oligosaccharides and tri- to tetra-galactooligosaccharides showed notable complement-activating activity; however, the activity of the intact ramified region was more potent than that of the isolated oligosaccharides. The ramified regions of the pectins were effective structures for activation of both the classical and alternative pathways. However, the galacturonan regions were suggested to modulate activation of the alternative pathway by the ramified regions, depending on their different methyl-ester distributions (Fig 2; Yamada and Kiyohara 1999).

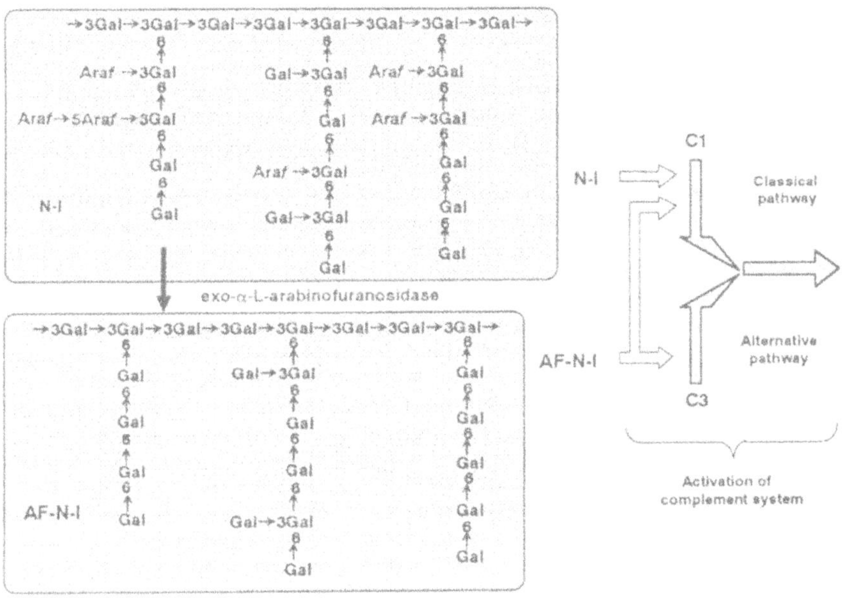

Figure 1. Enzymatic digestion of N-I and its complement activation.

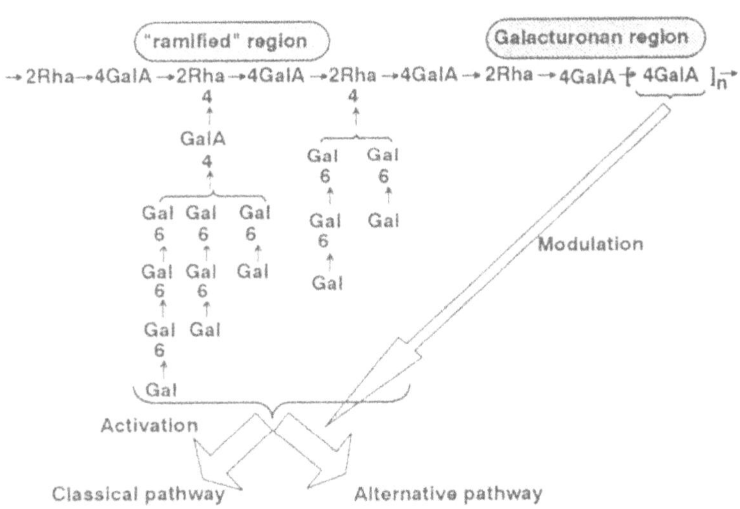

Figure 2. Structural requirements for complement activation by the active pectin from *Angelica acutiloba* Kitagawa.

2.2 An Arabinogalactan from Rhizomes of *Atractylodes lancea* DC that Acts Through the Intestinal Immune System to Increase Proliferation of Bone Marrow Cells

Because most herbal medicines have generally been taken orally, some active ingredients are absorbed from the intestine, but others may affect Peyer's patch cells and intraepithelial lymphocytes in gut-associated immune lymphoreticular tissues. Herbal components may also activate mucosal immune systems. Through these various mechanisms, herbal medicines may give several pharmacological activities. One of the Sino-Japanese herbal (Kampo) medicines, Juzen-taiho-to (Si-Quan-Da-Bu-Tang in Chinese), is known to modulate the systemic immune system. Juzen-taiho-to activates T cells in Peyer's patches and stimulates secretion of hematopoietic growth factors from T cells (Hong *et al* 1998). Recently, we found that rhizomes of *A. lencea* DC, one of the components of Juzen-taiho-to, contributes to the activity of the medicine mainly by stimulating Peyer's patch cells. An arabinogalactan was isolated from *A. lancea* DC rhizomes and was found to act through the intestinal immune system to increase proliferation of bone marrow cells (Yu *et al* 1998). The following methods were established for assaying this activity (Hong *et al* 1998). Peyer's patch cells were prepared from the small intestine of C3H/HeJ mice that had been administered test samples orally. The cells were cultured, and the resulting culture medium was tested by the Alamer blue reduction method for activity in stimulating the proliferation of bone marrow cells. Peyer's patch cells incubated with test samples were also used for an *in vitro* assay. The purified active polysaccharide, ALR-5IIa-1-1, had a molecular weight of 74,000 and consisted of about 92% carbohydrate with traces of uronic acid and protein. The major components of ALR-5IIa-1-1 were arabinose and galactose in a molar ratio of ~1.0:1.7, indicating that the active polysaccharide was a certain type of arabinogalactan molecule. Methylation analysis indicated that ALR-5IIa-1-1 was composed mainly of terminal Ara*f*, 4- or 5-linked Ara, 3,4- or 3,5-branched Ara, 4-linked Gal, 3-linked Gal and 3,6-branched Gal, suggesting that the structure was an arabino-3,6-galactan. The active polysaccharide, ALR-5IIa-1-1, strongly reacted with β-glucosyl-Yariv antigen, as does Type II arabinogalactan from Acacia, further suggesting the presence of an arabino-3,6-galactan moiety. Inactive polysaccharide fractions from *A. lancea* DC rhizomes did not react with Yariv antigen. When the active polysaccharide was digested with exo-α-L-arabinofuranosidase or exo-β-D-(1→3)-galactanase separately, the pharmacological activity was not changed, but when ALR-5IIa-1-1 was digested first with exo-α-L-arabinofuranosidase and then with exo-β-D-(1→3)-galactanase, the activity decreased significantly and in accordance with a reduction of the reactivity with β-glucosyl Yariv antigen (Fig 3; Yu *et al* 1998). Therefore, these results suggest that the arabino-3,6-galactan moiety in the active polysaccharide contributes to the expression of intestinal immune system modulating activity.

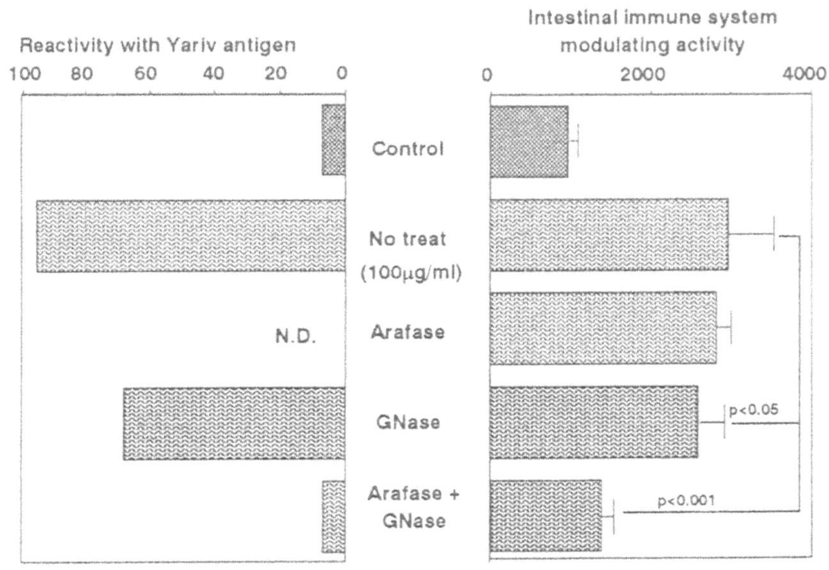

Figure 3. Effect of exo-α-L-arabinofuranosidase (Arafase) and exo-β-D-(1→3)-galactanase (GNase) digestions on reactivity with Yariv antigen and intestinal-immune-system-modulating activity of ALR-5IIa-1-1. N.D., not done.

2.3 Mitogenic Pectic Polysacharide from Roots of *Bupleurum falcatum*

The roots of *B. falcatum* have been used clinically in Sino-Japanese traditional herbal medicines for the treatment of chronic hepatitis, nephrotic syndrome and auto-immune diseases. This medicinal herb was found to contain an interesting bioactive pectin, bupleuran 2IIc, which has a molecular weight of 63,000 (Yamada *et al* 1991a). Approximately 86% of bupleuran 2IIc was galacturonan, consisting of 70% α(1→4) linked galacturonic acid and 30% carboxymethylated galacturonic acid and branched galacturonic acid. Bupleuran 2IIc also contained ramified pectin regions, which consisted of a rhamnogalacturonan core with several arabino- and galactooligosaccharide side chains attached to core 2-linked rhamnosyl residues either directly or through 4-linked galacturonic acid (Yamada *et al* 1991b). Bupleuran 2IIc also contained a minor region similar to the pectic RG-II region, which contains rare sugars such as 2-keto-3-deoxy-D-manno-octulosonic acid (KDO). Although bupleuran 2IIc had complement- activating and anti-ulcer activity (Yamada *et al* 1991a, Yamada 1994b), it also showed a potent mitogenic activity against mouse spleen cells and Peyer's patch cells of the small intestine *in vitro*. Citrus pectin did not show such activity (Fig 4; Sakurai *et al* 1999).

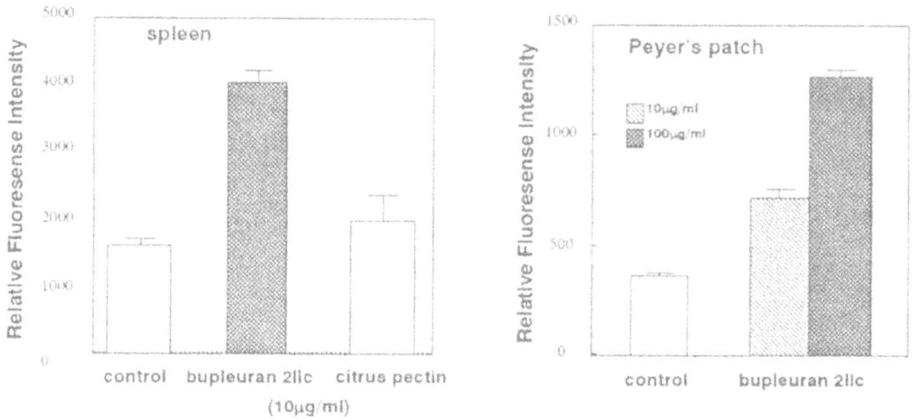

Figure 4. Mitogenic activity of bupleuran 2IIc on spleen cells and Peyer's patch cells from C3H/HeJ mice. Larger fluorescence intensity indicates larger mitogenic activity.

The population of B cells was increased in the presence of bupleuran 2IIc; however, that of T cells was not increased on flow cytometric analysis, suggesting that bupleuran 2IIc is a B cell-specific mitogen. Bupleuran 2IIc also showed *in vivo* mitogenic activity when orally administered to mice (Sakurai *et al* 1999). When adherent cells such as macrophages were removed from the population of spleen cells, enhancement of B cell proliferation by bupleuran 2IIc was not changed. When T cells were depleted from the population of spleen cells by treatment with anti-Thy 1.2 antibody and complement, the enhanced proliferation of B cells was also not changed. Therefore these results indicate that bupleuran 2IIc expresses mitogenic activity without the presence of adherent cells and T cells. Because bupleuran 2IIc consists of a galacturonan region, a ramified region PG-1 (which contains a rhamnogalacturonan core attached to neutral sugar chains) and a RG-II-like region, these structural fragments were obtained by endo-$\alpha(1\rightarrow4)$-polygalacturonase digestion, and their mitogenic activities were tested (Hirano *et al* 1994, Sakurai *et al* 1999). Among these fragments, only the ramified region PG-1 gave a potent mitogenic activity. Therefore, the ramified region may be involved in the activity. To analyze absorption and tissue distribution of the active polysaccharide after oral administrations to the animals, rabbit antipolysaccharide antibody was made against the ramified region (Sakurai *et al* 1996). When the pectic polysaccharide fraction BR-2, which contains bupleuran 2IIc, was administrated orally to mice, the ramified region of the polysaccharide was detected in the T cell area of follicles in Peyer's patches by immunohistochemical staining with the antipolysaccharide antibody (Sakurai *et al* 1996). A two-layer sandwich ELISA method was also established for detecting the ramified region of bupleuran 2IIc (Sakurai *et al* 1996). This assay system could measure the ramified region of bupleuran 2IIc with sensitivity more than one nanogram per well. By this sandwich ELISA method, the ramified region of bupleuran 2IIc was also detected in liver homogenate one week after oral administration. These results suggest that at least a part of bupleuran 2IIc can be absorbed in the body. To identify the antigenic epitopes of the antipolysaccharide antibody, the carbohydrate sequence in the ramified region was modified through application of specific carbohydrases such as exo-α-L-arabinofuranosidase, rhamnogalacturonase A, exo-β-(1→3)-galactanase, endo-β-(1→6) galactanase and β-glucuronidase (Sakurai *et al* 1998). The results suggest that the ramified region consists of a rhamnogalacturonan core with attached branched arabinofuranosyl chains and galactosyl chains (Fig 5). Within this structure, the antigenic epitope was suggested to be 6-linked galactosyl chains containing terminal glucuronic acid and 4-methylglucuronic acid attached to β-(1→3)-linked galactosyl chains (Sakurai *et al* 1998). The mitogenic activity of the ramified region was inhibited in the presence of the antipolysaccharide antibody. Although non-immune $(Fab')_2$ fragments showed no effect on the mitogenic activity of bupleuran 2IIc, antipolysaccharide antibody $(Fab')_2$ also reduced the mitogenic activity, suggesting that the antigenic epitope is involved in the mitogenic activity. The results of flow cytometric analyses indicated that bupleuran 2IIc affected the differentiation from pre B cells to mature B cells (Sakurai *et al* 1999). Bupleuran 2IIc also stimulated secretion of interleukin-6 (IL-6), and resulting IL-6 may contribute to induction of IgG and IgM antibody forming cells as an autocrine and/or paracrine mechanism (Fig 6; Sakurai *et al* 1999, Guo and others, unpublished data). Because the ramified region is involved in the activity, the result indicates that the same carbohydrate chains are recognized by the antipolysaccharide antibody. Therefore this antibody may be useful for further studies on the molecular mechanism of the activity and pharmacodynamics of bupleuran 2IIc.

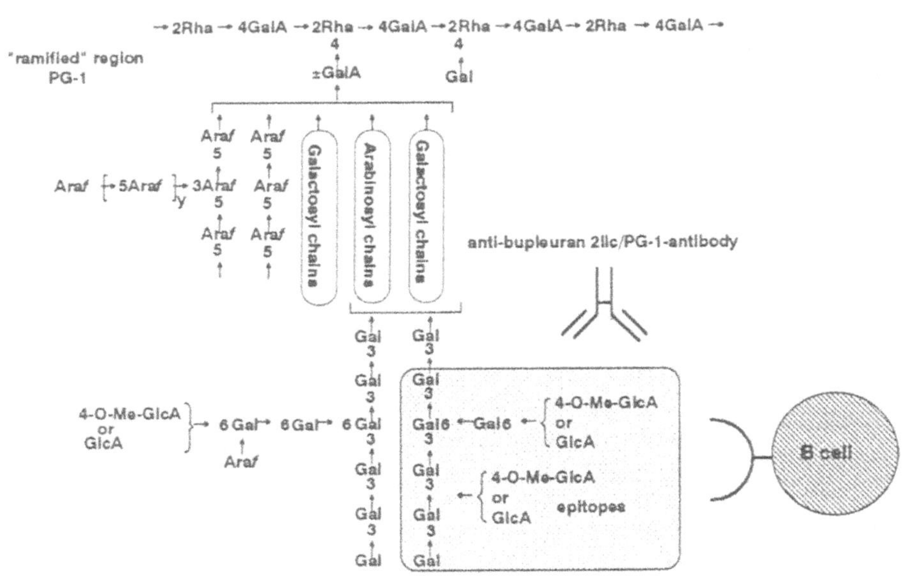

Figure 5. Proposed structure of ramified region of bupleuran 2IIc and its antigenic epitopes as the active site for mitogenic activity.

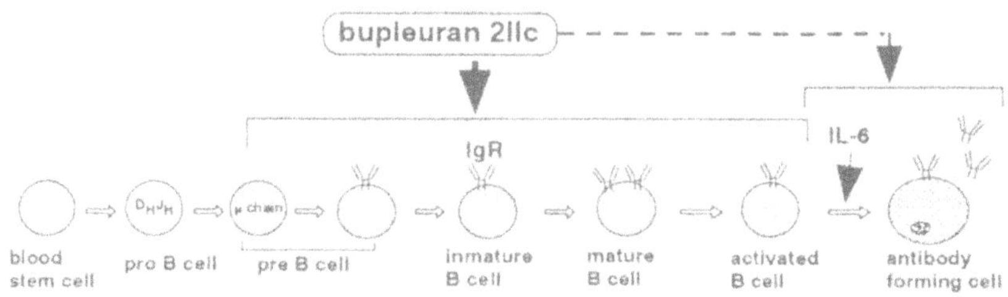

Figure 6. Effects of bupleuran 2IIc on spleen B cell differentiation. IL-6, interleukin-6.

3: CONCLUSION

Results presented herein indicate that arabino-3,6-galactan and 6-linked β-galactosyl side chains in the ramified region of the pectin play an important role for the expression of pharmacological activities of medicinal herbs. Each pharmacological activity of these pectic polysaccharides may depend on fine chemical structure. The present observation suggests that the application of arabinogalactans and related pectic polysaccharides for medicinal use can benefit humans.

ACKNOWLEDGEMENTS

The author is pleased to acknowledge all the work described in this paper that was carried out by his coworkers, H. Kiyohara, T. Matsumoto, M. H. Sakurai, T. Hong and K-S. Ra. The author also thanks Dr. Hashimoto (Saitama University) for kind gifts of exo-β-D-(1→3)-galactanase, endo-β-(1→6)galactanase and exo-β-D-glucuronidase.

REFERENCES

Hirano, M., Kiyohara, H., Matsumoto, T., and Yamada, H., 1994, Structural studies of endo-polygalacturonase-resistant fragments in anti-ulcer pectin from the roots of *Bupleurum falcatum* L., *Carbohyd. Res.* **251**: 145–162.

Hong, T., Matsumoto, T., Kiyohara, H., and Yamada, H., 1998, Enhanced production of hematopoietic growth factors through T cell activation in Peyer's patches by oral administration of Kampo (Japanese herbal) medicines, "Juzen-taiho-to", *Phytomedicine* **5**: 353–360.

Kiyohara, H., Cyong, J-C., and Yamada, H., 1988, Structure and anti-complementary activity of pectic polysaccharides isolated form the roots of *Angelica acutiloba* Kitagawa, *Carbohyd. Res.* **182**: 259–275.

Kiyohara, H., Zhang, Y-W., and Yamada, H., 1997, Effect of exo-β-D-(1→3)-galactanase digestion on complement activating activity of neutral arabinogalactan unit in a pectic arabinogalactan from roots of *Angelica acutiloba* Kitagawa, *Carbohyd. Polym.* **32**: 249–253.

Sakurai, M., Matsumoto, T., Kiyohara, H., and Yamada, H.,1996, Detection and tissue distribution of anti-ulcer polysaccharides from *Bupleurum falcatum* L. by polyclonal antibody, *Planta Med.* **62**: 341–346.

Sakurai, M., Kiyohara, H., Matsumoto, T., Tsumuraya, Y., Hashimoto, Y., and Yamada, H., 1998, Characterization of antigenic epitopes in anti-ulcer pectic polysaccharides from *Bupleurum falcatum* L. using several carbohydrases, *Carbohyd. Res.* **311**: 219–229.

Sakurai, M. H., Matsumoto, T., Kiyohara, H., and Yamada, H., 1999, B-cell proliferation activity of pectic polysaccharide from a medicinal herb, the roots of *Bupleurum falcatum* L. and its structural requirement, *Immunology* **97**: 540-547.

Yamada, H., 1994a, Modern scientific approaches to kampo medicine, *Asia Pacific J. Pharmacol.* **9**: 209–217.

Yamada, H., 1994b, Pectic polysaccharides from Chinese herbs: Structure and biological activity, *Carbohyd. Polym.*, **25**: 269–276.

Yamada, H., 1996, Contribution of pectins on health care, In *Pectin and Pectinase Progress in Biotechnology*, Vol. 4 (J. Visser and A. G. J. Voragen, eds.), Elsevier, Amsterdam, pp. 173–190.

Yamada. H., and Kiyohara, H., 1999, Complement-activating polysaccharides from medicinal herbs, In *Immunomodulatory Agents from Plants* (H. Wagner, ed.), Birkhauser Verlag, Basel, pp. 161–202.

Yamada, H., Kiyohara, H., Cyong, J-C., and Otsuka, Y., 1987, Structural characterization of an anti-complementary arabinogalactan form the roots of *Angelica acutiloba* Kitagawa, *Carbohyd. Res.* **159**: 275–291.

Yamada, H., Sun, X-B., Matsumoto, T., Ra, K-S., Hirano, M., and Kiyohara, H., 1991a, Purification of anti-ulcer polysaccharides from the roots of *Bupleurum falcatum* L., *Planta Med.* **57**: 555–559.

Yamada, H., Hirano, M., and Kiyohara, H., 1991b, Partial structure of an anti-ulcer pectic polysaccharide from the roots of *Bupleurum falcatum* L., *Carbohyd. Res.* **219**: 173–192.

Yu, K-W., Kiyohara, H., Matsumoto, T., Yang, H-C., and Yamada, H., 1998, Intestinal immune system modulating polysaccharides from rhizomes of *Atractylodes lancea*, *Planta Med.* **64**: 714–719.

Chapter 20

Uses of Gum Arabic (*Acacia* sp.) in the Food and Pharmaceutical Industries

Florian M. Ward
TIC Gums, Inc. Belcamp, MD 21017

1. INTRODUCTION

Gum arabic, also known as gum acacia, is defined in the National Formulary 18 (U.S. Pharmacopeial Convention, Inc. 1995) as the dried gummy exudate from the stems and branches of *Acacia senegal* (Linne) Willdenow and other related African species of Acacia (family Leguminosae). The specifications for gum arabic in commerce, as adopted by the U.S. Food and Nutrition Board (1996), are shown in Table 1.

Table 1. Gum arabic (Source: U.S. Food and Nutrition Board 1996)

Impurity/Property	Limit
Arsenic (As As)	Not more than 3 ppm
Ash (total)	Not more than 4%
Ash (acid insoluble)	Not more than 0.5%
Heavy metals (as Pb)	Not more than 20 ppm
Lead	Not more than 10 ppm
Insoluble matter	Not more than 1%
Loss on drying	Not more than 15%
Starch and dextrin	Passes test
Tannin-bearing gums	Passes test

A highly branched anionic polysaccharide, gum arabic consists mainly of three fractions: an arabinogalactan (AG) complex, an AG-protein complex (AGP) and a glycoprotein fraction (Williams *et al* 1990). Upon acidic hydrolysis it yields glucuronic acid and other sugars, including galactose, arabinose and rhamnose, which occur in various ratios depending on the type of Acacia species used. The protein moiety has an unusual amino acid composition rich in Hyp and Ser. The

Acacia species originating from various regions of Africa fall into a discrete cluster on the basis of chemometric analyses (Jurasek *et al* 1993). Jurasek *et al* (1993) noted that there were no significant differences in *A. senegal* from the main producing countries: Sudan, Niger, Nigeria, Uganda, Oman, Mauritania, Mali, Senegal, Ethiopia, and Kenya.

Gum arabic is reported to have an average molecular weight of 400,000. The high degree of branching and short spiral coils with side chains in the polysaccharide may account for the unusually low viscosity of gum arabic solutions. The gum is soluble up to 50% and forms a soft gel at this concentration. The high solubility and low viscosity of gum arabic make it ideal for use in confections and spray-dried flavors, and as a fiber source.

The maximum levels of gum arabic permitted in various food products, according to the U.S. Food and Drug Administration (1994) regulations, are shown in Table 2. A maximum usage level of 2% based on the weight of the finished product is permitted for beverages or drinks, and 4.0% is permitted for snack foods. For hard candy and cough drops, the level allowed is 46.5%, and as much as 85% is allowed for soft candy. In confections and frostings, where it is used for water binding, adhesive properties, and its ability to inhibit sugar crystallization, the maximum level of usage is 12.4%.

Table 2. Maximum usage levels permitted for gum arabic by Title 21 Code (U.S Food and Drug Administration 1994)

Food	Use Level, %
Beverages and bases	2.0
Chewing gum	5.6
Confections and frostings	12.4
Dairy product analogs	1.3
Fats and oils	1.5
Gelatins, puddings, fillings	2.5
Hard candy and cough drops	46.5
Nuts and nut products	8.3
Snack foods	4.0
Soft candy	85.0
All other food categories	1.0

2. SPRAY-DRIED GUM ARABIC: ESSENTIAL FEATURES

Spray-dried gum arabic (e.g., *Gum Arabic NF/FCC* from TIC Gums, Inc., MD, USA) is a high-purity, low-bacteria, cream-to-white powder, manufactured from selected species of gum arabic. It is processed by subjecting the crude gum arabic sorts to a series of purification and pasteurization steps, followed by spray drying. *Gum Arabic NF/FCC* has excellent functional properties, including film formation and improved protection of lipid from oxidation, increased solubility and dispersibility, and higher percentage purity over the traditional gum arabic. The advantages of using gum arabic over emulsifying starches include the following:

1. Multi-functional: good emulsifier, film-former, texturizer, and low-viscosity water binder and bulking agent; inhibits sugar crystallization
2. High source of fiber: contains not less than 85% soluble dietary fiber, which has been associated with reduction in high cholesterol levels
3. High percentage purity: no additives; free from sediment and impurities; has extremely low bacterial counts (less than 200 cfu/g)
4. Fast hydration and ease of dispersion: available in prehydrated or agglomerated form that achieves 80% hydration in the first few minutes of mixing with water
5. Cost-effective: *Gum Arabic NF/FCC* is competitive in price with modified starches
6. "Natural" labeling: Gum arabic is not chemically modified and qualifies for "natural" labeling or "no artificial additives" claims.

3. FORMULATIONS USING GUM ARABIC

3.1 Gum Arabic in Confections

Gum arabic is one of the essential ingredients in confectioneries because of its multiple functions (Ward and Andon 1997). It can retard sugar crystallization in gum drops and emulsify and distribute fat particles in caramel and toffee. Gum arabic also functions as a binder and structure builder in cough drops and lozenges. In panned sugar confections it serves as a coating agent and film-former in products including coated nuts, chewing gum, jelly beans, candy corn and many others. Gum arabic and other hydrocolloids also act as humectants to hold or bind water in formulations.

Candies that contain sugar in a non-crystalline, glass-like state include sour balls, butterscotch, etc. These products have moisture levels as low as 2%, as in hard candies. The chewy types, including caramel and taffy, contain 8–15% moisture, whereas gummy candies such as marshmallows, gum drops, and jellies have 15–22% moisture. The greater the percentage of gum arabic, the softer and more chewable the candy.

The hard candy formulation using gum arabic at a 5% level includes ingredients such as dextrose, sugar, and flavor (TIC Gums, Inc. 1996). Gum arabic binds the water and thus prevents sugar crystallization. A sugarless hard candy for dietary purposes uses sorbitol or mannitol rather than sucrose as the sweetener. A soft candy may contain 60% or more of gum arabic. The chewy centers of some candies may be prepared from coacervates between gelatin and gum arabic.

Gum arabic is used to prevent sugar crystallization and to emulsify fat or oil in caramels and nougats. Gum arabic also improves aeration characteristics and stabilizes foam in marshmallows. It is usually mixed with sucrose or corn syrup when used in panning confections. Small and round candies are glazed by placing the sugar centers in revolving heated pans, and a gum arabic syrup is sprayed into the pan. The uniform coating is achieved by gently tumbling the centers, applying the syrup solutions and drying these gently as the water is evaporated from the heated revolving pan. Chocolate-panned products are glazed or polished by spraying a solution of gum arabic into the pan after the application of the chocolate coating.

3.2 Gum Arabic Systems in Snack Foods and Bakery Products

Gum arabic is also recommended as a lubricant and binder in extruded snack cereals, at levels from 2–5%. It can be combined with cellulose gum and other hydrocolloids to improve "mouthfeel" and water-binding properties.

Gum arabic at 15–40% levels is typically used to provide adhesion of dry flavors in peanuts and similar products. The solution is sprayed onto the surface of the nuts, followed by drying at 100–170°C. Gum arabic solutions also add a glossy and shiny surface to the product.

Gum arabic is used as an adhesive and water-binding agent in the manufacture of granola from cereal grains (Ward and Andon 1993). A similar formulation can also be used for making rice cereal bars. A 12–15% gum solution can be sprayed on snack foods such as corn chips and tortilla chips at 10% of the weight of the product, followed by application or tumbling with the spices to be adhered. Alternatively, the spices and the powdered gum may be mixed to form a slurry and later applied to the surface using a spray device. A resilient film and sheen, as well as efficient spice adhesion, is achieved by replacing the oil partially or in full. Because of the unique effect of oil on flavor release, the formulator has the option of retaining 10% of the oil instead of total replacement with gum solution. In extrusion of cereals and snack foods, gum arabic and other hydrocolloids are added to act as a lubricant and binder as well as to give improved strength and texture to the finished product.

In the manufacture of bakery glazes and icings, the film-forming and adhesive properties of gum arabic are utilized. In low-fat cake and muffin mixes, gum arabic is used as a partial oil replacer as well as a moisture binder, which increases the shelf life of the bakery product. As a surface-active agent, gum arabic can also help stabilize foam, and hence it is used in whipped toppings and similar products.

3.3 Gum Arabic as a Flavor Encapsulating Agent

The retention of volatile flavor compounds is influenced by a variety of factors, including solids content and viscosity of the feed material, type and molecular weight of the carrier used, molecular weight and vapor pressure of the flavor component, and spray-drying equipment parameters. As a flavor-encapsulating agent, gum arabic has been the ideal carrier for the flavor industry because of its natural emulsifying and surface-active properties, good retention of volatile flavor components, high solubility in water (up to 50%), and acid stability. Additional advantages of gum arabic include its bland flavor, low viscosity, low hygroscopicity, good film-forming properties, and ability to protect flavors from oxidation. A typical spray-dried formulation would include 4 parts of gum arabic to 1 part of the flavor oil (Reineccius *et al* 1995). The average retention of strawberry flavor components (ethyl butyrate, benzaldehyde, methyl cinnamate, etc.) spray dried with gum arabic has been shown to be high, making it an excellent encapsulating agent.

3.4 Gum Arabic in Beverage Emulsions

The emulsifying properties of gum arabic have been attributed mainly to the presence of covalently linked protein moieties in the AGP and glycoprotein complexes found in gum arabic (Williams *et al* 1990). The processing conditions to which the gum arabic is subjected will affect its emulsifying properties, especially if the protein moieties are exposed to prolonged heat and low pH.

A typical beverage flavor emulsion (with weighting agents such as ester gum or brominated vegetable oils) is shown in Table 3. The stability of emulsions prepared with gum arabic can be evaluated by various methods, including the determination of the particle size distribution of the oil-in-water emulsion using a Coulter Counter device (Fig 1). A smaller particle size (less than 2 μm) indicates a higher stability or shelf life of the emulsion. To predict the stability of several emulsifying systems, the particle size distributions of the emulsions may be compared and correlated with other stability indices and sensory evaluation tests.

Table 3. Beverage emulsion composition and procedure using gum arabic

Ingredients	%
Water	70.54
Gum arabic	15.43
Sodium benzoate	0.08
Citric acid	0.32
Color solution	2.08
Oil phase/weighting agent	11.55

Procedure:

1. Add gum to water
2. Add sodium benzoate and citric acid
3. Mix for 1 hour
4. Add color solution and mix for 15 min
5. Add oil phase and mix for 10 min to form a coarse oil-in-water emulsion
6. Homogenize to desired particle size distribution

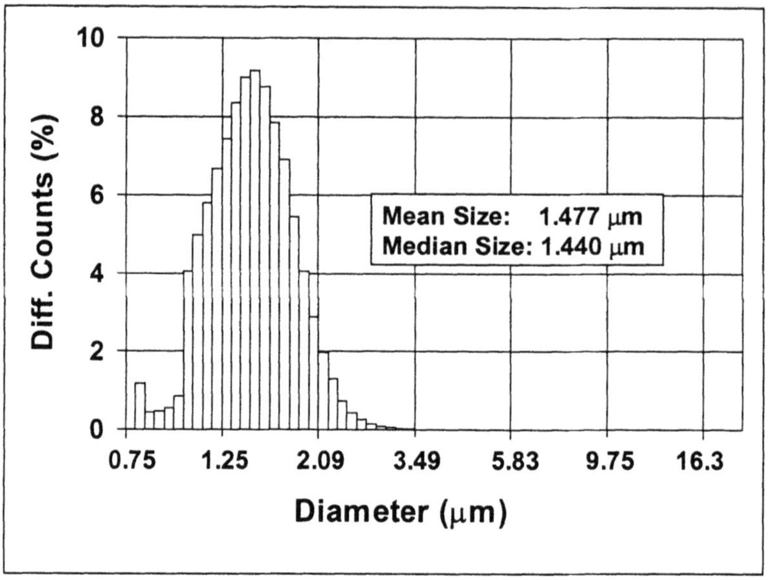

Figure 1. Emulsion size distribution (Coulter Counter) using *Arabic Spray Dry* as emulsifier.

3.5 Gum Arabic in Pharmaceutical Products

Gum arabic has been used traditionally as tablet excipient and binder because of its adhesive properties and non-hygroscopic nature. It is also used as a thickener, suspending agent, and texturizer in medicated syrup preparations. Gum arabic has been used for decades as an emulsifying agent for cod-liver oil and other therapeutic oil-in-water emulsions. It is also employed as an adhesive and texturizer in facial masks and skin lotions.

3.6 Gum Arabic in Nutraceuticals

Because of the high solubility, low viscosity, and high soluble dietary fiber content (80% minimum on an as is basis) of gum arabic, its use in meal replacers, nutritive drinks, and weight-loss products has increased steadily in the past decade. The sensory attributes of a fortified orange drink with and without gum arabic as a fiber source and texturizer is shown in Fig 2. Gum arabic contributes opacity and improves mouthfeel without increasing viscosity. A clinical study of the effects of dietary gum arabic has described its use to decrease the level of low-density lipoprotein serum cholesterol in clinically tested human subjects (McLean Ross et al 1983).

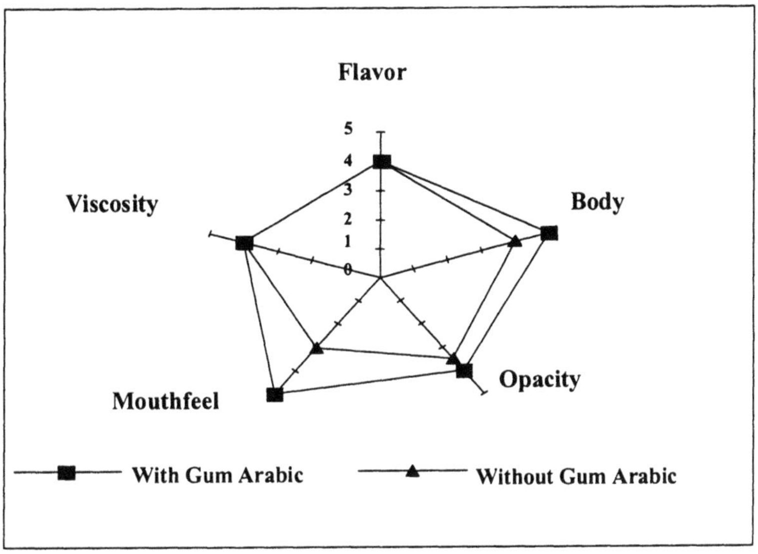

Figure 2. Sensory attributes of orange drink with and without gum arabic.

4. RECENT DEVELOPMENT IN PROCESSING: PREHYDRATED GUM ARABIC

Gum arabic in its spray-dried or crude form is difficult to disperse directly from its powdered form into water because of its slow hydration and lumping. Propylene glycol, premixing in oil slurries, increased temperature, and high-speed mixers must be used to increase hydration and avoid lumping.

All of these current processing problems are minimized by the use of prehydration or agglomeration techniques, whereby the powder is converted to aggregates with interstitial voids that allow water to be imbibed at a faster rate during the hydration step. Gum arabic is now available in a readily dispersible form that allows faster hydration within the first 10 min of dissolution in water (Fig 3). The use of Pre-Hydrated® *Gum Arabic NF/FCC* from TIC Gums, Inc., instead of regular spray-dried gum arabic, has been widely adopted in the food and pharmaceutical industries. It was initially accepted primarily by the process engineer and the plant personnel directly affected by the problems of dusting, slow hydration, and repeated filtration steps associated with the use of traditional gum arabic.

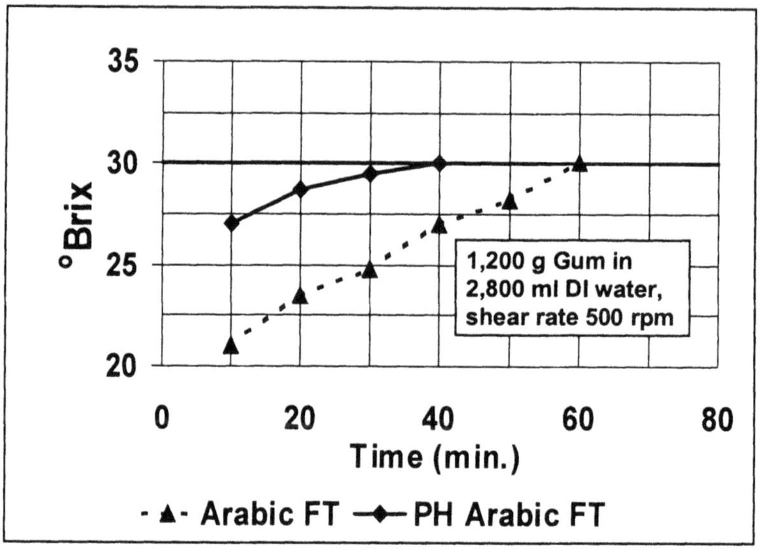

Figure 3. Hydration rates of *Arabic FT* and Pre-Hydrated® *Gum Arabic FT*, special spray-dried products manufactured by TIC Gums, Inc.

5. SUMMARY AND RECOMMENDATIONS

Gum arabic is widely used in the food and pharmaceutical industries because of its various functional properties: emulsifying and encapsulating agent, film former, adhesive, and foam stabilizer. It also prevents sugar crystallization, binds water, and provides structure and mouthfeel to the hard and soft candies and lozenges. In snack foods, gum arabic is used as an adhesive to replace oil in spice adhesion and to provide sheen or attractive glaze. It is also used as a lubricant and binder in extruded cereal products. As a binding agent, it may be used in the manufacture of snack bars from cereals, fruits, and nuts. The high percentage purity, low bacterial counts, and bland taste of spray-dried gum arabic make it ideal for use in the food and nutraceutical industries. The high soluble dietary fiber of a multi-functional natural product such as gum arabic, in addition to its reasonable cost relative to starches and other fiber sources, makes it an excellent food ingredient. It is recommended that food formulators carefully select the type of gum arabic for specific applications to assure the good quality and long shelf-life of the finished product.

ACKNOWLEDGEMENTS

The author acknowledges Stephen A. Andon and Chris E. Andon, TIC Gums, Inc.; Dr. Richard B. Ward, RBW Consulting, for assistance in preparing the typescript and the graphics; and Ken Kuschwara and Stuart Cantor, Food Technologists, TIC Gums, Inc., for their research and development assistance.

REFERENCES

Jurasek, P., Kosik, M., and Philips, G. O., 1993, A chemometric study of the Acacia (gum arabic) and related natural gums, *Food Hydrocoll.* **7**: 73–85.

McLean Ross, A. H., Eastwood, M. A., Anderson, J. R., and Anderson, D. M., 1983, A study of the effects of dietary gum arabic in humans, *Am. J. Clin. Nutr.* **37**: 368–375.

Reineccius, G. A., Ward, F. M., Whorton, C., and Andon, S. A., 1995, Developments in gum acacia for the encapsulation of flavors, *Am. Chem. Soc. Symp. Ser.* **590**: 161–168.

TIC Gums, Inc., 1996, *Technical Bulletins*, Belcamp, MD.

U.S. Food and Drug Administration, 1994, *Code of Federal Regulations*, Parts 170–199, Office of the Federal Register, National Archives and Records Administration, Washington, DC, p. 459.

U.S. Food and Nutrition Board, 1996, *Food Chemicals Codex*, Institute of Medicine, National Academy of Sciences, National Academy Press, Washington, DC., p. 9.

U.S. Pharmacopeial Convention, Inc., 1995, *National Formulary (NF 18) and United States Pharmacopeia (USP 23)*, Rockville, MD.

Ward, F. M., and Andon, S. A., 1993, Water-soluble gums used in snack foods and cereal products, *Cereal Foods World* **38**: 748–752.

Ward, F. M. and Andon, S. A., 1997, Gum acacia in confectioneries and snack foods, *Asia Pacific Food Ind.* **May**: 68–72.

Williams, P. A., Philips, G. O., and Randall, R. C., 1990. Structure-function relationship of gum arabic, In *Gums and Stabilizers for the Food Industry* (D. J. Wedlock and P. A. Williams, eds.), IRL Press, Oxford, UK, pp. 25–26.

Chapter 21

Structural Analysis of Gum from *Acacia senegal* (Gum Arabic)

P. A. Williams, O. H. M. Idris and G. O. Phillips
North East Wales Institute, Plas Coch, Mold Road, Wrexham, United Kingdom

1. INTRODUCTION

Gum arabic is an exudate obtained principally from the *Acacia senegal* tree, which grows widely across the Sahelian belt of Africa. The major producing country is Sudan, followed by Nigeria and Chad (Islam *et al* 1997). The gum oozes from the stems and branches of the tree when it is subjected to stress conditions such as drought and wounding. To stimulate production, sections of the bark are removed at particular periods during the year. This process is known as tapping. The gum is produced in the tree within a zone between the inner bark and the cambial zone (Joseleau and Ullmann 1990). The trees must be 5 years of age or more for gum production to occur, suggesting, therefore, that gummosis is in direct competition with tree growth. The sticky gummy substance dries on the branches to form hard nodules, which are picked from the tree by hand and are sorted according to color and size. The gum is used to stabilize flavor oils for soft drinks, as an encapsulator for spray dried flavors for use in dried food mixes, and in the production of certain confectionery, notably fruit gums and pastilles.

A considerable body of literature exists concerning the structure of the gum (Stephen and Churms 1995, Islam *et al* 1997). The gum is a highly heterogeneous complex polysaccharide consisting of galactose (~44%), arabinose (~27%), rhamnose (~13%), glucuronic acid (~14.5%) and 4-O-methylglucuronic acid (~1.5%). The carbohydrate compositions can vary, depending on the location, the age of the tree and site of tapping, and from season to season (Idris *et al* 1998). Analysis by Smith degradation indicates that the gum has a highly branched carbohydrate structure (Street and Anderson 1983, Churms *et al* 1983). Churms *et al* (1983) have suggested that the molecules have regularly repeating blocks of 8 kDa consisting of a (1→3)-β-D-galactan core with extensive branching through 3- and 6-linked galactose and 3-linked arabinose. Rhamnose and glucuronic acid are positioned at the periphery of the molecule, where they terminate some of the branches.

The gum also contains a small amount of proteinaceous material (~2%) as an integral part of the structure (Islam *et al* 1997). It has been described as an arabinogalactan-protein (AGP) complex since it interacts with Yariv antigen (Fincher *et al* 1983, Akiyama *et al* 1984). A high molecular mass fraction of the gum can be degraded by pronase enzyme (Connolly *et al* 1988, Randall *et al* 1988, Osman *et al* 1993), and hence it has been proposed that the gum has a "wattle-blossom"-type structure in keeping with other AGPs (Fincher *et al* 1983).

Considerable progress has been made in recent years in gaining a more definitive description of the structure of the gum through characterization of fractions isolated by various separation procedures, including gel permeation chromatography (GPC) (Vandevelde and Fenyo 1985, Randall *et al* 1988, Qi *et al* 1991, Idris *et al* 1998), hydrophobic interaction chromatography (HIC), (Randall *et al* 1989, Osman *et al* 1993) and ion exchange chromatography (Osman *et al* 1995). This paper describes further work we have undertaken with the use of HIC to fractionate two gum samples obtained from different regions in Sudan.

2. MATERIALS AND METHODS

Two *A. senegal* gum samples were collected by one of the authors from different regions in Sudan. Sample E was obtained from the El Daley area in East Sudan, which has clay soils, and sample W was obtained from the Kordofan area in West Sudan, which has sandy soils. Both samples were from 15-year-old trees.

2.1 Fractionation

The samples were fractionated by HIC (Randall *et al* 1989). A total of 100 g of phenyl Sepharose CL-4B hydrophobic gel (Pharmacia) in 20% ethanol was filtered through a 120-μm sintered glass funnel. The gel was extensively washed with distilled water, followed by several washings with 4.2 M NaCl and finally suspended in 250 ml of 4.2 M NaCl. The suspension was poured into a glass column of dimensions 26 mm × 700 mm (Pharmacia) and allowed to settle overnight at room temperature. The excess eluent was allowed to drain out under gravity with use of a valve attached to the base of the column.

Gum solutions were prepared by dissolving 35 g (dry weight) of the gum nodules in 200 ml of 4.2 M NaCl. The gum solution was filtered through a Whatman filter paper No. 54 with use of a water pump. The filtrate was loaded onto the phenyl Sepharose CL-4B column and was eluted with 4.2 M NaCl at a flow rate of 1 ml/min controlled by a peristaltic pump (Pharmacia). Fractions (~10 ml) were collected with use of an LKB fraction collector, and their absorbance was measured at 214 and 280 nm with use of a Perkin-Elmer Lambda 5 spectrophotometer equipped with a 1.0-mm path length cuvette.

Most of the gum sample did not appreciably adsorb to the phenyl Sepharose CL-4B gel and thus quickly passed through the column. Fractions of the gum that had adsorbed to the gel were sequentially desorbed by passing 2 M NaCl down the column, followed by distilled water. The fractions were extensively dialysed at 4°C against distilled water until the dialysate was free from chloride ions as indicated by the lack of a cloudy precipitate on adding silver nitrate. The fractions were freeze dried, and the weight of each recovered fraction was determined.

2.2 Molecular Mass Distribution

The molecular mass distributions of the whole gums and fractions were determined by GPC with use of either the Pharmacia Fast Protein Liquid Chromatography system fitted with a 10 mm × 30 mm Superose 6 column or high pressure PLGFC 4000, 1000 and 3000 columns (300 mm × 7.5 mm) in series (Polymer Laboratories, UK). Gum solutions (1% w/w) were prepared in 0.5 M NaCl, filtered through a 0.45-μm filter and injected onto the column. The flow rate was set at 0.5 ml/min. The Pharmacia system detection was by ultraviolet absorbance (UV, wavelength 206 nm, LKB 2238 Uvicord S11), refractive index (RI, R401 Waters Millipore) and photon correlation spectroscopy (Oros 801 molecular size detector, Oros Instruments, Cambridge, MA). For the high-pressure columns, detection was by multiangle laser light scattering (DAWN, Wyatt Technology, Santa Barbara, CA) and RI (Optilab 903 interferometer, Wyatt Technology).

2.3 Sugar Composition

Samples were hydrolyzed in 4% sulfuric acid at 100°C for 4 hours, and the solution was neutralized by adding solid barium carbonate. Neutral sugars were determined by injecting 60 μl of the filtered solution onto a Waters Carbohydrate Analysis 125A HPLC column (3.9 mm × 300 mm) with 75/25 acetonitrile/water as eluant at a flow rate of 0.8 ml/min. Detection was by RI.

Glucuronic acid contents were determined by Chembiotech Laboratories (University of Birmingham) with use of HPLC.

2.4 Amino Acid Composition

The amino acid composition, and hence total protein content, was determined by Alto Bioscience (University of Birmingham) with use of ion exchange chromatography.

3. RESULTS

The GPC elution profiles showing the molecular mass distribution of the whole gum samples are given in Fig 1. Figure 1 (left) gives the RI profiles and the molecular mass of the eluting species as determined by multiangle laser light scattering. The RI elution curves indicate the presence of at least two molecular mass species. The major peak (peak 1) corresponds to material of ~250 kDa, whereas the minor peak (peak 2) corresponds to material of ~1,500 kDa. Figure 1 (right) shows the UV elution profiles, which indicate that at least three molecular mass species are present. The additional species (peak 3) corresponds to material of ~200 kDa. The differences in the intensities of the three peaks observed by RI and UV are due to the fact that RI is sensitive to the concentrations of materials eluting, whereas UV is also sensitive to their chemical nature. We have shown previously (Randall *et al* 1988 1989, Osman *et al* 1993) that peak 1 corresponds to polysaccharide molecules with a low protein content, whereas peaks 2 and 3 correspond to polysaccharide molecules with a significantly higher protein content. It is the high protein content that is responsible for the higher UV absorbance values obtained. Information regarding molecular shapes is provided by Fig 2 (left), which shows the radius of gyration (Rg) and the hydrodynamic radius (Rh) as a function of elution volume, and Fig 2 (right), which shows the ratio of these two parameters. The Rg values for

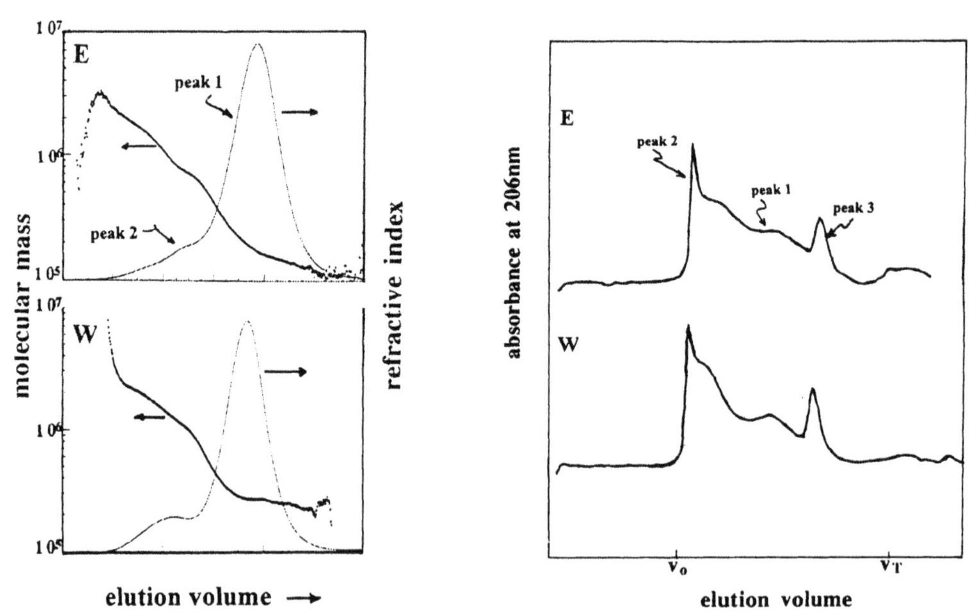

Figure 1. GPC elution profiles of gum arabic samples. (Left) RI and molecular mass of eluted species determined by multiangle laser light scattering detection; (right) UV absorbance at a wavelength of 206 nm detection. E, sample from East Sudan; W, sample from West Sudan.

the molecular species corresponding to peaks 1, 2 and 3 are ~10, 35 and 10 nm respectively, whereas for Rh the values are ~15, 45 and 15 nm respectively. Interestingly, the value of the ratio Rg/Rh remains constant at ~0.8 ± 0.2 for all molecular species. This ratio is a function of molecular shape and indicates that all of the species have a compact globular structure consistent with the high degree of branching in the carbohydrate chains (Idris *et al* 1998).

The two gum samples were fractionated by HIC. The elution profiles obtained by monitoring the UV absorbance of the eluate from the HIC column are given in Fig 3. The two profiles appear to be very similar to each other and to previously published data (Randall *et al* 1989, Osman *et al* 1993). Each profile shows at least eight peaks indicating the presence of a spectrum of molecular species. Previous fractionation studies using HIC (Randall *et al* 1989, Osman *et al* 1993, Ray *et al* 1995) have been undertaken, but only three or four fractions were isolated.

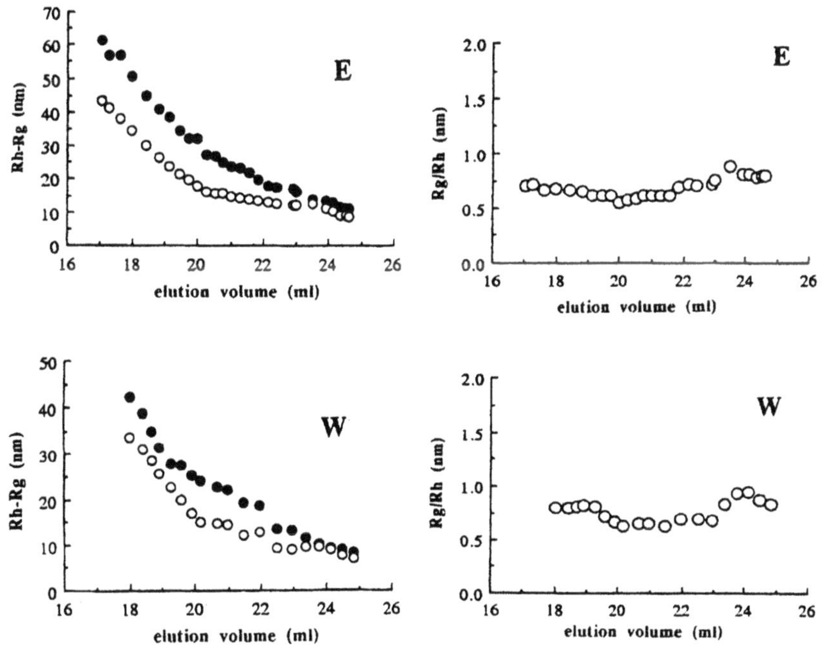

Figure 2. (Left) Hydrodynamic radii (Rh, closed circles) and radii of gyration (Rg, open circles), and (right) ratio of radii of gyration to hydrodynamic radii of gum arabic molecules eluting from a GPC column.

The percentage recovery for each fraction is given in Table 1. Similar proportions of the various fractions were obtained for both gum samples. Fraction 1A, which passed through the HIC column without adsorbing, represents more than 95% of the total weight of both gum samples, in reasonable agreement with previous findings (Randall *et al* 1989, Osman *et al* 1993, Ray *et al* 1995).

Table 1. Percentage recovery of HIC fractions

Fraction Sample	1A	1B	1C	1D	2A	2B	3A	3B
E	96.2	0.2	1.1	0.3	1.2	0.6	0.3	0.1
W	95.8	0.5	1.5	0.4	1.1	0.3	0.3	0.1

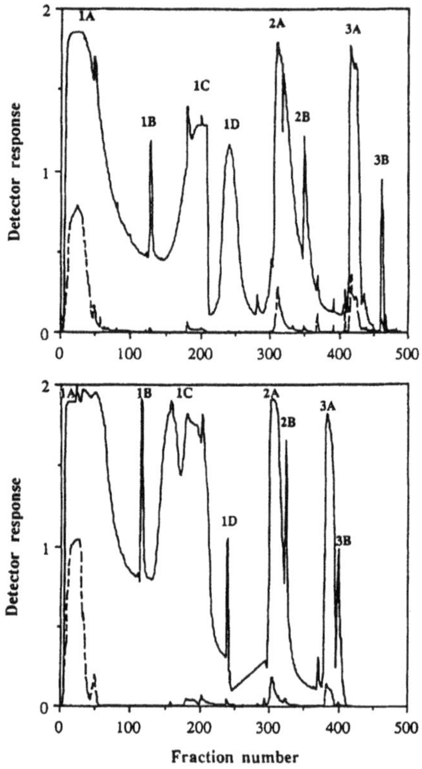

Figure 3. HIC elution profiles of gum arabic samples (top sample E, bottom sample W) using UV absorbance detection at wavelengths of 214 nm (solid line) and 280 nm (broken line). Fractions 1A, 1B, 1C and 1D were eluted with 4.2 M NaCl, 2A and 2B with 2 M NaCl, and 3A and 3B with distilled water.

The molecular mass distributions of the fractions as monitored by RI and UV absorption (214 nm) are shown in Fig 4. All fractions are seen to be polydisperse, containing species of two or more molecular mass sizes. The RI elution profile for fraction 1A corresponds mainly to peak 1 for the whole gum, with some contribution from the high molecular mass peak (peak 2). The lower proportion of the high molecular mass fraction (peak 2) is more evident in the UV elution profile, which in addition shows that the molecular mass component corresponding to peak 3 is absent. Other fractions in both gum samples were found to have high molecular mass material corresponding mainly to peak 2. As can be seen from the RI profiles, the material corresponding to peak 3 is present mainly in fractions 1C and 1D for both samples. Fractions 3A and 3B of both samples have an additional peak close to the salt peak at Vt, which was not detected in the GPC profiles of the whole gums.

The carbohydrate and protein compositions of both gum samples and their fractions are shown in Table 2. The data show that all fractions contain galactose, arabinose, rhamnose and glucuronic acid. Fraction 1A, as might be expected, has a very similar composition to the whole gum. The other fractions, apart from fraction 1B of sample W, have lower galactose contents. The most significant difference between the fractions

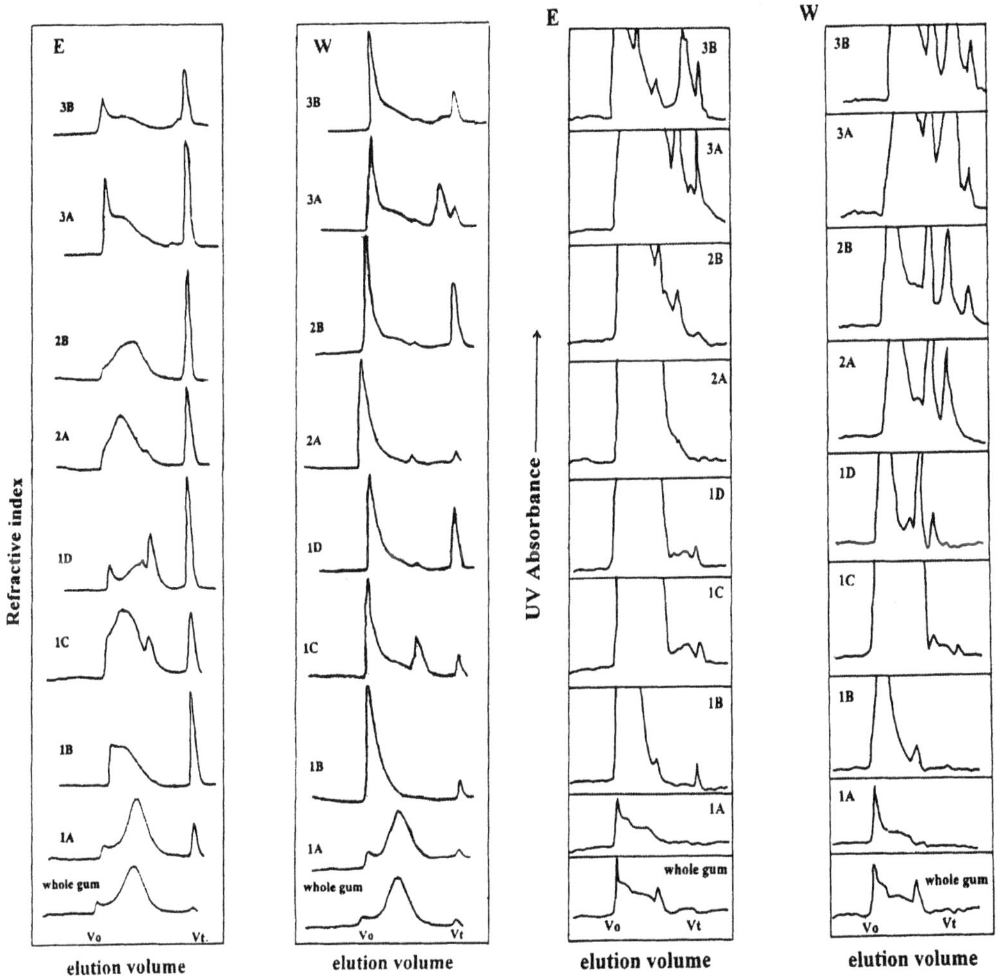

Figure 4. GPC elution profiles for fractions of gum arabic samples obtained by HIC (Fig 3). (Left) RI detection; (right) UV absorbance at 206 nm detection.

however, is their glucuronic acid content, with significantly lower values being obtained for fractions 3A of both samples and fraction 1D of sample E. These results for carbohydrate composition are consistent with the work of Osman *et al* (1993). Previous ^{1}H-NMR and methylation studies by Williams *et al* (1990) on isolated fractions from another *A. senegal* gum sample originating from Sudan found that HIC fractions all had similar branched carbohydrate structures based on a (1→3)-β-D-galactan core.

Further structural similarities between the same HIC fractions were revealed by immunological studies with anti-AGP antibodies directed against carbohydrate epitopes (Osman *et al* 1993). The present data (Table 2) also show that the proportion of protein in the fractions varies significantly and that the major fraction 1A has a much lower protein content than do the other fractions, which is in agreement with previous findings (Randall *et al* 1989, Osman *et al* 1993).

Table 2. Percentage monosaccharide compositions and protein contents of the whole gums and their HIC fractions (Fig 3)

	Whole gum	1A	1B	1C	1D	2A	2B	3A	3B
Sample E									
Galactose	40	43	34	33	30	33	32	29	n.d.[a]
Arabinose	27	24	26	21	27	26	22	23	n.d.
Rhamnose	14	16	12	11	11	11	12	8	n.d.
Glucuronic acid	13.3	15.3	n.d.	10.2	1.8	13.9	12.0	4.0	n.d.
Protein	2.4	n.d.	n.d.	n.d.	n.d.	n.d.	n.d.	n.d.	n.d.
Sample W									
Galactose	42	43	40	32	30	36	29	29	n.d.
Arabinose	26	25	28	19	20	28	21	24	n.d.
Rhamnose	13	15	13	10	13	12	13	10	n.d.
Glucuronic acid	n.d.	n.d.	n.d.	9.4	9.2	15.2	n.d.	3.9	n.d.
Protein	2.0	1.2	7.1	26.4	10.5	14.4	12.8	36.4	26.5

[a]n.d., not determined

Table 3. Amino acid compositions of sample W and its HIC fractions (residues/1000 residues)

	Whole gum	1A	1B	1C	1D	2A	2B	3A	3B
Hyp	252	455	357	61	179	237	220	68	74
Asp	57	20	34	126	87	70	74	116	116
Thr	76	79	89	47	67	70	69	54	60
Ser	139	129	155	84	120	119	115	95	92
Glu	40	11	26	98	69	49	53	54	65
Pro	63	59	63	61	65	66	58	51	51
Gly	61	45	51	89	83	66	68	82	88
Ala	30	16	19	47	47	32	40	48	65
Cys	0	0	0	0	0	0	0	0	0
Val	44	16	24	84	53	45	47	68	60
Met	1	1	<1	<1	3	<1	3	5	6
Ile	16	6	7	22	19	14	17	34	32
Leu	82	62	70	84	80	79	73	99	92
Tyr	5	13	11	27	23	19	28	37	34
Phe	35	11	22	89	53	46	39	48	51
His	56	57	54	31	47	53	51	44	39
Lys	30	15	14	38	28	27	31	68	51
Arg	12	6	5	12	11	7	15	30	25

The amino acid compositions of sample W and its fractions are reported in Table 3. The major amino acids in the whole gum and fractions 1A, 1B, 1D, 2A, and 2B are Hyp, Ser, Thr, Leu, Pro, and Gly in the approximate ratio of 4:2:1:1:1:1, in keeping with the findings for AGPs generally (Fincher *et al* 1983). For fractions 1C, 3A and 3B, however, the predominant amino acids are Asp, Ser, Leu, Gly, Ala, Val and Lys. Randall *et al* (1989), Osman *et al* (1993) and Ray *et al* (1995) also reported that HIC fractions 3A and 3B had amino acid compositions that differed from those of the other fractions.

4. DISCUSSION

The GPC experiments with a range of detectors applied to the whole gum samples indicate that the gum is principally composed of three molecular mass species. The RI and light-scattering data indicate that peak 1, which corresponds to the bulk of the gum (~90% by comparison of the area of peaks 1 and 2), is ~250 kDa, and the Rg and Rh data suggest that the molecules are globular and highly branched. The data also indicate that peak 2, which corresponds to about 10% of the total, is ~1,500 kDa and is also highly branched. The UV elution profiles show that peak 1 has a low molar absorptivity, consistent with the molecules having a low protein content. However, peak 2 has a high molar absorptivity, consistent with the molecules having a high protein content. It has been shown previously that this fraction can be degraded by proteolytic enzymes to give molecules with molecular mass similar to the bulk of the gum, and hence it has been suggested that this fraction has a wattle-blossom-type structure, where (approximately five) blocks of carbohydrate of ~250 kDa are attached to a common polypeptide chain (Connolly *et al* 1988, Osman *et al* 1993). Qi *et al* (1991) isolated the molecular species corresponding to peak 2 by GPC fractionation, and following hydrogen fluoride–deglycosylation, they concluded that the polypeptide chain consisted of ~400 amino acid residues with a simple empirical formula [Hyp_4, Ser_2, Thr, Pro, Gly, Leu, His] consistent with our findings. They identified galactosyl-O-Hyp as the glycopeptide linkage. Akiyama *et al* (1984) had previously shown that arabinosyl-O-Hyp and glycosyl-O-Ser linkages were present in studies on the whole gum. Qi *et al* (1991) reported that the isolated fraction was only 220 kDa, which is about one-sixth of the value found by us and other workers (Vandevelde and Fenyo 1985, Randall *et al* 1988). Qi *et al* (1991) suggested that the molecules were rod-like and resembled a "twisted hairy rope" with small blocks of polysaccharide (~30 residues) attached to the peptide chain. This model was supported by electron microscopy images, which showed filamentous rod-like molecules ~150 nm in length. This model conflicts with the wattle-blossom model accepted by many workers, and we have argued against it previously (Osman *et al* 1993).

The third molecular mass species corresponds to peak 3 and is only detectable by UV absorbance, confirming its presence in only small quantities (<1%) and suggesting that it has a high protein content. This fraction was found to be ~200 kDa, and the Rg and Rh data again indicate a highly compact structure. Interestingly, even though this fraction does have a very high protein content it is not degraded by proteolytic enzymes (Osman *et al* 1993), suggesting that the polypeptide is within the center of the molecule and inaccessible to attack. This resistance to proteolytic enzymes is common for many AGPs (Fincher *et al* 1983).

Fractionation of the gum by HIC reveals an even more complex picture of the molecular composition of the gum. The HIC elution profiles show multiple peaks, indicating that many molecular fractions are present. The elution profiles have been shown to be very similar for the two samples included in this study and also for a range of samples studied previously (Osman *et al* 1993). Molecules fractionate in HIC according to their degree of hydrophobic character, which depends on not only the chemical nature of the moieties within the molecules but also the accessibility of the hydrophobic sites for interaction with the column packing. The least hydrophobic molecules are expected to be those with the highest glucuronic acid and lowest protein and rhamnose contents. This expectation is consistent with the observations that fraction 1A (high glucuronic acid and low protein) does not appreciably adsorb onto the column matrix, but fractions 3A and 3B (low glucuronic acid and high protein) do strongly adsorb onto the column. As noted above, the proteinaceous component within fractions 3A and 3B may be within the core of the molecule and inaccessible for interaction with the column. The low glucuronic acid content, therefore, may be an important factor controlling the interaction of this species with the column packing.

The fact that all of the HIC fractions are polydisperse containing varying molecular mass species clearly demonstrates the extreme heterogeneous nature of the gum. Further insights into the complexities of the molecular structure of the gum can best be gained by combining two or more separation procedures to obtain discrete fractions.

REFERENCES

Akiyama, Y., Eda, S., and Kato, K., 1984, Gum arabic is a kind of arabinogalactan-protein, *Agric. Biol. Chem.* **48**: 235–237.

Churms, S. C., Merrifield, E. H., and Stephen, A. M., 1983, Some new aspects of the molecular structure of *Acacia senegal* gum (gum arabic), *Carbohyd. Res.* **123**: 267–279.

Connolly, S., Fenyo, J.-C., and Vandevelde, M.-C., 1988, Effect of a proteinase on the macromolecular distribution of *Acacia senegal* gum, *Carbohyd. Polym.* **8**: 23–32.

Fincher, G. B., Stone, B. A., and Clarke, A. E., 1983, Arabinogalactan proteins: structure, biosynthesis, and function, *Annu. Rev. Plant Physiol.* **34**: 47–70.

Idris, O. H. M., Williams, P. A., and Phillips, G. O., 1998, Characterisation of gum from *Acacia senegal* trees of different age and location using multidetection gel permeation chromatography, *Food Hydrocoll.* **12**: 379–388.

Islam, A. M., Phillips, G. O., Sljivo, A., Snowden, M. J., and Williams, P. A., 1997, A review of recent developments on the regulatory, structural and functional aspects of gum arabic, *Food Hydrocoll.* **11**: 493–505.

Joseleau, J.-P., and Ullmann, G., 1990, Biochemical evidence for the site of formation of gum arabic in *Acacia senegal*, *Phytochemistry* **11**: 3401–3405.

Randall, R. C., Phillips, G. O., and Williams, P. A., 1988, The role of the proteinaceous component on the emulsifying properties of gum arabic, *Food Hydrocoll.* **2**: 131–140.

Randall, R. C., Phillips, G. O., and Williams, P. A., 1989, Fractionation and characterisation of gum from *Acacia senegal*, *Food Hydrocoll.* **3**: 65–75.

Osman, M. E., Menzies, A. R. Williams, P. A., Phillips, G. O., and Baldwin, T. C., 1993, The molecular characterisation of the polysaccharide gum from *Acacia senegal*, *Carbohyd. Res.* **246**: 303–318.

Osman, M. E., Menzies, A.R., Martin, B. A., Williams, P. A., Phillips, G. O., and Baldwin, T. C., 1995, Characterization of gum arabic fractions obtained by anion-exchange chromatography, *Phytochemistry* **38**: 409–417.

Qi, W., Fong, C., and Lamport, D. T. A., 1991, Gum arabic glycoprotein is a twisted hairy rope, *Plant Physiol.* **96**: 848–855.

Ray, A. K., Bird, P. B., Iacobucci, G. A., and Clark, Jr., B. C., 1995, Functionality of gum arabic. Fractionation, characterisation and evaluation of gum fractions in citrus oil emulsions and model beverages, *Food Hydrocoll.* **9**: 123–131.

Stephen, A. M., and Churms, S. C., 1995, Gums and mucilages, In *Food Polysaccharides and their Applications* (A.M. Stephen, ed.), Marcel Dekker, New York, pp. 337–440.

Street, C. A., and Anderson, D. M. W., 1983, Refinement of structures previously proposed for gum arabic and other *Acacia* gum exudates, *Talanta* **30**: 887-893.

Vandevelde, M.-C., and Fenyo, J.-C., 1985, Macromolecular distribution of *Acacia senegal* gum (gum arabic) by size exclusion chromatography, *Carbohyd. Polym.* **5**: 251–273.

Williams, P. A., Phillips, G. O., and Stephen, A. M., 1990, Spectroscopic and molecular comparisons of three fractions from *Acacia senegal* gum, *Food Hydrocoll.* **4**: 305–311.

Chapter 22

Promising Gums from Sources other than *Acacia senegal*

Gladys León de Pinto
Centro de Investigaciones en Química de los Productos Naturales, Facultad de Humanidades and Educación, La Universidad del Zulia, Apartado 526, Maracaibo, Venezuela

1. INTRODUCTION

Many trees, widely disseminated in Venezuela, produce gum exudates of potential economic interest. Investigations have involved gums from species belonging to the Meliaceae (León de Pinto *et al* 1996a), Anacardiaceae (León de Pinto *et al* 1995) and Mimosaceae (León de Pinto and Corredor 1986, León de Pinto *et al* 1993, Martínez *et al* 1996).

The proven presence of indigenous Mimosaceae species such as *Acacia tortuosa, Acacia macracantha* and *Enterolobium cyclocarpum*, among others, adapted to the precise ecological conditions and capable of yielding good quality gum, comparable in composition and structure to *Acacia senegal* gum, offers an opportunity for increasing agroforestry development within Venezuela. Furthermore, the climatological conditions of hot arid areas in Venezuela are suitable for commercial gum production in high yield.

Gum arabic (*A. senegal*) remains an important international commodity, despite falling demand, over the last 20 years, resulting from drought-induced severe shortages and increased availability of a wide range of extremely price-competitive modified starches. Series Gummiferae Benth. (e.g., *Acacia seyal*) have been of minor commercial interest in comparison with gums from the series Vulgares Benth. (Anderson *et al* 1991).

Analytical characterizations of the gums from *A. tortuosa*, *A. macracantha* and *E. cyclocarpum*, together with relevant structural features of the gum polysaccharides from *A. tortuosa*, are presented in this work.

2. EXPERIMENTAL

2.1 Origin and Purification of Gum Samples

Gums from *A. tortuosa*, *A. macracantha* and *E. cyclocarpum* were collected from different locations in Zulia State, Venezuela, South America. The identification of voucher specimens was confirmed by Dr. Lourdes Cárdenas de Guevara, a botanical taxonomist of the Universidad Central de Venezuela. The voucher specimens are deposited at the Botanical Garden, Maracaibo, Venezuela.

The gum exudates dissolved readily in water, and solutions were filtered through muslin and then through Whatman N° 1 and N° 42 filter papers. The final filtrates were dialyzed against running tap water for 2 days, and then the gum was recovered by freeze drying.

2.2 General Methods

Standard methods for gum analysis were used (Anderson and Bell 1975). Paper chromatography was carried out on Whatman N° 1 and 3 MM papers with the following solvent systems (v/v): (a) benzene:butan-1-ol:pyridine:water (1:5:3:3), (upper layer); (b) acetic acid:ethyl acetate:formic acid:water (3:18:3:4); (c) acetic acid:ethyl acetate:formic acid:water (3:18:3:9); (d) butan-1-ol:ethanol:water (4:1:5); (e) ammonia (0.88):butan-2-one:water (1:200:17); and (f) butan-1-ol:ethanol:0.1 M hydrochloric acid (1:10:5). Before solvent (f) was used, the paper was pretreated with 0.3 M sodium dihydrogen phosphate solution and allowed to dry.

Gas-liquid chromatography of the methyl glycosides was performed with a Varian 2700 instrument fitted with a flame-ionization detector. The chromatograph was operated at 190°C with nitrogen flow rates of about 40 ml/min. The glass column (166 × 0.57 cm) packing was 10% by weight of polyethylene glycol adipate on Chromosorb WHP (Supelco, Bellefonte, PA). Retention times were referenced to that of methyl 2,3,4,6-tetra-O-methyl-β-D-glucopyranoside for the methyl ethers. Optical rotations were measured with a Polax-D polarimeter (Atago Co., Tokyo, Japan) at 30°C. The ^{13}C-NMR spectra were recorded in D_2O with a Bruker AM-300 spectrometer, and the experimental conditions were as described previously (León de Pinto *et al* 1994).

Neutral sugars were determined by the phenol-sulphuric acid method (Dubois *et al* 1956), and uronic acids were measured by direct titration with standard sodium hydroxide solution on exhaustively electrodialyzed samples. The procedures for partial hydrolysis, isolation and identification of neutral and acidic components, and methylation of the oligosaccharides and polysaccharides have been reported (León de Pinto 1991).

The nitrogen and tannin contents were determined by the Kjeldahl and Porter methods (Porter *et al* 1989), respectively.

2.3 Preparation and Examination of Degraded Gums A and B

Unless otherwise stated, the experimental procedures used for the preparation and examination of degraded gums A and B from the original polysaccharide isolated from *A. tortuosa* were the same as described previously (León de Pinto 1991). In brief, purified gum (4 g) was partially hydrolyzed with 5 mM H_2SO_4 for 96 hours at 100°C and then neutralized and dialyzed to yield degraded gum A (1.54 g). Degraded gum B was prepared by treating degraded gum A with 0.25 M $NaIO_4$ for 120 hours.

2.4 Smith-Degradation Studies

Three sequential Smith-degradations were performed with the pure gum of *A. tortuosa* as the starting material to obtain polysaccharide I (30% yield from the pure gum), polysaccharide II (38% yield from polysaccharide I) and polysaccharide III (20% yield from polysaccharide II). The experimental conditions for these degradations were, in general, as described previously (León de Pinto 1991). The preparation of polysaccharides I–III was repeated to check the yields and to have enough sample to complete the Smith-degradation process.

3. RESULTS AND DISCUSSION

A. macracantha, A. tortuosa and *E. cyclocarpum* (Mimosaceae) are all widely disseminated in Venezuela and produce good yields of exudate gums that are very soluble in water at room temperature (25°C). The solubilities of these two Gummiferae *Acacia* gums (>50% w/v) are comparable to that reported for *A. senegal* gum and are higher than that exhibited by *E. cyclocarpum* gum, although the solubility of this gum increases when the samples are collected as soon as the gum is exuded (León de Pinto *et al* 1996b).

The analytical data obtained for the gum exudates from the three Mimosaceae species (Table 1) show interesting features. The gum from *A. macracantha* and *A. tortuosa* are levorotatory, although many Gummiferae gums have strongly positive optical rotations (Anderson and Karamalla 1966, Anderson *et al* 1984, Churms *et al* 1986). The nitrogen contents are relatively high in comparison with the values reported for other Gummiferae gums (Anderson and Karamalla 1966, Anderson and Farquhar 1979, Anderson *et al* 1984) but lower than that of *A. hebeclada* gum, the highest recorded for an *Acacia* gum (Anderson and Farquhar 1979). The limiting viscosity number exhibited for these gums is comparable with that reported for *A. ehrenbergiana* (Anderson *et al* 1984), *A. kirkii* (Anderson and Farquhar 1979), *A. nubica* and *A. nilotica* (Anderson and Karamalla 1966) gum exudates. The ratio of galactose to arabinose is in the range published for other

Gummiferae spp. (Anderson and Farquhar 1979, Anderson *et al* 1984). The low content of rhamnose in *A. macracantha* gum is not unusual in Gummiferae spp.; the absence of this sugar and the presence of traces of xylose in *A. tortuosa* gum are good evidence to distinguish these very close *Acacia* native species.

Table 1. Analytical data of gum exudates from Mimosaceae

	A. senegal[a]	*A. tortuosa*[b]	*A. macracantha*	*E. cyclocarpum*
Yield (g/specimen/week)	39	n.d.[c]	n.d.	36
Moisture (%)	13	13.3	11.10	13.12
Ash (%)	3.6	3.16	3.86	5.62
Intrinsic viscosity (ml/g)	16	8.16	13	93.2
Specific rotation (°).	−30	−27	−12	+6.5
Nitrogen (%)	0.34	6.04	4.54	0.23
Hence protein (%)	2.1	37.75	28.37	1.44
Equivalent weight (g).	1050	1130	840	873
Hence uronic acids (%)	17	17	21	21
Neutral sugars composition, after hydrolysis (%)				
Galactose	44	63	42	49
Arabinose	25	20	34	18
Rhamnose	14	0	3	12

[a]The sample was obtained from Sudan.
[b]*A. tortuosa* gum contained traces of xylose.
[c]n.d., not determined.

Gum is exuded from *E. cyclocarpum* within 3 days after the incision is made at the trunk level. The average yield is very high (36 g/specimen/week) (León de Pinto *et al* 1996b). The intrinsic viscosity limit (~100 ml/g) is higher than that of the *Acacia* gums studied so far (Anderson and Farquhar 1979, Anderson *et al* 1983) but comparable to some *Combretum* gums (Anderson and Bell 1977). The sugar composition of *E. cyclocarpum* is similar to that of *A. senegal* gum (Anderson *et al* 1991, Osman *et al* 1993).

Comparison of analytical data from *E. cyclocarpum* gums in different conditions (crude, pure, stored for 6 months and heated at 100°C) does not show any significant differences (León de Pinto *et al* 1996b).

The properties of *A. tortuosa*, *A. macracantha*, and *E. cyclocarpum* gums indicate that they may be new sources of gums for industrial applications. A comparative study of the rheological properties of *E. cyclocarpum* gum with that of commercially important *A. senegal* gum showed the former to be a pseudo-plastic material. That gum has been used as stabilizer agent in emulsifying systems in pharmaceutical preparations (Avila *et al* 1994).

Preliminary toxicological studies of these Mimosaceae gums have been done, but further work in this area is needed with regard to the new international specifications for food additives (Anderson *et al* 1991). In addition, it is important to study the behavior of these gums in the manufacture of confections and in other technological applications.

The polysaccharide from *A. tortuosa* contains galactose, arabinose, xylose (as traces) and uronic acid residues (Table 2). Partial acid hydrolysis, followed by paper chromatography, led to the isolation of 6-O-(β-D-glucopyranosyluronic acid)-D-galactose, which was characterized through chromatography, hydrolysis, and methylation analysis. Neutral oligosaccharides of galactose were not detected, which may indicate a low abundance of long chains of galactose in the structure of the polysaccharide of *A. tortuosa*.

Table 2. Sugar composition of *A. tortuosa* gum and its degradation products

		Sugars, %[a,b]		
Polymer	Yield, %	Galactose	Arabinose	Uronic acids
Original	–	69	13	18
Degraded gum A	–	72	7	21
Degraded gum B	45	62	9	29
Polysaccharide I	30	58	20	22
Polysaccharide II	38	63	17	20
Polysaccharide III	20	60	19	21

[a]Corrected for moisture.
[b]The presence of xylose (<1%) was corroborated in the original polysaccharide.

Methylation analysis of the original polysaccharide showed the presence of 3-, 6- and 3,6-linked galactosyl residues; terminal and 3-linked L-arabinofuranosyl and 3-linked L-arabinopyranosyl residues, together with terminal glucuronosyl residues.

During the preparation of degraded gum A, portions of the galactose, arabinose, xylose and glucuronic acid were removed. The unusual presence of xylose in *Acacia* gums appeared to occur in terminal positions as indicated by its isolation during the preparation of this polymer.

Degraded gum B, obtained from degraded gum A, is an acidic branched (1→3)-β-galactan; arabinose and uronic acids were not totally removed from the core. This finding, observed in *A. macracantha* (Martínez *et al* 1996) and *E. cyclocarpum* (León de Pinto *et al* 1994) gums, may be related to steric hindrance due to hydrogen bonding between those sugars, thus working against their removal by periodate oxidation.

The *A. tortuosa* gum was subjected to three successive Smith-degradations. The high yields obtained for the polysaccharides I–III (Table 2) may be related to the incomplete removal of neutral and acidic components.

Signal assignments of the ^{13}C-NMR spectra have been made on the basis of chemical evidence and previous studies (León de Pinto 1991, León de Pinto *et al* 1994) (Tables 3–5; Figs 1, 2).

The ^{13}C-NMR spectra of the original polysaccharide and its degradation products showed unequivocal signals due to 3- and 6-linked galactosyl residues, β-D-glucuronosyl residues and their 4-O-methyl analogue, terminal and 3-linked α-L-arabinofuranosyl residues and 3-linked β-L-arabinopyranosyl residues.

Table 3. ^{13}C-NMR data[a] of ß-D-galactopyranosyl residues in the spectrum of degraded gum B[b] of *A. tortuosa* gum.

Type of linkage	C-1	C-2	C-3	C-4	C-5	C-6
→3)ß-D-Gal*p*(1→	103.47	71.84	82.03	69.28	75.08	60.48
		72.05	82.36	69.87	75.15	61.00
Model compound[c]	103.90	71.15	82.07	68.85	74.95	61.00
→6)ß-D-Gal*p*(1→	102.59	70.51	72.59	[d]	73.01	68.62
Model compound[c]	103.30	70.40	72.80	67.80	73.00	68.10

[a]Values relative to the signal of 1,4-dioxane (δ 66.67 ppm).
[b]The same signals were observed in the spectra of the original gum and the other degraded products.
[c]León de Pinto *et al* 1994.
[d]This resonance was overlapped with the signal of 1,4-dioxane.

Table 4. ^{13}C-NMR data[a] of uronosyl residues in the spectrum of degraded gum B[b,c] of *A. tortuosa* gum.

Type of linkage	C-1	C-2	C-3	C-4	C-5	C-6
ß-D-GlcA (1→	104.30	75.15	76.17	73.65	76.74	175.66
						176.80
Model compound[d]	104.70	75.50	77.10	73.30	77.50	177.50
4-OMe-α-D-GlcA(1→	99.62	72.05	73.65	83.36	70.73	59.76
					70.92	59.92
						60.07
Model compound[d]	99.10	71.80	73.30	82.56	70.77	59.94

[a]Values relative to the signal of 1,4-dioxane (δ 66.67 ppm).
[b]The same signals were observed in the spectra of the original gum and the other degraded products.
[c]The spectrum of polysaccharide III shows signals at 181.45 and 181.51 ppm due to C-6 of α-D-glucuronic acid residues substituted by metals.
[d]León de Pinto *et al* 1994.

Table 5. ^{13}C-NMR data[a] of 3-O-β-L-arabinopyranosyl residues in the spectra of degraded gums A and B and polysaccharide I of *A. tortuosa* gum.

Polymer	C-1[b]	C-2	C-3	C-4	C-5
Degraded gum B	101.20	–	75.08	65.26	62.63
Degraded gum A	101.00	67.00	75.08	65.00	62.67
Polysaccharide I	101.20	66.72	75.07	–	62.10
Model compound[c]	101.00	67.60	74.50	65.90	62.80

[a]Values relative to the signal of 1,4-dioxane (δ 66.67 ppm).
[b]The signal due to C-1 of α-L-arabinofuranosyl residues appears in the spectrum of the original gum.
[c]León de Pinto *et al* 1994.

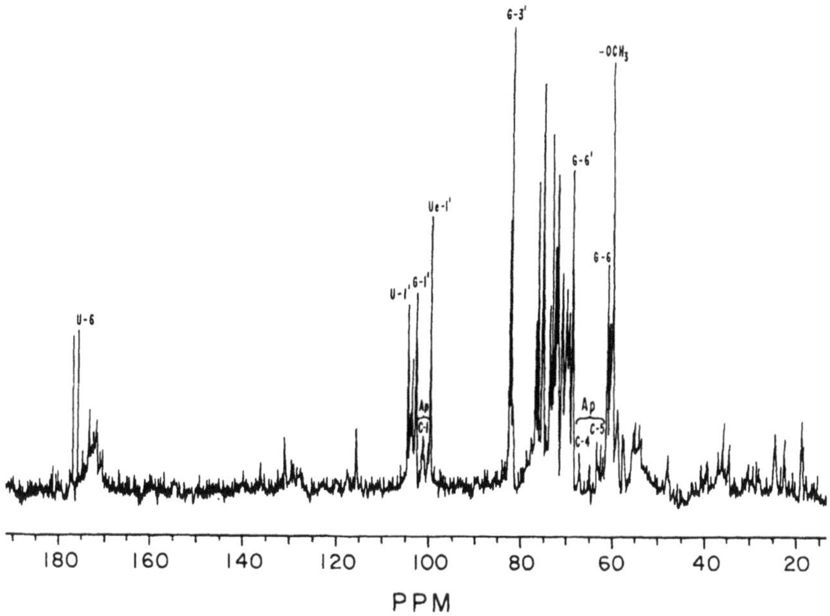

Figure 1. ¹³C-NMR spectrum of degraded gum A of *A. tortuosa* gum.

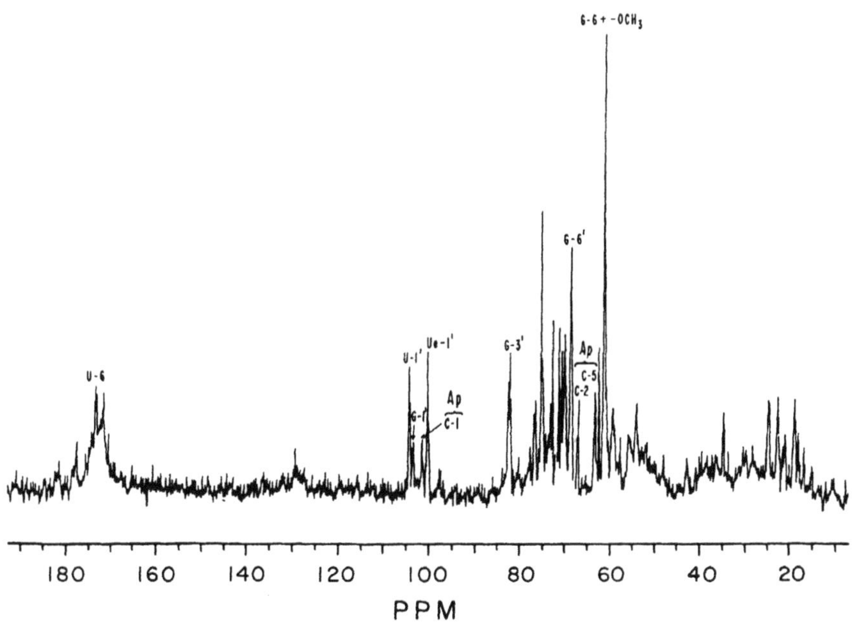

Figure 2. ¹³C-NMR spectrum of polysaccharide I of *A. tortuosa* gum.

Chemical and spectroscopic evidence support the existence of an acidic branched (1→3)-β-galactan as the backbone of the structure of the polysaccharide gum of *A. tortuosa*. Removal of uronic acid residues was difficult. Arabinose occurs as terminal and 3-linked furanosyl and pyranosyl residues. Many structural features of the polysaccharide from *A. tortuosa* gum are comparable to those reported for *E. cyclocarpum* (León de Pinto *et al* 1994), *A. macracantha* (Martínez *et al* 1996) and *A. senegal* gums (Anderson and Street 1983, Osman *et al* 1993).

REFERENCES

Anderson, D. M. W., and Bell, P. C., 1975, Structural analysis of the gum polysaccharide from *Anacardium occidentale, Anal. Chim. Acta* **79**: 185–197.

Anderson, D. M. W., and Bell, P. C., 1977, The composition of the gum exudates from some *Combretum* species; the botanical nomenclature and sytematics of the Combretaceae, *Carbohydr. Res.* **57**: 215–221.

Anderson, D. M. W., and Farquhar, J. K. G., 1979, The composition of eight *Acacia* gum exudates from the series Gummiferae and Vulgares, *Phytochemistry* **18**: 609–610.

Anderson, D. M. W., and Karamalla. K. A., 1966, Studies on uronic acid materials, *Carbohydr. Res.* **2**: 403–410.

Anderson, D. M. W., and Street, C. A., 1983, Refinement of structures previously proposed for gum arabic and other *Acacia* gum exudates, *Talanta* **30**: 887–893.

Anderson, D. M. W., Farquhar, J. K. G., and McNab, C. G. A., 1983, Some highly proteinaceous *Acacia* gum exudates of the subseries Juliflorae, *Phytochemistry* **22**: 2481–2484.

Anderson, D. M. W., Bridgeman, M. M. E., and Pinto, G., 1984, Acacia gum exudates from species of the series Gummiferae, *Phytochemistry* **23**: 575–577.

Anderson, D. M. W., Millar, J. R. A., and Weiping, W., 1991, Gum arabic (*Acacia senegal*) from Niger—Comparison with other sources and potential agroforestry development, *Biochem. Syst. Ecol.* **19**: 447–452.

Avila de Avila, G., Attias de Galindez, D., y León de Pinto, G, 1994, Propiedades fisicoquimicas del exudado gomoso de *Enterolobium cyclocarpum* y su aplicación en la industria farmacéutica, *Acta Científica Venezolana* **45**: 71–74.

Churms, S. C., Stephen, A. M., and Steyn, Ch. B., 1986, Analytical comparison of gums from *Acacia hebeclada* and other Gummiferae species, *Phytochemistry* **25**, 2807–2809.

Dubois, M., Gilles, K. A., Hamilton, J. K., Rebers, P. A., and Smith, F., 1956, Colorimetric method for determination of sugars and related substances, *Anal. Chem.* **28**: 350–356.

León de Pinto, G., 1991, Carbon-13 NMR spectral study of *Acacia xanthophloea* gum and its degradation products, *Carbohydr. Res.* **220**: 229–242.

León de Pinto, G., y Ludovic de Corredor, A., 1986, Estudio analítico de los exudados gomosos provenientes de la especie *Enterolobium cyclocarpum, Acta Científica Venezolana* **37**: 92–93.

León de Pinto, G., Martínez, M., Ortega, S., Villavicencio, N., and Borjas, L., 1993, Comparison of gum specimens from *Acacia tortuosa* and other Gummiferae species, *Biochem. Syst. Ecol.* **21**: 795–797.

León de Pinto, G., Martínez, M., Ludovic de Corredor, A., Rivas, C., and Ocando, E., 1994, Chemical and ^{13}C-NMR studies of *Enterolobium cyclocarpum* gum and its degradation products, *Phytochemistry* **37**: 1311–1315.

León de Pinto, G., Martínez, M, Mendoza, J. A., Ocando, E., and Rivas, C., 1995, Comparison of three Anacardiaceae gum exudates, *Biochem. Syst. Ecol.* 23, 151–156.

León de Pinto, G., González de Troconis, N., Martínez, M., Clamens, C., Vera, A., Rivas, C., and Ocando, E., 1996a, Composition of three Meliaceae gum exudates, *Ciencia* **4**: 47–52.

León de Pinto, G., .Martínez, M., Clamens, C., y Vera, A., 1996b, Uso potencial de la goma de *Enterolobium cyclocarpum*, *Rev. Facult. Agron.*, *La Plata* **101**: 51–56.

Martínez, M., León de Pinto, G., Rivas, C. and Ocando, E., 1996, Chemical and spectroscopic studies of the gum polysaccharide from *Acacia macracantha*, *Carbohydr. Polym.* **29**: 247–252.

Osman, M. E., Menzies, A. R., Williams, P. A., Phillips, G. O., Baldwin, T. C., 1993, The molecular characterization of the polysaccharide gum from *Acacia senegal*, *Carbohydr. Res.* **246**: 306–318.

Porter, L. J., 1989, Tannin, In *Methods in Plant Biochemistry (1):Plant Phenols* (P. M. Dey and J. B. Harborne, eds.), Academic Press, London, UK, pp 389–420.

Chapter 23

Immunochemical, Structural and Functional Properties of Mesquite Gum Compared with Gum Arabic

F. M. Goycoolea, A. M. Calderón de la Barca, G. Hernández, J. R. Valenzuela and J. R. Balderrama

Centro de Investigación en Alimentación y Desarrollo, A.C. (C.I.A.D., A.C.), Biopolymers Laboratory, P.O. Box 1735, Hermosillo, Sonora 83000, Mexico

1. INTRODUCTION

It is well documented that the bark of mesquite (*Prosopis spp.*) exudes a proteoglycan mucilage (Whistler and Smart 1953) as a response to damage to the cambium by direct means, such as insect attack or wounding, or by physiological stresses that indirectly affect the vascular cambium, such as severe water and heat stresses (Greenwood and Morey 1979). The axial xylem parenchyma and ray parenchyma at sites of cambial wounding appear to synthesize massive amounts of gum which is excreted into the wound area where it may prevent undue desiccation of the tissues, thereby enabling formation of wound parenchyma, and may also prevent the ingress of pathogens, as is shown for gum deposition in other plants (Banko and Helton 1972). Indeed, response to wounding is one of the hypotheses for the origin of gum arabic (Joseleau and Ullman 1990).

Prosopis exudate gum can be regarded as an arabinogalactan-protein (AGP) or a related member of this class of proteoglycans, which includes gum arabic and is defined (Fincher *et al* 1983) on the basis of its macromolecular structure, namely, a Type II arabinogalactan polysaccharide covalently attached to protein. *Prosopsis* gum has been suggested as a possible substitute for gum arabic from *Acacia sengal* (L.) Willd., a far more studied and industrialized AGP. Although *Prosopsis* and *Acacia* gums have been documented to have closely related chemical compositions, the uses of mesquite gum have been confined to few regions of Mexico, mostly in households (e.g., in Sonora it is chewed as a candy) and in the soft drink, ink, textile, glue and other local industries. In

Cell and Developmental Biology of Arabinogalactan-Proteins
Edited by Nothnagel *et al.*, Kluwer Academic/Plenum Publishers, 2000

the past, the Seri Indians of Sonora (Mexico) used mesquite gum to prepare eye drops (Felger and Moser 1974). Presently, the gum, which is known locally in Sonora as *chúcata*, is also used for several household and folk medicine purposes. The collection of mesquite gum for commercial purposes currently accounts for only a few tons per year (Orozco *et al* 1997).

The basic chemistry of the polysaccharide components of gum mesquite has been established for several decades (Whistler and Smart 1953, Aspinall and Whitehead 1970a b, Anderson and Farquhar 1982): a central backbone chain comprised of β 1,3- and 1,6-linked D-galactose to which branches attach at O(6) on some of the 1,3- linked residues. These branches consist of single residues or oligosaccharide chains of D-galactose, L-arabinose (which can occur in both pyranose and furanose forms), D-glucuronic acid, 4-O-methyl D-glucuronic acid and occasionally L-rhamnose. A very similar primary structure has also been described for the carbohydrate components of gum arabic (Street and Anderson 1983, Defaye and Wong 1986, Stephen 1990). This highly branched structure on both polysaccharides drives the molecules into a compact amorphous macromolecular conformation, thus effectively resulting in high solubility in water (~50%). Although mesquite gum and gum arabic have very similar overall compositions of their carbohydrate domains, the two can be distinguished by different sets of polyclonal isoantibodies, which specifically recognize non-reducing terminal sugars of the peripheral chains (Miskiel and Pazur 1991).

A "wattle blossom" macromolecular assembly has been proposed to describe the tertiary structure of the highest molecular-mass proteoglycan fraction of gum arabic (Connolly *et al* 1987 1988, Randall *et al* 1989). In this model, several polysaccharide domains of $M_w \sim 2 \times 10^5$ are held together by a short peptide chain. The occurrence of this type of assembly in the structure of mesquite gum AGP has not been demonstrated. Recent studies, however, have shown subtle macromolecular differences between mesquite gum and gum arabic (Goycoolea *et al* 1995).

Mesquite gum forms and stabilizes oil-in-water emulsions and has the ability to encapsulate orange citrus oil during spray drying, and both properties compare well to those of gum arabic (Beristain *et al* 1996, Vernon-Carter *et al* 1996, Goycoolea *et al* 1997). Despite these similar functional properties, gum arabic is a firmly established food additive with "ADI not specified" and GRAS status, respectively, by JECFA (European Community, official additive number E 414) and the U.S. FDA (Sec. 184.1330), but mesquite gum and gum exudates from *Acacia* species other than *A. senegal* (Anderson and Weiping 1989) are not permitted as food additives. Nevertheless, the Ministry of Health of Mexico has recently granted authorization for use of mesquite gum in stabilizing soft drinks, coating tablets and manufacturing feeds. The leading argument to ban the use of exudate gums other than *A. senegal* in food in Western Europe is the lack of toxicological tests to affirm their safety. Moreover, tannin compounds, as well as high Mn levels (up to 1680 ppm), have been identified in gum samples from *Prosopis* spp. of varying botanical origins and from Mexican commercial suppliers (Anderson and Weiping 1989). These two associated factors have been related to potential mutagenic effects documented for tannic acid (Stich and Powrie 1982). The presence of high tannin levels (~1.9%) has recently been confirmed consistently in

Prosopis exudate gum samples, particularly in dark gum nodules from well-identified individual trees (Goycoolea *et al* 1997). The chemical nature of these polyphenols and the AGP fractions with which they associate remain poorly understood. We present an overview of recent developments regarding the immunochemical, structural, and functional properties of native and purified low-tannin mesquite AGP from *Prosopis* species indigenous to the drylands of Mexico. These properties are compared with those of gum arabic from *A. senegal* as the standard product, so that some of the underlying barriers that have jeopardized the industrial use of mesquite gum can be identified.

2. MATERIALS AND METHODS

The *Prosopis* gum samples used in the present and previous studies carried out in our laboratory have been collected mostly in the plains of the Sonoran Desert in the late spring season (early June). *Prosopis velutina* botanical vouchers of sampled trees were lodged at the herbarium of the University of Arizona. The average mesquite gum yield per tree, sampled in the native form (i.e., without incisions), was 81.6 g (n = 24; 95% std. error = 17.7 g). Commercial mesquite gum (nodules) from Northwestern Mexico was purchased from a local retailer. Commercial spray-dried mesquite gum from the central highlands (San Luis Potosí state), presumably from *P. laevigata* (Rodríguez and Maldonado 1996), was obtained from Makimat S.A. de C.V. Commercial spray-dried gum arabic was from *Acacia* trees and supplied by Sigma Chemicals Co. (St Louis, MO, USA), and an authenticated sample of gum nodules of *A. senegal* (from Sudan) was a gift from Prof. Peter A. Williams of NEWI (Wales, UK). A sample of cold-pressed orange peel essential oil (5x) was from Esencítricos S. de R.L. (Tlalnepantla, Mexico). Details of most of the experimental methods followed to obtain our results are found in the cited references. Emulsion properties were determined according to the methods described by James and Patel (1988) and Chikamai *et al* (1996): 2 ml of a 5% (w/v) gum solution were vigorously mixed with 0.5 ml of orange peel essential oil with an Ultraturrax T25 homogenizer (15,000 rpm; 60 s). The optical density (λ = 500 nm) of a 100-fold dilution was defined as emulsion activity, and the optical density (λ = 500 nm) of a subsequent 100-fold dilution of the bottom layer after 30-min standing, expressed as a percentage of emulsion activity value, was defined as emulsion stability.

3. RESULTS

3.1 Immunological Relationships Between Arabic and Mesquite Gums

Figure 1 shows two sets of crossed immunoelectrophoresis gels obtained for *Acacia* and *Prosopis* native AGPs. Plates 1 and 3 present each gum electrophoresed in the presence of polyclonal antibodies prepared by injecting a rabbit with that gum. In plate 2, mesquite gum produced a new peak (b), partially fused with peak a, and an increased right shoulder of peak a. Thus, two molecular species in mesquite gum are related to arabic gum and give rise to two peaks, the first one presenting partial, and the second one total immunological identity. However, anti-mesquite gum antibodies (plates 3 and 4) showed only one molecular species of partial identity between both gums. One of the three peaks in plate 3 (peak c) is partially fused with the new peak d produced by gum arabic in plate 4. In conclusion, only one molecular species in each gum exhibits a total immune relationship; the unique species in arabic gum produces peaks a and d, and the unique species in mesquite gum produces peaks b and c. This observation confirmed previous evidence that part of the anti-arabic gum antibodies reacted with mesquite gum, and part of the anti-mesquite antibodies reacted with arabic gum (Goycoolea *et al* 1997). Crossed electrophoresis with Yariv reagent embedded in the second dimension has been used successfully to differentiate AGPs from gum arabic and various organs of wild tomato (van Holst and Clarke 1986). In an analogous manner, crossed electrophoresis with polyclonal antibodies in the second dimension appears to be adequate to distinguish exquisitely between the various components comprising AGPs from *Acacia* and *Prosopis* exudates.

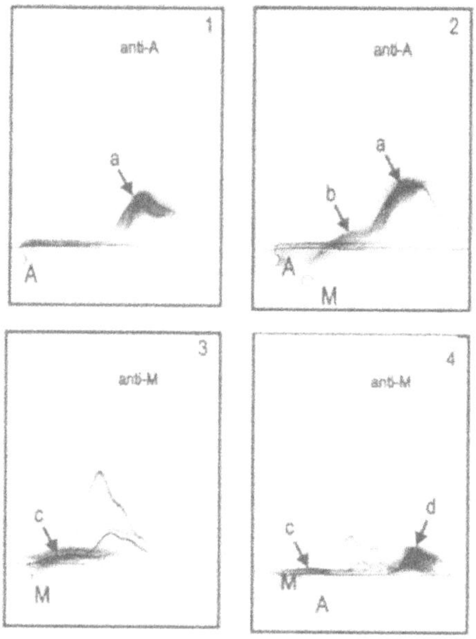

Figure 1. Crossed immunoelectrophoresis of mesquite gum and gum arabic. Samples of 5 μl of arabic (A) or mesquite (M) gums were electrophoresed in 1% agarose gels in the first (horizontal) dimension. Plates 1 and 2 contained 250 μl of anti-A antibodies and plates 3 and 4, 250 μl of anti-M antibodies, mixed with the agarose gel for the second (vertical) dimension.

3.2 Separation and Fractionation of Mesquite Gum Glycoproteins

Acacia and *Prosopis* gums have been separated on a T-Gel column (β-mercapto-ethanol coupled to divinylsulfone-activated agarose) on the basis of their relative protein content, yielding glycoprotein fractions of 52 and 36% protein content for arabic and mesquite gum, respectively (Goycoolea *et al* 1997). Subsequent analysis by SDS-PAGE showed these fractions to be polydisperse, containing at least three putative AGPs under 35 kDa for both *Acacia* and *Prosopis*. Two of these bands in each preparation shared similar molecular weight (17 and 31 kDa). In addition, a similar AGP component of 66 kDa was detected in both preparations. Distinctive bands for *Acacia* AGPs were identified at 21 kDa (a major band) and 35 kDa, whereas *Prosopis* had a principal AGP band at 23 kDa. Blotting and immunodetection of these high-protein AGPs showed cross reactivity between the antisera produced against the 21- and 35-kDa components of gum *Acacia* and against the 23- and 66-kDa components of mesquite gum.

3.3 Molecular Size and Structure

The existence of "wattle blossom" molecular assemblies in gum mesquite, similar to those postulated to exist in gum arabic AGPs (Connolly *et al* 1988), has not yet been assessed experimentally. However, the observed mean hydrodynamic radius Rh of mesquite gum (~8.2–10.6 nm) spans the mean value of Rh = 9.2 nm reported for the protease insensitive Fraction 1 of gum arabic as shown in Table 1 (Randall *et al* 1989). Weight average molecular weight values for whole mesquite gum (~3.4×10^5) and Fraction 1 of gum arabic (M_w ~3.8×10^5) are also similar, although the radius of gyration Rg and molecular mass (M_w and M_n) values of mesquite gum appear notably lower than the corresponding values for whole gum arabic or for the "wattle blossom" assembly (Fraction 2). Also noteworthy is the lower polydispersity index (M_w/M_n) observed for mesquite gum as compared to that of gum arabic, a result consistent with differences in size of the discrete molecular fractions present in each AGP preparation. Rheological measurements in concentrated solutions of both gums (Goycoolea *et al* 1995) indicated, however, that the increase in viscosity with increasing concentration above the overlap parameter $c[\eta] \sim 1$, where molecules start to touch each other, was steeper for mesquite gum than for gum arabic, indicating a more compact structure with greater resistance to forced contraction. The evidence is consistent with the suggestion that mesquite AGPs may be similar to the isolated carbohydrate domains that comprise the "wattle blossom" AGP assemblies of *A. senegal* gum. Clearly, a detailed structural analysis of mesquite gum fractions, including their susceptibility to proteolysis, would be timely.

Table 1. Molecular size dimensions for whole gum mesquite in comparison with whole and fractionated gum arabic

Material	Rh (nm)	Rg (nm)	M_w	M_n	M_w/M_n	[η] (ml/g)
Mesquite gum	8.5[a]	11.6[b]	3.4×10^{5}[b]	2.8×10^{5}[b]	1.2[b]	11.0[a]
Gum arabic: "whole" gum	14.1[c]	13.3[b]	5.5×10^{5}[b]	3.0×10^{5}[b]	1.8[b]	19.0[a]
Fraction 1	9.2[c]		3.8×10^{5}[c]			
Fraction 2	22.8[c]	25.5[c]	14.5×10^{5}[c]			

[a]At 20°C in 0.1 M NaCl (from Goycoolea *et al* 1995).
[b]Obtained by GPC-multiangle laser light scattering in gum exudates from *A. senegal* and *P. velutina* by kind help of Professor Marguerite Rinaudo of CERMAV (Grenoble, France).
[c]From Randall *et al* (1989).

3.4 Protein and Emulsifying Properties

Table 2 shows results for the protein content analysis for mesquite and arabic gums as well as for their emulsifying properties. The mean protein contents (~4–4.8%) for mesquite gum exudates appear to be consistently greater than the typical average protein contents (~1.79–2.58%) reported for gum arabic (Phillips and Williams 1994). No straightforward correlation is evident between the net protein contents and the emulsifying properties of *A. senegal* and *Prosopis* gums. Indeed, the emulsifying behaviors of both gums were very comparable, but the overall protein content of *Acacia* gum was lower than that for mesquite gum. No simple relationship between the net protein content and emulsifying behavior has been found in gum arabic (Dickinson *et al* 1988), arguing in favor of a role of structural characteristics in determining functional properties. This hypothesis is consistent with the expected structural differences in AGP fractions between the two materials, particularly those that account for the protein-rich components. In the case of mesquite gum, the importance of low-molecular-mass glycoprotein fractions in long-term emulsion stabilization is clearly demonstrated below. In gum arabic, it has been suggested that the good balance of AGPs of varying molecular size dictates the capacity to lower the oil–water interfacial tension, and thus effectively to create the emulsifying capacity and stability (Dickinson *et al* 1991a b, Ray *et al* 1995). In a recent study, Vernon-Carter *et al* (1996) found lower mean droplet size values and coalescence rates in oleoresin-in-water emulsions with mesquite gum than in those with gum arabic. Although these observations suggest that molecular size and average protein contents might influence the emulsifying behavior of exudate AGPs, it can be expected that emulsification might also be influenced by other factors such as the fine details of the aminoacyl and glycosyl compositions.

Table 2. Protein contents and emulsifying properties of mesquite gum and gum arabic

Source	Total N (%)	Protein (%)[a]	Emulsion activity[c]	Emulsion stability[d]
P. velutina (raw nodules)	0.61	4.0[a]	2.5	83.2
Commercial mesquite gum (spray dried)				
Northwest Mexico	—	5.1[b]	2.6	99.4
Central Mexico	—	5.0[b]	2.5	96.1
Commercial gum arabic (spray dried)	0.31	2.4[a]	2.7	98.2
A. senegal (raw nodules) (Sudan)	0.40	2.7[a]	2.6	98.3

[a]N conversion factors: *Prosopis* spp. gums 6.53 (Anderson and Weiping 1989), *A. senegal* gums 6.64 (Anderson *et al* 1990).
[b]Determined by Lowry *et al* (1951) assay.
[c]Optical density measurements at $\lambda = 500$ nm (as described in Materials and Methods).
[d]% of original optical density readings ($\lambda = 500$ nm) after 30 min standing and diluting.

3.5 Tannins

Polyphenols (tannins) have been positively detected in Mexican mesquite gum samples (Goycoolea *et al* 1998). The mean concentration (expressed as tannic acid percentage) for *P. velutina* nodules of red-amber color was 0.46 ± 0.04%. This content seems to be identical to the 0.40 ± 0.04% present in gum arabic from *A. senegal* (from Sudan). However, individual dark brown nodules of mesquite gum had much greater tannin content (1.9 ± 0.06%) than the light "tears" (0.43 ± 0.03%) sampled from the same tree and season. The tannin content of such dark "tears" was similar to that of commercial batches of mesquite gum (2.2 ± 0.03%) and even greater than the 1.6 ± 0.5% in the brown spray-dried sample from the highlands of Mexico. These latter figures agree well with previously reported values for commercial mesquite gum of Mexican origin which, as explained earlier, were indicated as being potentially hazardous (Anderson and Weiping 1989). A detailed study of gummosis in mesquite has shown that colorless to amber gum is produced in the phloem cavities, whereas dark brown materials arise from the xylem, parenchyma and vascular cambium (Greenwood and Morey 1979). It can be argued that impurities present in mesquite gum collected in wild conditions contribute substantially to the high tannin content found particularly in commercial batches of mesquite gum. Controlling the depth of tapping (i.e., to induce gummosis as in *A. senegal* trees) may be a route to obtain mesquite gum of low tannin content. The detailed chemical nature of the polyphenolic compounds, their specific binding sites and the possible biological significance of a putative interaction between tannins and mesquite AGPs remain unknown. The gross distribution of tannin compounds associated with mesquite gum AGPs is addressed below.

3.6 Low-Tannin Mesquite Gum

3.6.1 Ultradiafiltration of Mesquite Gum

The use of ultrafiltration and cross-flow microfiltration has been reported to clarify gum arabic solutions (Decloux *et al* 1996). This technology was tested for efficacy in reducing tannin contents of commercial mesquite gum (Goycoolea *et al* 1998). These trials were conducted on a benchtop ultrafiltration unit (20 l) fitted with hollow fiber polysulfone cartridge membranes of 10- and 50-kDa molecular weight cut-off (MWCO). Samples from the retentate and the permeate lines were collected at short time intervals. The effects of the process on the immunochemical and functional behavior of mesquite gum are discussed next.

3.6.2 Tannin and Protein Elimination

Figure 2a shows tannin concentrations of retentate samples of mesquite gum as a function of overall ultradiafiltration processing time (Goycoolea *et al* 1998). Inspection of the curves for the two processing regimes with membranes of different MWCO shows that, in both cases, tannins are removed as monotonic exponential functions of time, with no signs of discontinuities in the curves at the changes between the ultrafiltration and diafiltration washing processes. In both cases a steady value is reached, beyond which tannins are no longer removed by diafiltration. During ultrafiltration with the 50-kDa MWCO membrane, the rate of tannin removal is faster, as evidenced by the steeper slope than that observed during ultrafiltration with the 10-kDa membrane. It is also evident that the overall decrease from initial tannin content to final limiting value is greater for the 50-kDa MWCO membrane (2.11% initial, 0.64% final) than for the 10-kDa membrane (2.27% initial, 0.85% final). Notice that with the cartridge of smaller pore size most of the tannins are removed during ultrafiltration operation, with a slight further decrease achieved during diafiltration. With the larger cutoff size, however, at least one diafiltration washing cycle seems necessary to reduce the tannin content to close to the final steady value corresponding to removal of ~70% of the initial tannins. Quantitative analysis of removed tannins indicates that 30% of the total amount of such polyphenolic compounds originally present in mesquite gum are associated with gum fractions of equivalent size greater than 50 kDa, 62% with fractions smaller than 10 kDa and the remaining 8% with fractions between 10 and 50 kDa.

Figure 2b illustrates the retention patterns of glycoproteins for the two membranes tested. From the 10-kDa MWCO trace, notice a progressive but very subtle loss of low-molecular-weight AGPs as a function of processing time. By contrast, with the 50-kDa membrane the protein content of the retentate initially decreases to ~4% during ultrafiltration, but then the protein content decreases drastically as soon as the diafiltration washing cycle is applied.

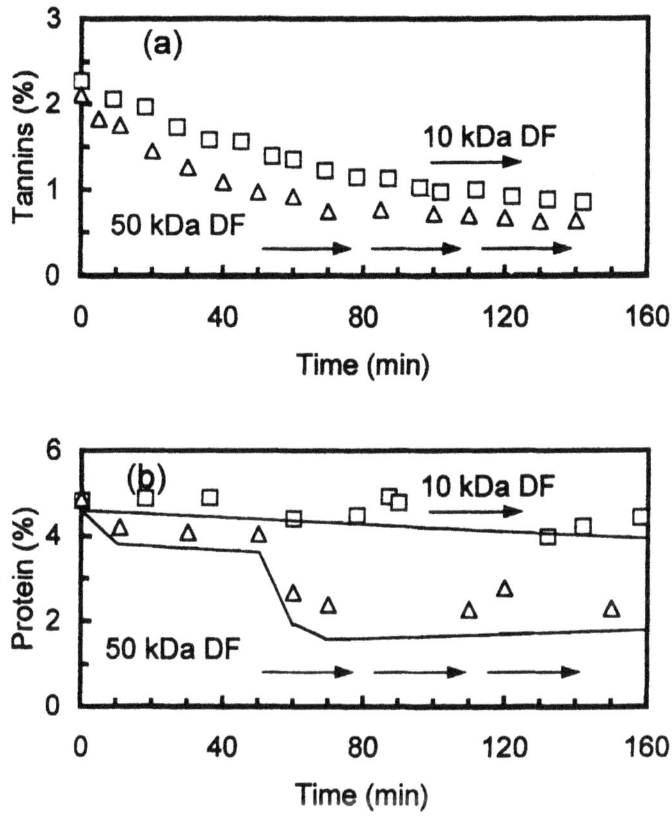

Figure 2. Profiles of tannin (a) and protein (b) retention as a function of processing time during ultrafiltration and diafiltration (DF) operations (labelled with arrows) of mesquite gum in hollow fiber membrane cartridges of 10-kDa (□) and 50-kDa (Δ) molecular weight cut-off (MWCO) (initial mesquite gum concentration 4.5% [w/w]; 55°C; membrane area 0.45 m^2). From Goycoolea *et al* (1998), with kind permission from The Royal Society of Chemistry.

3.6.3 Functionality

Goycoolea *et al* (1998) studied the capacity for the retention of citrus oil in microcapsules formed by spray drying oil-in-water emulsions prepared with mesquite or arabic gums. Gum arabic and native mesquite gum both had greater oil-retention capacity than did low-tannin mesquite gum prepared by ultradiafiltration. Moreover, low-tannin mesquite gum prepared with the 50-kDa MWCO membrane had lower oil retention capacity than did gum prepared with the 10-kDa MWCO membrane. These results point to the importance of the low-molecular-mass (10 kDa < M_W < 50 kDa) glycoprotein fractions. Such fractions, previously identified by SDS-PAGE electrophoresis (Goycoolea *et al* 1997), are likely to be lost in the permeate during ultradiafiltration. These glycoproteins might facilitate the rapid migration and adsorption of gum fractions into the oil-in-water interface and hence the formation of a uniform interface before spray drying (Anderson and Weiping 1989).

To test this hypothesis, the change in emulsion properties (emulsion activity and emulsion stability indices) was assessed. Figure 3a indicates that the capacity of mesquite gum to form an emulsion is not affected by the loss of AGP fractions of $M_w < 50$ kDa.

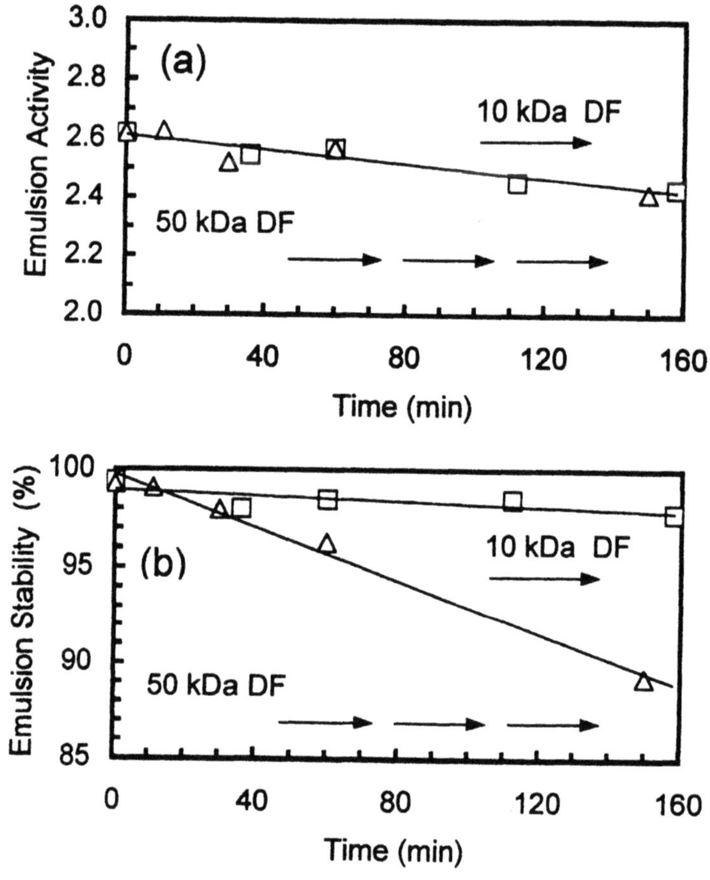

Figure 3. Profiles of (a) emulsion activity (as optical density at $\lambda = 500$ nm) and (b) % emulsion stability after 30 min as a function of processing time during ultrafiltration and diafiltration (DF) operations (labeled with arrows) of mesquite gum in hollow fiber membrane cartridges of 10-kDa (□) and 50-kDa (Δ) MWCO (same conditions as in Fig 2).

The gradual decrease in emulsion stability as increasing amounts of the different glycoprotein fractions ($M_w < 50$ kDa) were removed from native mesquite gum (Fig. 3b) might be explained in terms of the disruption of the fine balance of the different molecular fractions varying in molecular mass and N-content. Previous results obtained on different fractions of gum arabic (Dickinson *et al* 1991b, Ray *et al* 1995) showed that a mixture of low- and high-molecular-mass glycoprotein fractions of gum arabic made better emulsions than any single fraction (Ray *et al* 1995). The role of low-molecular-weight, fast-adsorbing, glycoprotein fractions of gum arabic in producing fine emulsions with good stability with respect to creaming and Ostwald ripening has been demonstrated by Dickinson *et al* (1988). Ostwald ripening is the predominant instability mechanism operating in oil-in-water emulsions made with D-limonene or orange oil, because of the appreciable solubility of these oils in water (Dickinson 1986). Vernon-Carter *et al* (1996) observed better emulsification properties with gum mesquite than with gum arabic in oleoresin-in-water emulsions, as judged by the initial droplet size and coalescence kinetics. These authors argue that the high-molecular-weight proteinaceous fraction of gum mesquite has a larger size than the corresponding one in gum arabic, leading to better emulsifying behavior. Our results clearly show that in addition to the high-molecular-weight factions involved in steric stabilization of the dispersed oil droplets, the fine proportion and nature of the low-molecular-weight glycoprotein fractions is also important with regard to rapid migration and adsorption into the oil-in-water interface and its further stabilization.

3.6.4 Immune-related Molecular Species Found after Ultradiafiltration

Tandem immuno-electrophoresis with a mixture of antibodies against both of the gums (arabic and mesquite) embedded in the second dimension gel is shown in Fig 4. As seen in plate 1, three principal peaks were produced by precipitation of the two whole gums before ultradiafiltration. The upper plates in Fig 4 (2 and 3) correspond to, respectively, permeate and retentate of mesquite gum ultradiafiltrated with the 50-kDa MWCO membrane and show that all three molecular species that produced the precipitation peaks were modified by the ultradiafiltration process. First, the right shoulder of peak a disappeared in permeate (plate 2) and was concentrated in the retentate (plate 3). Second, peak b showed a new shoulder, especially concentrated in retentate, that was closer to the application well than in whole mesquite gum (plate 1). The last peak, peak c, was present only in permeate (plate 2) and disappeared in retentate (plate 3). The behavior of these three molecular species after ultradiafiltration through the membrane of 10-kDa MWCO (plates 4 and 5) was very similar to that observed for the 50-kDa MWCO membrane. However, comparing plates in Fig 4 with those in Fig 1, it is possible to conclude that peak a was obtained by the cross-reactivity of anti-A antibodies against one immune-identical molecular species in each gum. Anti-M antibodies also reacted against the same molecular species in each gum, but the patterns in Fig 1 show a partial identity between them.

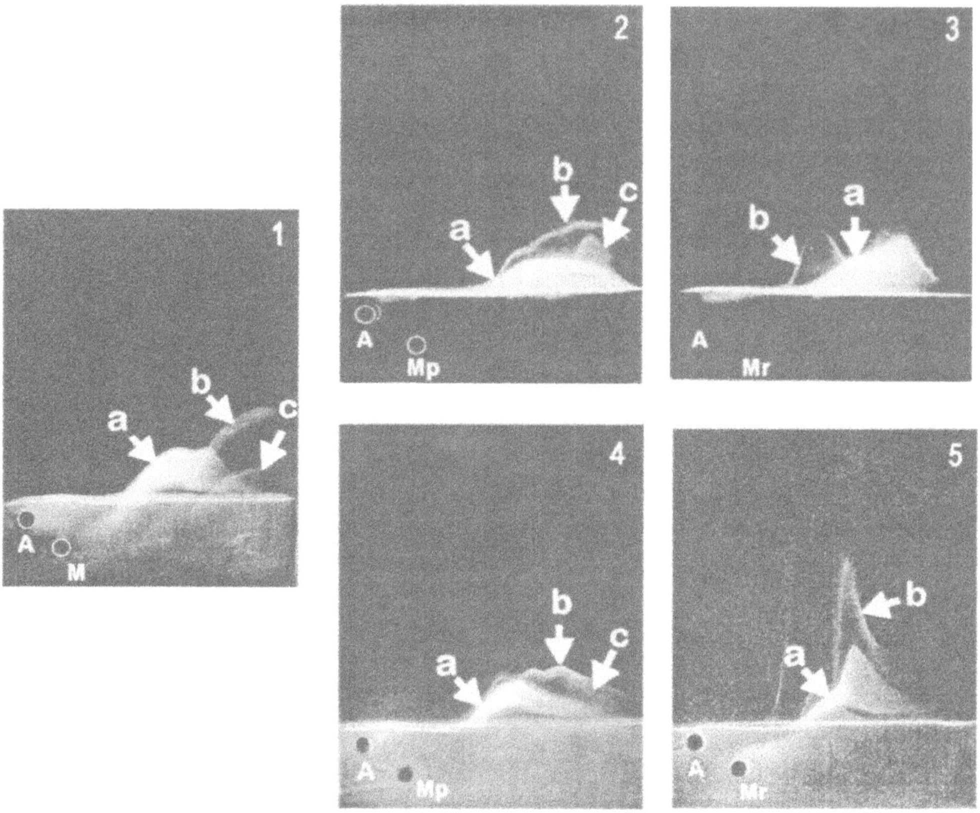

Figure 4. Crossed immunoelectrophoresis of mesquite gum (M) and its fractions obtained by ultradiafiltration (M_p = permeate, M_r = retentate) as compared to gum arabic (A). Samples of 7 μl of the gum or fraction were electrophoresed in 1% agarose gel in the first dimension (horizontal). All of the plates contained 125 μl of anti-A + 125 μl of anti-M antibodies embedded in the agarose gels in the second dimension (vertical). Tandem electrophoresis of both gum samples run before (plate 1) ultradiafiltration and of M_p and M_r fractions of mesquite gum along with whole gum arabic after ultradiafiltration on 50- and 10 kDa MWCO membranes are shown, respectively, in the upper (2 and 3) and lower part (4 and 5). Plates 1, 2 and 3 are reproduced from Goycoolea *et al* (1998), with permission.

In terms of the functional performance, the molecular species in peak a in Fig 4 could be related to the encapsulation capacity of the mesquite gum. Whole gum produced a bigger precipitation area of peak a (plate 1) and the best emulsification capacity, whereas the ultrafiltration retentate, with a minor peak a area (plates 3 and 5), presented a lower capacity. In addition, the 50-kDa ultrafiltration retentate (plate 3), which lost part of the peak a in permeate (plate 2), had less emulsification capacity than the 10-kDa retentate (plate 5), which contained almost the entire peak a (the contribution of M_p to peak a is almost negligible in plate 4). From these trials, it appears reasonable to hypothesize that the low-molecular-weight glycoprotein components of arabic and mesquite gums, which appear identical at this level of analysis, are responsible for the long-term stabilization of oil-in-water emulsions.

Ultrafiltration seems to be a feasible technology to reduce high tannin levels in gum exudates by as much as 60%. Although use of a large-pore, hollow fiber membrane (i.e., 50-kDa MWCO) leads to a rapid reduction of the original tannin contents, considerable loss of glycoprotein components also occurs and compromises gum functionality (Fig 3b). Using a small-pore membrane (i.e., ~10-kDa MWCO) and combining ultrafiltration with one diafiltration washing cycle seems adequate to eliminate up to 62% of polyphenolic compounds. These latter processing conditions do not significantly reduce the amount of small glycoprotein components, and the oil emulsion and encapsulation properties of mesquite gum are retained.

Ultrafiltration, combined with adequate hand sorting to eliminate high-tannin "dark tears" from mesquite gum collected from the Sonoran Desert, seems to be a suitable technological approach to produce mesquite gum of sufficiently high quality for clearance for use in food.

REFERENCES

Anderson, D. M. W., and Farquhar, J. G. K., 1982, Gum exudates from the genus *Prosopis, Int. Tree Crops J.* **2**: 15–24.

Anderson, D. M. W., and Weiping, W., 1989, The characterization of proteinaceous *Prosopis* mesquite gums which are not permitted food additives, *Food Hydrocoll.* **3**: 235–242.

Anderson, D. M. W., Brown Douglas, D. M., Morrison, N. A., and Weiping, W., 1990, Specifications for gum arabic (*Acacia senegal*); analytical data for samples collected between 1904 and 1989, *Food Add. Contam.* **7**: 303–321.

Aspinall, G. O., and Whitehead, C. C., 1970a, Mesquite gum. I. The 4-*O*-methylglucurono-galactan core, *Can. J. Chem.* **48**: 3840–3849.

Aspinall, G. O., and Whitehead, C. C., 1970b, Mesquite gum. II. The arabinan peripheral chains, *Can. J. Chem.* **48**: 3850–3855.

Banko, T. J., and Helton, A. W., 1972, Cytospora-induced changes in stems of *Prunus persica, Phytopatology* **64**: 899–901.

Beristain, C. I., Azuara, E., García, H. S., and Vernon-Carter, E. J., 1996, Kinetic model for water/oil absorption of mesquite gum (*Prosopis juliflora*) and gum arabic (*Acacia senegal*), *Int. J. Food Sci. Tech.* **31**: 379–386.

Connolly, S., Fenyo, J. C., and Vandevelde, M. C., 1987, Heterogeneity and homogeneity of an arabinogalactan-protein: *Acacia senegal* gum, *Food Hydrocoll.* **1**: 477–480.

Connolly, S., Fenyo, J. C., and Vandevelde, M. C., 1988, Effect of a proteinase on the macromolecular distribution of *Acacia senegal* gum, *Carbohydr. Polym.* **8**: 23–32.

Chikamai, B. N., Banks, W. B., Anderson, D. M. W., and Weiping, W., 1996, Processing of gum arabic and some new opportunities, *Food Hydrocoll.* **10**: 309–316.

Decloux, M., Dornier, M., and Gratius, I., 1996, Crossflow microfiltration of gum arabic solutions: Comparison of the classical system with the co-current permeate flow system. *Int. J. Food Sci. Technol.* **31**: 153–166.

Defaye, J., and Wong, E., 1986, Structural studies of gum arabic. The exudate polysaccharides from *Acacia senegal, Carbohydr. Res.* **150**: 221–231.

Dickinson, E., 1986, Emulsions, *Annu. Rep. Prog. Chem., Sect. C, Phys. Chem.* **83**: 31–58.

Dickinson, E., Murray, B. S., Stainsby, G., and Anderson, D. M. W., 1988, Surface activity and emulsifier behaviour of some *Acacia* gums, *Food Hydrocoll.* **2**: 477–490.

Dickinson, E., Galazka, V. B., and Anderson, D. M. W., 1991a, Emulsifying behaviour of *Acacia* gums. Part 2: Effect of the nature of the oil phase on the emulsion droplet-size distribution, *Carbohydr. Polym.* **14**: 373–383.

Dickinson, E., Galazka, V. B., and Anderson, D. M. W., 1991b, Emulsifying behaviour of *Acacia* gums. Part 2: Effect of the gum molecular weight on the emulsion droplet-size distribution,

Carbohydr. Polym. **14**: 385–392.
Felger, R. S., and Moser, M. B., 1974, The traditional pharmacopoeia of the Seri Indians of Sonora, Mexico, *Econ. Bot.* **28**: 414–436.
Fincher, G. B., Stone, B. A., and Clarke, A. E., 1983, Arabinogalactan-proteins: structure, biosynthesis and function, *Annu. Rev. Plant Physiol.* **34**: 47–70.
Goycoolea, F. M., Morris, E. R., Richardson, R. K., and Bell, A. E., 1995, Solution rheology of mesquite gum in comparison with gum arabic, *Carbohydr. Polym.* **27**: 37–45
Goycoolea, F. M., Calderón de la Barca, A. M., Balderrama, J. R., and Valenzuela, J. R., 1997, Immunological and functional properties of the exudate gum from northwestern Mexican mesquite (*Prosopis* spp.) in comparison with gum arabic, *Int. J. Biol. Macromol.* **21**: 29–36.
Goycoolea, F. M., Calderón de la Barca, A. M., Balderrama, J. R., Valenzuela, J. R., and Hernández, G., 1998, Processing and functional behaviour of low-tannin mesquite gum, In *Gums and Stabilisers for the Food Industry 9* (P. A. Williams and G. O. Phillips, eds.), Royal Society of Chemistry, Cambridge, pp. 305–313.
Greenwood, C., and Morey, P.,1979, Gummosis in honey mesquite, *Bot. Gaz.* **140**: 32–38.
James, M. J., and Patel, P. D., 1988, *Development of a Standard Oil-in-Water Emulsion Test for Proteins,* Leatherhead Food Res. Association, Leatherhead, UK.
Joseleau, J. P., and Ullman, G., 1990, Biochemical evidence for the site of formation of gum arabic in *Acacia senegal, Phytochemistry* **29**: 3401–3405.
Lowry, O. H., Rosebrough, N. J., Farr, A. L., and Randall, R. J., 1951, Protein measurement with the Folin-phenol reagent, *J. Biol. Chem.* **193**: 265–275.
Miskiel, F. J., and Pazur, J. H., 1991, The preparation and characterization of antibodies with specificity for the carbohydrate units of gum arabic and gum mesquite, *Carbohydr. Polym.* **16**: 17–35.
Orozco, J., Meráz, S., Lechuga, J. A., Cruz, F, and Vernon-Carter, E. J., 1997, Aplicaciones del mezquite (*Prosopis leavigata*) un recurso de norteamérica, In *Productos Naturales Vol. 3: Perspectivas Biotecnológicas* (J. A. Peña, J. A. Lechuga, and F. Cruz, eds.), Universidad Autónoma Metropolitana, Iztapalapa, México, pp. 43–47.
Phillips, G. O., and Williams, P. A., 1994, The specification of gum arabic in commerce, In *Food Hydrocolloids: Structures, Properties and Functions* (K. Nishinari and E. Doi, ed.), Plenum Press, New York, pp. 45–61.
Randall, R. C., Phillips, G. O., and Williams, P. A., 1989, Fractionation and characterization of gum from *Acacia senegal, Food Hydrocoll.* **3**: 65–75.
Ray, A. K., Bird, P. B., Iacobucci, G. A., and Clark, B. C., 1995, Functionality of gum arabic: fractionation, characterization and evaluation of gum fractions in citrus oil emulsions and model beverages, *Food Hydrocoll.* **9**: 123–131.
Rodríguez, C., and Maldonado, L. J., 1996, Overview of past, current and potential uses of mesquite in Mexico, In *Prosopis: Semiarid Fuelwood and Forage Tree Building Consensus for the Disenfranchised.* (P. Felfer and J. Moss, ed.), Center for Semi-arid Forest Resources, Kingsville, TX, pp. 6.41–6.52.
Stephen, A. M., 1990, Structure and properties of the exudate gums, In *Gums and Stabilisers for the Food Industry* 5 (G. O. Phillips, P. A. Williams and D. J. Wedlock, eds.), IRL Press at Oxford Univ. Press, Oxford, pp. 3–16.
Stich, H. F., and Powrie, W. D., 1982, Plant phenolics as genotoxic agents and as modulators for the mutagenicity of other food components, In *Carcinogens and Mutagens in the Environment, Vol. 1, Food Products* (H. F. Stich, ed.), CRC Press, Boca Raton, FL, pp. 135–145.
Street, C. A, and Anderson, D. M. W., 1983, Refinement of structures previously proposed for gum arabic and other Acacia gum exudates, *Talanta* **30**: 887–893.
van Holst, G. J., and Clarke, A. E., 1986, Organ-specific arabinogalactan-proteins of *Lycopersicon peruvianum* Mill. demonstrated by crossed electrophoresis, *Plant Physiol.* **80**: 786–789.
Vernon-Carter, E. J., Gómez, S. A., and Beristain, C. I., 1996, Color degradation and coalescence kinetics of Aztec marigold oleoresin-in-water emulsions stabilized by mesquite or arabic gums and their blends. *J. Text. Stud.* **27**: 625–641.
Whistler, R. L., and Smart, C. L., 1953, *Polysaccharide Chemistry*, Academic Press Inc., New York, pp. 315–317.

Structure of an Arabinogalactan-Protein Glycosylphosphatidylinositol Anchor

David Oxley, Joelian Youl, and Antony Bacic
Plant Cell Biology Research Centre, School of Botany, University of Melbourne, Parkville, Victoria 3052, Australia

The observation that two AGPs (Na1 from *Nicotiana alata* and Pc1 from pear) were present in extracellular secretions, when the respective cDNAs indicated that they both contained a C-terminal transmembrane domain, prompted us to examine these AGPs for evidence of C-terminal processing. The AGPs Na1 and Pc1 were purified and deglycosylated with anhydrous hydrogen fluoride. Their protein backbones and respective peptides produced by chemical and enzymic cleavage were analyzed by electrospray-ionization mass-spectrometry, protein sequencing and amino acid analysis. The results showed that the entire C-terminal hydrophobic domain was absent from both proteins and that they each contained an ethanolamine residue on the C-terminal amino acid (Youl *et al* 1998). This finding is a very good indication that both AGPs contained glycosylphosphatidylinositol (GPI) membrane anchors, although, interestingly, neither contained lipid. The GPI-glycan moiety was obtained, however, after dephosphorylation of AGP Pc1, suggesting that the lipid had been removed by a phospholipase. A lipid-anchored form of AGP with a glycan moiety identical to that found in the secreted Pc1 was purified from detergent extracts of suspension-cultured pear cells. The complete structure of the GPI anchor has been determined (Oxley and Bacic 1999). We are currently investigating the mechanism by which GPI-AGPs are released from the plasma membrane.

REFERENCES

Oxley, D., and Bacic, A., 1999, Structure of the glycosylphosphatidylinositol anchor of an arabinogalactan-protein from *Pyrus communis* suspension-cultured cells, *Proc. Natl. Acad. Sci. USA* **96**: 14246-14251.

Youl, J. J., Bacic, A., and Oxley, D., 1998, Arabinogalactan-proteins from *Nicotiana alata* and *Pyrus communis* contain glycosylphosphatidylinositol membrane anchors, *Proc. Natl. Acad. Sci. USA* **95**: 7921-7926.

Glycosylphosphatidylinositol Ceramide Lipid Anchor on Rose Arabinogalactan-Proteins

Jelka Svetek, Madhav P. Yadav, and Eugene A. Nothnagel
Department of Botany and Plant Sciences, University of California, Riverside, CA 92521

Arabinogalactan-proteins (AGPs) are cell-surface proteoglycans that are widely found throughout the plant kingdom and are present in leaves, roots, floral parts, and seeds. The functions of AGPs are unknown but appear to involve fundamental processes in plant development including cell proliferation, cell expansion, cell wall assembly, cell identity marking, and cell death. Characterization of AGPs from the plasma membrane, cell wall, and culture medium of suspension-cultured rose cells has revealed structural similarities between certain AGPs found in these cell surface domains. The present investigation focused on the question: What is the mechanism through which some AGPs associate with the plasma membrane while other AGPs with very similar aminoacyl and glycosyl compositions are soluble molecules in the aqueous phases of the cell wall and extracellular spaces? The results show that some plasma membrane AGPs, as well as a small proportion of culture medium AGPs, carry a ceramide-class glycosylphosphatidylinositol (GPI) lipid anchor (Svetek *et al* 1999). The evidence supporting this conclusion includes: (a) the release of AGPs from plasma membrane vesicles *in vitro* by treatment with exogenous phosphatidylinositol-specific phospholipase C (PI-PLC); (b) the shifting of elution characteristics from hydrophobic to hydrophilic when certain AGPs are treated with exogenous PI-PLC and analyzed by reverse-phase chromatography; (c) the presence of characteristic GPI-linker oligosaccharide components including inositol, glucosamine, and mannose; and (d) the presence of ceramide components, i.e. long-chain fatty acid and long-chain base, in hydrophobic AGPs that are sensitive to PI-PLC. The predominant lipid components are tetracosanoic acid and phytosphingosine. Plasma membrane vesicles readily shed AGPs by an inherent mechanism that appears to involve a phospholipase. The presence of GPI lipid anchors may be a common feature of rose AGPs. This finding has important implications towards understanding the biosynthesis, the subcellular localization, and the biological functions of AGPs, and towards stimulating searches for other GPI-linked proteins in higher plants.

ACKNOWLEDGEMENTS

Supported by U. S. Department of Agriculture NRICGP grant 95-37304-2292

REFERENCE

Svetek, J., Yadav, M. P., and Nothnagel, E. A., 1999, Presence of a glycosylphosphatidylinositol lipid anchor on rose arabinogalactan proteins, *J. Biol. Chem.* **274**: 14724-14733.

Characterization of Arabinogalactan-Proteins Secreted by Suspension Cells and Protoplasts of Sugar Beet

Anna Majewska-Sawka,[1] Rafal Butowt,[1] and Eugene A. Nothnagel[2]
[1]Institute for Plant Breeding and Acclimatization, 85-950 Bydgoszcz, Poland, [2]Department of Botany and Plant Sciences, University of California, Riverside, CA 92521

Arabinogalactan-proteins (AGPs) - the widespread components of plant plasma membranes and cell walls - are also abundantly produced by cells cultured *in vitro*. These proteoglycans are postulated to have important functions in cell proliferation, expansion and differentiation, and their activity in developmental processes seems to be related to the structure and concentration of the molecules in the cellular environment.

In the present study the AGPs which are synthesized and secreted by non-embryogenic suspension cells and suspension-derived protoplasts of sugar beet have been characterized. Arabinogalactan-proteins were extracted from the cells with 0.02 M phosphate buffer, precipitated with $(\beta\text{-D-Glc})_3$ Yariv phenylglycoside, and purified by the dimethyl sulfoxide-acetone method, whereas AGPs from the culture medium were precipitated by adding $(\beta\text{-D-Glc})_3$ directly to the culture supernatant. All samples were subsequently analyzed by gas chromatography for the glycosyl composition of the carbohydrate moiety. Amounts of AGPs obtained from the cells and from the culture supernatant were estimated by rocket electrophoresis, and the charge/size characteristics of the molecules were examined by crossed electrophoresis. The subcellular locations of two carbohydrate epitopes of AGPs were determined for both suspension cells and regenerating protoplasts by immunocytochemical detection of JIM13 and JIM8 binding sites (Butowt *et al* 1999). Microscopic observations revealed that labeling was present in the cell walls of most suspension cells and also in incipients of cell walls synthesized around the protoplasts. The labeling around protoplasts became much more intense as rebuilding of cell walls progressed during culture. Relatively weaker labeling was associated with the vacuolar compartment.

These results document the presence of JIM13 and JIM8 carbohydrate epitopes in sugar beet cells and suggest their involvement in early stages of cell wall formation, as well as their role as structural components of fully developed walls in established suspension cells.

ACKNOWLEDGEMENTS

Supported by a Fulbright Senior Research Grant and by the Polish Committee for Scientific Research - grant 6 PO4C 087 12.

REFERENCE

Butowt, R., Niklas, A., Rodriguez-Garcia, M. I., and Majewska-Sawka, A., 1999, Involvement of JIM13- and JIM8-responsive carbohydrate epitopes in early stages of cell wall formation, *J. Plant Res.* **112**: 107-116.

A Proteoglycan from Saffron Corm (*Crocus sativus* L.) Inhibits Root Elongation of *Nicotiana tabacum* Seedlings and is Highly Cytotoxic on Tobacco Cells and Protoplasts

José Antonio Fernández,[1] Abel Piqueras,[2] Joaquín Medina,[3] and Julio Escribano[1]
[1]Genetics Group, ETSIA, University of Castilla-La Mancha, Albacete, Spain, [2]Center of Soil Science and Applied Biology, CSIC, E-30080 Murcia, Spain, [3]Department of Crop Protection, Research and Technology, INIA, Madrid, Spain

Recently we have characterized a proteoglycan from saffron corms that is cytolytic on human tumoral cells (Escribano *et al* 1999). Herein we explore the biological function of this molecule. To study the effect of the proteoglycan on root growth, seedlings of *Nicotiana tabacum* were grown for up to 30 days on agar medium with proteoglycan concentrations ranging from 0.1 to 100 µg/ml. Seed germination did not appear to be affected by the treatment. The proteoglycan did produce a significant dose-dependent reduction of root growth. Roots of seedlings grown in 100 µg/ml proteoglycan were one-eighth the length of those of untreated seedlings. Concentrations of 0.1 µg/ml proteoglycan did not alter overall root elongation, but analysis of roots by microscopy revealed epidermal cell bulging. At proteoglycan concentrations greater than 10 µg/ml, roots exhibited a complete absence of root hairs. Seedlings grown on medium containing 100 µg/ml proteoglycan exhibited enlargement of root vascular tissues, apparently due to thickening of cell walls. We also found a strong cytotoxic effect on isolated tobacco cells and protoplasts where the proteoglycan concentrations inducing 50% cytotoxicity were 0.5 and 2 µg/ml, respectively. Microscopic changes induced by the exposure to proteoglycan included a decrease in cell size, loss of regular cell morphology, and evident cytoplasmic collapse. Release of intracellular proteins after proteoglycan treatment demonstrated that this compound was cytolytic on both tobacco cells and protoplasts. It is proposed that alteration of plasma membrane permeability and cytolysis are responsible for the morphological changes observed in root epidermal cells.

REFERENCE

Escribano, J., Rios, I., and Fernández, J. A., 1999, Isolation and cytotoxic properties of a novel glycoconjugate from corms of saffron plant (*Crocus sativus* L.), *Biochim. Biophys. Acta* **1426**: 217-222.

Arabinogalactan-Protein Epitopes Are Host-Derived in *Frankia-Alnus* Symbiosis

A. M. Berry,[1] U. Rasmussen,[2] K. Bateman,[2] S. Lindwall,[2] K. Huss-Danell,[3] and B. Bergman[2]

[1]*Department of Environmental Horticulture, University of California, Davis, CA 95616*
[2]*Department of Botany, Stockholm University, Stockholm, Sweden,* [3]*Department of Crop Science, Swedish University of Agricultural Sciences, Umeå, Sweden.*

Because of the potentially important role of arabinogalactan-proteins (AGPs) in facilitating cell-cell interactions, we investigated the origin and occurrence of AGP epitopes in a *Frankia-Alnus* root nodule symbiosis by the use of anti-AGP monoclonal antibodies (MAC207, JIM4, JIM13). In western blots, AGP epitopes were detected in protein extracts from symbiotic *Frankia* vesicle clusters purified from nodule preparations, but not in extracts from free-living *Frankia* cells, or from *Frankia* culture supernatant. Immunogold localization and transmission electron microscopy of *Alnus* root nodules revealed distinct patterns of cellular distribution of JIM4 and JIM13 epitopes within the nodule tissue. Similar patterns were observed in young root tips of *Alnus*. Our findings indicate that cellular expression of AGPs detected in *Alnus* root nodule symbiosis is consistent with root tissue expression patterns. In contrast to the *Gunnera-Nostoc* symbiosis, wherein both partners secrete AGP-like compounds (Rasmussen *et al* 1996), *Alnus* synthesizes AGPs but *Frankia* apparently does not.

REFERENCE

Rasmussen, U., Johansson, C., Renglin, A., Petersson, C., and Bergman, B., 1996, A molecular characterization of the *Gunnera-Nostoc* symbiosis: Comparison with *Rhizobium-* and *Agrobacterium*-plant interactions, *New Phytol.* **133**: 391-398.

The Role of Chitinases, Arabinogalactan-Proteins and Nodulation Factors in the Regulation of Somatic Embryogenesis in Norway Spruce

Malgorzata Wiweger,[1] Ulrika Egertsdotter,[2] and Sara von Arnold[1]
[1]*Uppsala Genetic Centre, Department of Forest Genetics, SLU, Box 7027, S-750 07 Uppsala, Sweden,*
[2]*NISK, Hogskoleveien 12, N-1432, Norway*

As a part of our research on somatic embryogenesis in Norway spruce (*Picea abies* (L.) Karst), we are studying the regulation of early embryo development.

Proliferating embryogenic cultures consist of single cells, proembryogenic masses and somatic embryos. We are studying cell lines capable of normal maturation and lines that are blocked at an early developmental stage. The amount and composition of extracellular arabinogalactan-proteins (AGPs) change in time, reflecting the developmental stage attained during somatic embryo development (Egertsdotter and von Arnold 1995). The composition of extracellular chitinases also varies with the stage of embryo development (Egertsdotter and von Arnold 1998). Furthermore, AGPs, chitinases and rhizobial nodulation factors can influence embryogenesis in Norway spruce (Egertsdotter and von Arnold 1998, Dyachok *et al* 2000). Similar results were obtained in the carrot system (De Jong *et al* 1992, Kreuger and van Holst 1993, Kragh *et al* 1996).

Our results indicate that some chitinases, AGPs and nodulation factors interact biochemically to exert their regulatory effect on spruce embryogenesis. We have isolated and expressed a putative class IV chitinase (*Chi4-Pa1*), and a bioassay is being developed to study the regulatory function of chitinases.

REFERENCES

De Jong, A. J., Cordewener, J., Lo Schiavo, F., Terzi, M., Vandekerckhove, J., Van Kammen, A., and De Vries, S. C., 1992, A carrot somatic embryo mutant is rescued by chitinase, *Plant Cell* **4**: 425-433.

Dyachok, J. V., Tobin, A. E., Price, N. P. J., and Von Arnold, S., 2000, Rhizobial Nod factors stimulate somatic embryo development in *Picea abies, Plant Cell Reports* **19**: 290-297.

Egertsdotter, U., and Von Arnold, S., 1995, Importance of arabinogalactan proteins for the development of somatic embryos of Norway spruce (*Picea abies*), *Physiol. Plant.* **93**: 334-345.
Egertsdotter, U., and Von Arnold, S., 1998, Development of somatic embryos in Norway spruce. *J. Exp. Bot.* **49**: 155-162.
Kragh, K. M., Hendriks, T., De Jong, A. J., Lo Schiavo, F., Bucherna, N., Hojrup, P., Mikkelsen, J. D., and De Vries, S. C., 1996, Characterization of chitinases able to rescue somatic embryos of the temperature-sensitive carrot variant ts11, *Plant Mol. Biol.* **31**: 631-645.
Kreuger, M., and Van Holst, G.-J., 1993, Arabinogalactan proteins are essential in somatic embryogenesis of *Daucus carota* L., *Planta* **189**: 243-248.

The Class III Pistil-Specific Extensin-Like Proteins of *Nicotiana tabacum* Show Arabinogalactan-Protein-Like Characteristics and are Non-Specifically Translocated Through Pollen Tube Walls *In Vivo*

Barend H. J. De Graaf, Bart Knuiman, Maurice J. M. Bosch, and Celestina Mariani
Graduate School of Experimental Plant Science, Laboratory of Plant Cell Biology, University of Nijmegen, Toernooiveld 1, 6525 ED Nijmegen, The Netherlands

Sexual plant reproduction initially involves a unique interaction between fast-growing pollen tubes and the complex fluid matrixes produced by the female reproductive organ during flower development. After pollen lands on a compatible stigma, pollen grains rapidly hydrate, germinate and form a pollen tube that grows by polarized tip extension. The secretory cells and intercellular matrix of the stigma are in continuum with the cells and intercellular matrix of the transmitting tissue present in the center of the style. Both the architecture of these tissues and the presence of specific components in the intercellular matrixes are thought to be directly involved in promoting an optimal and directional pollen tube growth during the progamic phase of compatible matings. The most abundant proteins present in the intercellular matrix are glycoproteins that share similarities with one or more classes of the traditional Hyp-rich glycoproteins (HRGPs) (Sommer-Knudsen *et al* 1998).

The class III pistil-specific extensin-like proteins (PELPs) (Goldman *et al* 1992) are some of the intercellular matrix proteins that tobacco pollen tubes come across during their tip growth through the pistil transmitting tissue in *Nicotiana tabacum*. Polyclonal antibodies specific for PELPs were produced against a recombinant PELP that was synthesized in *E. coli*. When used to detect native PELPs after electrophoretic separation, these antibodies label components in the range of 120-140 kDa, which contrasts with the 42 kDa size predicted from the deduced amino acid sequence of the encoding region of MG15, a PELP cDNA. The purified native PELPs show a reaction with Yariv reagent, indicating that these proteins have AGP-like characteristics. Further analysis of the biochemical characteristics of the class III PELPs is in progress. After pollination, the PELP-epitopes are detected in the pollen tube walls. Despite their large size, the PELPs are probably directly translocated from the intercellular matrix into the pollen tube wall. Furthermore, in the non-growing pollen tube parts behind the actively growing tips, the PELP-epitopes are translocated farther inwards, towards the pollen tube membrane. These results suggest that the walls of *in vivo*-growing pollen tubes in the transmitting tract of *N. tabacum* styles are more permeable than assumed. Furthermore, from studies of interspecific crosses, the translocation of PELPs through pollen tube walls could be a non-specific phenomenon. In conclusion, our results suggest that the PELPs are different from other stylar Pro-HRGPs, and that the function(s) of PELPs during pistil development and/or during pollen-pistil interaction is presumably determined by the specific strategies of post-translational modifications of these proteins by the tobacco transmitting tissue cells.

REFERENCES

Goldman, M. H. D. S., Pezzotti, M., Seurinck, J., Mariani, C., 1992, Developmental expression of tobacco pistil-specific genes encoding novel extensin-like proteins, *Plant Cell* **4**: 1041-1051.

Sommer-Knudsen, J., Clarke, A. E., and Bacic, A., 1997, Proline- and hydroxyproline-rich gene products in the sexual tissues of flowers. *Sex. Plant Reprod.* **10**: 253-260.

Arabinogalactan-Proteins, Pollen Tube Growth and Effect of Yariv Phenylglycoside

Jean-Claude Mollet, Sunran Kim, Guang-Yuh Jauh,[1] and Elizabeth M. Lord

Department of Botany and Plant Sciences, University of California, Riverside, CA 92521, [1]present address: Institute of Botany, Academia Sinica, Taipei 11529, Taiwan

Arabinogalactan-proteins (AGPs) are complex macromolecules involved in plant growth and reproductive and vegetative development. Yariv phenylglycoside (β-D-Glc)$_3$, a synthetic probe known to precipitate AGPs, is added to the germination medium to investigate the effect on the pollen tube growth of different plant families (Jauh and Lord 1996, Roy *et al* 1998). Among the six species tested, *Lilium longiflorum, Zea mays* and *Annona cherimoya* pollen tubes clearly exhibit inhibited growth within 1-2 hours of Yariv phenylglycoside treatment, *Lycopersicum esculentum* and *Aquilegia eximia* pollen tubes are not arrested in tip growth, and *Nicotiana tabacum* pollen tubes might be slowed in growth. Interestingly, the species showing inhibition of pollen tube growth with Yariv phenylglycoside also exhibit evidence of AGP secretion at the tip during normal development, as revealed by a strong labeling with JIM13 monoclonal antibody. Lily pollen tubes, upon removal of Yariv phenylglycoside from the germination medium, regenerate new tube tips within 4 hours (56% of the population). These new sites of polarized growth emerge from the flanks of the tips arrested by the Yariv phenylglycoside. This tip regeneration provides us with a system to study initial events in polarized growth. The location of the new emerging tips can be anticipated by localizing AGPs with JIM13. Within 1 hour after the removal of Yariv phenylglycoside from the germination medium, 49% of the pollen tubes show evidence of JIM13 labeling as a faint line on the flanks near the arrested tip or as a strong fluorescence in a new, just-emerging tip. Within 4 hours after the removal of Yariv phenylglycoside, only 17% of the pollen tubes reveal this pattern, whereas 39% already have a new, well-developed tip that strongly labels with JIM13. Since secretion of AGPs toward the new emerging tip is one of the later stages in the polarization event, we are looking at earlier markers. Preliminary results concerning the re-arrangement of actin filaments using confocal microscopy did not reveal a clear pattern. We are currently investigating several other intracellular probes (protein kinases) to point out the early molecular changes involved in the determination of the new growth sites.

REFERENCES

Jauh, G. Y., and Lord, E. M., 1996, Localization of pectins and arabinogalactan-proteins in lily (*Lilium longiflorum* L.) pollen tube and style, and their possible roles in pollination, *Planta* **199**: 251-261.

Roy, S., Jauh, G. Y., Hepler, P. K., and Lord, E. M., 1998, Effects of Yariv phenylglycoside on cell wall assembly in the lily pollen tube, *Planta* **204**: 450-458.

Pollination in *Arabidopsis thaliana*: Cell-Cell Interaction During Pollen Tube Growth

K. A. Lennon,[1] S. Roy,[2] P. K. Hepler,[2] and E. M. Lord[1]

[1]*Department of Botany and Plant Sciences, University of California, Riverside, CA 92521,*

[2]*Department of Biology, University of Massachusetts, Amherst, MA 01003*

Pollination and the guidance of the pollen tube through the transmitting tract of the gynoecium to the ovule involve a series of complex cell communication events. In this study, we demonstrate that while within the central transmitting tract of the gynoecium, these pollen tubes are tightly adherent to the transmitting tract cells. Once these tubes leave the extracellular matrix (ECM) of the central septum, they adhere to and travel along the ECM of the septum epidermis. The epidermal ECM, which surrounds the pollen tubes, is secreted through breaks in the cuticle of the septum surface. The transmitting tract ECM stains strongly for acidic polysaccharides and stains lightly with the (β-D-Glc)$_3$ Yariv reagent. Monoclonal antibodies to esterified and low-esterified homogalacturonans do not bind to this ECM. Monoclonal antibodies to arabinogalactan-proteins (AGPs) provide evidence that AGPs may be a component of the septum epidermal ECM (Lennon *et al* 1998). Preliminary data indicate that *in vitro* pollen tube growth is inhibited by the addition of the (β-D-Glc)$_3$ Yariv reagent to the germination medium, though little binding is seen between monoclonal antibodies to AGPs and pollen tubes.

REFERENCE

Lennon, K. A., Roy, S., Hepler, P. K., and Lord, E. M., 1998, The structure of the transmitting tissue of *Arabidopsis thaliana* (L.) and the path of pollen tube growth, *Sex. Plant Reprod.* **11**: 49-59.

Adhesion Molecules in Lily Pollination

S.-Y. Park, J.-C. Mollet, G.-Y. Jauh,[1] K. J. Eckard, E. A. Nothnagel, L. L. Walling and E. M. Lord

Department of Botany and Plant Sciences, University of California, Riverside, CA 92521, [1]present address: Institute of Botany, Academia Sinica, Taipei 11529, Taiwan

Molecules involved in adhesion of the pollen tube to the transmitting tract cells in lily have been studied. At least two molecules in the stylar extracellular matrix (ECM) are involved in pollen tube adhesion. One of the molecules has been purified from the gynoecium using an *in vitro* adhesion assay. The cDNA corresponding to the protein has been cloned (Park *et al* 2000). Analysis of the cDNA indicated that this molecule is highly homologous to nonspecific lipid transfer proteins (nsLTPs) with eight cysteine residues, which is typical in known nsLTPs in plants. The LTP purified from the gynoecium is not adhesive by itself, but adhesion of pollen tubes to an artificial matrix *in vitro* (Jauh *et al* 1997) occurs when the LTP is combined with leaf or petal homogenate, or with a style wall fragment extract which has the sugar composition of an AGP. These homogenates and extract are not adhesive in the absence of the LTP. Plant nsLTPs are so named because of their ability to transfer lipids from one membrane to another *in vitro*, but the fact that they are secreted molecules localized to the ECM, instead of to the cytoplasm as initially expected, means their function is as yet unknown in plants. Here we demonstrate a function for LTPs in pollen tube adhesion to the stigma/stylar transmitting tract ECMs in the lily gynoecium.

REFERENCES

Jauh, G. Y., Eckard, K. J., Nothnagel, E. A., and Lord, E. M., 1997, Adhesion of lily pollen tubes on an artificial matrix, *Sex. Plant Reprod.* **10**: 173-180.

Park, S.-Y., Jauh, G.-Y., Mollet, J.-C., Eckard, K. J., Nothnagel, E. A., Walling, L. L., and Lord, E. M. 2000, A lipid transfer-like protein is necessary for lily pollen tube adhesion to an in vitro stylar matrix, *Plant Cell* **12**: 151-163.

A Role for Arabinogalactan-Proteins in Root Growth

Clare G. Steele-King and J. Paul Knox

Centre for Plant Sciences, University of Leeds, Leeds, LS2 9JT, U. K.

The synthetic phenylglycoside, β-glucosyl Yariv reagent (βGlcY), binds specifically to a subset of AGPs in *Arabidopsis thaliana* roots. When wild-type seedlings are germinated in media containing 30 μM βGlcY, a dramatic reduction in root and shoot growth, compared to control seedlings, is observed (Willats and Knox 1996). Short roots are a result of reduced cell elongation rather than reduced cell proliferation. We have used this βGlcY-phenotype in wild-type plants as a reference point in the screening of T-DNA and EMS-generated mutants for seedlings that are supersensitive or insensitive to βGlcY, or which show an altered pattern of βGlcY binding. The EMS-generated mutant, YSS1, has an apparently wild-type phenotype at 14 days. However, when YSS1 is germinated in the presence of βGlcY, the growth of the seedling is retarded to a much greater extent than wild-type seedlings, and the root is less than a millimetre in length at 14 days. This result with YSS1 is in contrast to wild-type seedlings grown in 30 μM βGlcY, where maximum root length is 8 mm. When YSS1 is transferred to medium without βGlcY, the mutant appears to recover a wild-type phenotype. This mutant, YSS1, is a very recent finding in our laboratory, and results are preliminary. It is possible, however, that YSS1 is a mutant that is supersensitive to βGlcY treatment. If this is the case, molecular studies of YSS1 are likely to yield important data regarding the βGlcY-binding subset of AGPs and their role in cell expansion.

REFERENCE

Willats, W. G. T., and Knox, J. P., 1996, A role for arabinogalactan-proteins in plant cell expansion: Evidence from studies on the interaction of beta-glucosyl Yariv reagent with seedlings of *Arabidopsis thaliana, Plant J.* **9**: 919-925.

Cytochemical Analysis of Cell Wall Composition in Non-Articulated Laticifers

Marcelo D. Serpe
Department of Biology, Boise State University, Boise, ID 83725

Non-articulated laticifers are long, coenocytic cells that contain latex, the conspicuous and often white fluid that oozes from many plants upon injury. As a first step to analyze the composition of laticifer walls, I have developed procedures to identify laticifers in tissue sections and to isolate laticifer fragments from *Euphorbia heterophylla* hypocotyls. Laticifers can be identified using nile red, a fluorescent dye that binds to neutral lipids present in the latex. Staining of hypocotyl cross sections with nile red revealed that laticifers are mainly present in the cortex, where they are distributed in a ring pattern. Various procedures were tested for their effectiveness in separating laticifer fragments from other cells. Of these procedures, the most successful one involved incubation of hypocotyls in a saturated solution of picric acid and subsequent separation of laticifers by micro-dissection. With hypocotyl sections and isolated laticifer fragments, I have begun the analysis of laticifer wall composition using various fluorescent dyes and colored stains that are often employed in the study of cell wall cytochemistry. Alcian blue, a colored stain used with bright-field microscopy to detect polyanions such as pectins, stained laticifer walls. Furthermore, after incubation with a fluorescein-conjugated pectate probe, laticifer walls were intensely fluorescent, indicating the presence of unesterified pectins. Laticifer walls also fluoresced with Calcofluor, but not with aniline-blue fluorochrome. These results indicate that β-D-glucans, probably cellulose, are abundant in laticifer walls, but (1→3)-β-D-glucans such as callose are absent. Staining with $(\beta\text{-D-Glc})_3$ Yariv phenylglycoside suggests that arabinogalactan-proteins are not abundant in laticifer walls. Hypocotyl sections and isolated laticifer fragments were incubated in 0.2% $(\beta\text{-D-Glc})_3$ Yariv phenylglycoside for 24 hours. After this incubation, no red coloration was observed in laticifer walls or other walls of the hypocotyl except for those of xylem vessels (Serpe and Nothnagel 1999). Lignin or other phenolics were not detected in laticifer walls, indicating that these walls are entirely primary.

REFERENCE

Serpe, M. D., and Nothnagel, E. A., 1999, Arabinogalactan-proteins in the multiple domains of the plant cell surface, *Adv. Bot. Res.* **30**: 207-289.

A Glycoconjugate Isolated from the Saffron Plant (*Crocus sativus* L.) is Cytolytic Against Tumoral Cells and Activates Macrophages *In Vitro*

Julio Escribano,[1] M. José M. Díaz-Guerra,[2] Hans H. Riese,[3] Alberto Alvarez,[2] Remedios Proenza,[3] Damián García-Olmo,[4] Dolores C. García-Olmo,[4] Jesús Ontañón[4] and José Antonio Fernández[1]

[1]Section of Biotechnology, IDR, Universidad de Castilla-La Mancha, Campus Universitario s/n, E-02071 Albacete, Spain, [2]Department of Biochemistry and Molecular Biology, Instituto de Bioquímica CSIC-UCM, Universidad Complutense, Av. Ramón y Cajal s/n, E-28040 Madrid, Spain, [3]Pharmacia and Upjohn, Department of Immunology and Oncology, Centro Nacional de Biotecnologia, Campus Universidad Autónoma, E-28049 Madrid, Spain, [4]Unit of Clinical and Experimental Research, Hospital General de Albacete, c/Hermanos Falcó s/n, E-02006 Albacete, Spain

Saffron corms contain a toxic glycoconjugate composed of 94.5% carbohydrate and 5.5% protein. We undertook to study the specificity and molecular events involved in cytotoxicity, as well as the possible immunomodulatory properties of this compound (Escribano *et al* 1999). Studies of intracellular calcium fluctuations, transmembrane potential and release of lactate dehydrogenase showed that this compound caused plasma membrane damage, allowing movements of both calcium and macromolecules, and leading to cell lysis. Microscopy of treated cells revealed cellular swelling and cytoplasmic collapse. This molecule was active *in vitro* against human tumoral cells derived from fibrosarcoma, cervical epithelioid carcinoma and breast carcinoma, the calculated IC_{50} values being 7, 9 and 22 μg/ml, respectively. Comparison of IC_{50} data for fibroblasts (60 μg/ml) and fibrosarcoma cells revealed that the glycoconjugate was nearly 8 times more cytotoxic for malignant cells than for their normal counterparts. A 100 μg/ml concentration of the compound produced 50% lysis of normal human erythrocytes, while 320 μg/ml were unable to induce 50% cell death of cultured human hair follicles. Non-cytotoxic concentrations promoted a significant macrophage activation, as detected by the release of nitric oxide, but did not stimulate either cAMP generation or mitogen-activated protein kinases. Activation of protein kinase C and transcriptional activator NF-κB was also observed. The activation of macrophages suggests a potential immuno-stimulating capacity of this compound.

REFERENCE

Escribano, J., Rios, I., and Fernández, J. A., 1999, Isolation and cytotoxic properties of a novel glycoconjugate from corms of saffron plant (*Crocus sativus* L.), *Biochim. Biophys. Acta* **1426**: 217-222.

Production of Arabinogalactan-Proteins in *Beta vulgaris* Cell Suspension Cultures: A Response to Hydrodynamic Stress

Mario Rodriguez-Monroy[1] and Enrique Galindo[2]
[1]*Centro de Desarrollo de Productos Bióticos, IPN. Apdo. Postal 24, Yautepec, Morelos, 62730 Mexico*
[2]*Instituto de Biotecnologia, UNAM, Apdo. Postal 510-3, Cuernavaca, Morelos, 62250 Mexico*

INTRODUCTION

The isolation of arabinogalactan-proteins (AGPs) from plant cell suspension cultures has been reported for several different species (Nothnagel 1997). However, the role of these molecules has not been established. In this work, the accumulation of AGPs in *B. vulgaris* cultures growing in a fermentor was associated with a cellular response to hydrodynamic stress conditions.

METHODS

The conditions for growing *B. vulgaris* suspension cultures were as described (Rodriguez-Monroy and Galindo 1999). Two systems with different conditions of hydrodynamic stress were used. Erlenmeyer flasks (125 ml) were agitated at 100 revolutions per minute, and a fermentor (2000 ml) was operated at a tip speed of 95.3 cm/second. Under these conditions, the power consumption in the fermentor was one order of magnitude higher than in the shake flasks.

Samples of *B. vulgaris* at the stationary phase of growth were used when harvesting the extracellular compounds. High molecular mass material was precipitated from cell-free medium with ethanol during overnight incubation at 4°C. The macromolecules were collected by centrifugation at 13,000 *g* and then dialyzed against water. Precipitated material was assayed with the Yariv reagent using the methodology reported by Van Holst and Clarke (1985). Molecular weights of proteins were estimated by electrophoretic analysis in a polyacrylamide gel (8%) under denaturing conditions with sodium dodecyl sulfate. Isoelectric points were determined by electrophoresis under non-dissociating conditions.

RESULTS AND DISCUSSION

The accumulation of extracellular material in the broth of the fermentor was six-fold higher than that obtained in the cultures grown in shake flasks (Table 1). Because a cell viability of 80% was observed in both culture conditions, cell lysis was not the source of extracellular material. Dissolved oxygen tension was not a limiting factor since the cells grew at practically the same rate in shake flasks and in the fermentor (in which dissolved oxygen tension was above 10%). Thus, the increased production of extracellular material by cells in the fermentor was interpreted as a response of the cells to the high hydrodynamic stress. The biochemical analysis of the precipitated material with the Yariv assay was positive. In addition, the precipitated material had a high content of hydroxyproline and an isoelectric point of 4 (Table 1). These characteristics indicated that the extracellular material contained AGP. Different molecular weights, however, were observed for the two preparations: two bands were identified in the product obtained from shake flasks (45 and 70 kDa), whereas the precipitated material obtained from the fermentor showed only one band (97 kDa). A previous work (Rodriguez-Monroy and Galindo 1999) reported a change in the rheology of the cell-free medium: it was Newtonian for cultures grown in shake flasks and pseudoplastic for cultures grown in the fermentor. It is postulated that this rheology change reduces the hydrodynamic turbulence in the cultures developed in the fermentor.

Table 1. Biochemical analysis of *B. vulgaris* extracellular material produced in shake flasks and in a fermentor.

	Yield (mg pp[a]/g cells)	Yariv test	Hydroxyproline (μg/mg pp)	Molecular weight (kDa)	Isoelectric pH
Shake flasks	1.0	+	4.70 ± 0.66	70, 45	ND[b]
Fermentor	6.2	+	5.27 ± 0.85	97	4

[a]pp, precipitated polymer
[b]ND, not determined

CONCLUSION

The cells of *B. vulgaris* secrete AGPs as a response to hydrodynamic stress.

ACKNOWLEDGEMENTS

This work was financed by CONACyT (grants 25165B and 26405-N). M. Rodriguez-Monroy is indebted to COFFA-IPN and CONACyT for the doctoral fellowship awarded.

REFERENCES

Nothnagel, E. A., 1997, Proteoglycans and related components in plant cells, *Int. Rev. Cytol.* **174**: 195-291.

Rodriguez-Monroy, M., and Galindo, E., 1999, Broth rheology, growth and metabolite production of *Beta vulgaris* suspension culture: A comparative study between cultures grown in shake flasks and in a stirred tank, *Enzyme Microb. Technol.* **24**: 687-693.

Van Holst, G.-J., and Clarke, A. E., 1985, Quantification of arabinogalactan-protein in plant extracts by single radial gel diffusion, *Anal. Biochem.* **148**: 446-450.

Quantitative and Qualitative Study of Arabinogalactan-Peptide During Bread Making

Anne-Marie A. Loosveld,[1] Caroline Maes,[1] Piet J. Grobet,[2] and Jan A. Delcour[1]
[1]*Laboratory of Food Chemistry, Katholieke Universiteit Leuven,* [2] *Center for Surface Chemistry and Catalysis, Katholieke Universiteit Leuven, Kardinaal Mercierlaan 92, B-3001 Heverlee, Belgium*

The quantitative and qualitative changes of water-extractable arabinogalactan-peptide (WE-AGP) and water-extractable arabinoxylan (WE-AX) in the flour from the two wheat varieties Skirlou and Soissons were investigated during a straight dough wheat bread-making process. The extractability of WE-AGP remained constant or increased slightly as a result of the bread-making process (increase: 12% for Skirlou and 10% for Soissons, expressed relative to the amount of WE-AGP initially present). This slight increase contrasts with the great increase in extractability of the WE-AX (29% for Skirlou and 77% for Soissons). No changes in arabinose-to-xylose ratio of WE-AX (0.47-0.50 (w:w) for Skirlou, 0.52-0.54 for Soissons) and arabinose-to-galactose ratio of WE-AGP (0.66-0.70 (w:w) for Skirlou, 0.69-0.72 for Soissons) occurred during the bread-making process. In contrast to differences found in the molecular weight distribution of WE-AX, no differences in the molecular weight distribution of WE-AGP were observed as a result of processing. Apparent molecular weights of 24,000 and 22,000 were found for WE-AGP samples of flour and doughs at different stages of the bread-making process for Skirlou and Soissons, respectively. The ^{1}H-NMR spectra (300 MHz, D_2O, 85°C) of the WE-AGP isolated from flour, dough and bread fractions were comparable. Taken together, the above data suggest that, unlike what can be observed in gluten agglomeration in dilute systems (Roels *et al* 1998), no specific interaction between WE-AGP and gluten proteins occurs in a straight dough bread-making process (Loosveld *et al* 1998).

REFERENCES

Loosveld, A.-M. A., Maes, C., Grobet, P. J., and Delcour, J. A., 1998, Quantitative and qualitative study of arabinogalactan-peptide during bread making, *J. Agric. Food Chem.* **46**: 5026-5030.

Roels, S. P., Grobet, P. J., and Delcour, J. A., 1998, Distribution of carbohydrates in gluten fractions isolated from European wheats (*Triticum aestivum* L.) in a batter system, *J. Agric. Food Chem.* **46**: 1334-1343.

The Effect of Larch Arabinogalactan on Mixing Characteristics of Wheat Flour Dough

Anne-Marie A. Loosveld and Jan A. Delcour
Laboratory of Food Chemistry, Katholieke Universiteit Leuven, Kardinaal Mercierlaan 92, B-3001 Heverlee, Belgium

The effect of larch arabinogalactan on mixing characteristics of wheat flour dough was investigated. Wheat flour was substituted with 1.0% and 2.0% larch arabinogalactan, and the influence on farinograph characteristics (Duedahl-Olesen *et al* 1999) and mixing time was evaluated. The farinograph absorption, based on dough consistency at the 500-Brabender units line, was lower for larch arabinogalactan substitution (57.6% and 55.7% farinograph absorption for 1.0% and 2.0% substitution, respectively) than for the reference (60.4%). Dough stability was likewise decreased for dough containing larch arabinogalactan (6.9 min and 2.2 min stability for 1.0% and 2.0% substitution, respectively) compared with the doughs of unsubstituted wheat flour (10.8 min). Optimal mixing time was determined with a 10-g mixograph on the basis of the 500 Brabender units farinograph absorption value. Longer mixing times for an optimal dough development were required in the case of larch arabinogalactan substitution (240 seconds and 270 seconds for 1.0% and 2.0% substitution, respectively) compared to the peak time of 180 seconds for the unsubstituted flour. The profiles were a little broader in the case of larch arabinogalactan substitution.

REFERENCE

Duedahl-Olesen, L., Zimmermann, W., and Delcour, J. A., 1999, Effects of low molecular weight carbohydrates on farinograph characteristics and staling endotherms of wheat flour-water doughs, *Cereal Chem.* **76**: 227-230.

Index

GPSR Compliance
The European Union's (EU) General Product Safety Regulation (GPSR) is a set of rules that requires consumer products to be safe and our obligations to ensure this.

If you have any concerns about our products, you can contact us on

ProductSafety@springernature.com

In case Publisher is established outside the EU, the EU authorized representative is:

Springer Nature Customer Service Center GmbH
Europaplatz 3
69115 Heidelberg, Germany

www.ingramcontent.com/pod-product-compliance
Ingram Content Group UK Ltd.
Pitfield, Milton Keynes, MK11 3LW, UK
UKHW052101011125
464545UK00008B/253
* 9 7 8 0 3 0 6 4 6 4 6 9 0 *